METALLOCENE-CATALYZED POLYMERS

Materials, Properties, Processing & Markets

George M. Benedikt

Brian L. Goodall

Society of Plastics Engineers

Plastics Design Library

TABLE OF CONTENTS

FOREWORD

INTRODUCTORY REMARKS

This book represents a collection of papers describing advances made in the area of metallocene catalysts for olefin polymerization over the last 3-4 years. There are contributions from most of the major industrial companies active in the field as well as from a number of academic groups. In this introductory chapter we attempt to "set the scene" for the technical papers which comprise the text by giving a brief overview of the "Metallocene Revolution", putting metallocenes into a historical perspective in the development of olefin polymerization catalysts and finally we attempt to assist the interested readers in finding their way to the various contributions of particular interest to them.

The origins of Ziegler catalysis and Ziegler-Natta catalysts can be traced back to the mid-1950's. First, there was Karl Ziegler's discovery that zirconium and titanium salts, in combination with aluminum alkyl co-catalysts, are able to polymerize ethylene to high molecular weight, linear polyethylene at low pressures followed less than a year later by Giulio Natta's finding that propylene can be polymerized to form crystalline, isotactic polypropylene using certain crystalline modifications of titanium trichloride. These landmark inventions resulted in the commercialization of the world's biggest thermoplastics (HDPE, LLDPE and polypropylene) and in Ziegler and Natta sharing the Nobel Prize for chemistry.

While the vast majority of the world's capacity for both polyethylene and polypropylene is still based on heterogeneous titanium catalysts directly derived from those first systems, the phrase "Metallocene Revolution" has become a ubiquitous expression within the polyolefins industry over the last decade.

THE HISTORICAL ORIGINS OF METALLOCENE CATALYSTS

The phrase "Metallocene Revolution" was coined to explain the enormous changes brought about to polyolefin preparation and design as a result of the discovery and development of the metallocene catalysts. Initially the excitement in metallocenes was limited to their ability to homopolymerize ethylene at unprecedented rates, but over the last several years it has become overly clear that metallocene catalysts (by virtue of their well-characterized, homogeneous nature) offer the ability for rational design and tailoring of the catalysts. Even after more than 40 years of study and development of the commercial heterogeneous catalysts this is simply not possible with the titanium chloride catalysts due to their ill-defined, heterogeneous nature. Indeed the difference between the two classes of catalysts is perhaps best explained by another ubiquitous phrase used to define metallocene systems - they are "single-site catalysts" whereas the heterogeneous systems, by their very nature comprise a multiplicity of active sites with differing activity, selectivity, hydrogen (molecular weight regulator) response, etc. The ability to rationally design the metallocenes results in unprecedented control over polymer microstructure (isotactic, syndiotactic, atactic, stereo-block), unprecedented control over molecular weight and polydispersity as well as some additional flexibility in terms of co-polymerizing "difficult to polymerize monomers" such as certain cyclic olefins (e.g., cyclopentene and norbornenes).

The origins of metallocene catalysts can also be traced back to the 1950's when Natta and Breslow (of Hercules, Inc.) both discovered that titanocene dichloride in combination with aluminum alkyls could homopolymerize ethylene. In 1973 Breslow found that the addition of water resulted in increased activity. In the later 1970's attention turned from titanocenes to zirconocenes due to the work of Professors Sinn and Kaminsky at the University of Hamburg (patents to BASF and Hoechst). During this period Sinn and Kaminsky reported that methaluminoxane (the product of the controlled addition of water to trimethylaluminum) was a unique cocatalyst with regard to it's ability in the activation of zirconocene dichloride. During this period (and subsequently) much effort was expended in finding the most effective route to generate methaluminoxane (for example by reacting trimethylaluminum with the water of crystallization of aluminum sulfate octadecahydrate), defining it's catalytic role(s) and elucidating it's composition and structure. The resulting catalyst systems showed unprecedented reaction rates in the polymerization of ethylene.

THE METALLOCENE "REVOLUTION"

The use of zirconocene dichloride in the polymerization of propylene inevitably gave rise to atactic polypropylene of low molecular weight. Breakthroughs were reported in 1984 when John Ewen of Exxon (using titanocenes) and Walter Kaminsky (using zirconocenes) showed that the use of chiral ansa-metallocenes (first prepared by Hans Brintzinger of the University of Konstanz and reported in 1982) resulted in control of the stereostructure of the polypropylene produced. Ewen described blocky, partially isotactic PP using titanium while Kaminsky reported the first ever example of highly isotactic polypropylene using a homogeneous catalyst (with zirconium). In retrospect, the history of metallocene development until this point (1955-1984) was probably evolution rather than revolution. However, the reports of Ewen and Kaminsky opened the floodgates in terms of research expenditure in the area and the true metallocene revolution was started.

Other key metallocene catalyst contributions over the last decade include the use of a titanocene catalyst to generate highly syndiotactic polystyrene (Ishihara of Idemitsu Kosan) in 1985, highly syndiotactic polypropylene (Ewen and Razavi of Fina) in 1988, the first "single component" zirconocene catalysts (Hlatky and Turner of Exxon, Jordan of the University of Iowa) in 1988/9 (using fluorinated phenylboron counter-anions), "Constrained Geometry" catalysts (Bercaw of CalTech, Stevens of Dow) in 1990, sophisticated ansa-zirconocene catalysts representing the first metallocenes capable of generating commercially viable (high MW and very highly stereospecific) isotactic polypropylene (Spaleck of Hoechst) in 1993 and "oscillating" metallocene catalysts (no bridge between the two cyclopentadiene rings) which generate elastomeric, stereoblock polypropylene (Waymouth of Stanford, Moore of Amoco) in 1995.

COMMERCIAL APPLICATIONS OF METALLOCENE CATALYSTS

Up to the present time there are fewer commercial milestones in the metallocene development story than catalyst/technology milestones but this will surely change over the next decade as metallocenes gradually replace more conventional heterogeneous titanium catalysts. Indeed, at present, the volume of commercial polyolefins generated using metallocenes totals only a few percent (less than 10%) of total polyolefin production capacity. Considering the huge investments that have been made in metallocene research and development, it is safe to conclude that this situation will change, but it is likely that some of the first products will be specialty polyolefins that cannot be manufactured with proven, existing catalysts. Commercial milestones to date include Dow and Idemitsu beginning joint development of syndiotactic polystyrene (1989), Exxon announcing the first commercial metallocene-made resin ("Plastomer" in 1991), Dow announcing plans for commercial LLDPE production with "Constrained Geometry" catalysts (1995) and Hoechst and Mitsui test marketing "Topas", a copolymer of ethylene and norbornene (1996).

AN OVERVIEW OF THIS BOOK'S CONTENTS

This book includes a collection of papers presented at the recent meetings Polyolefins IX and X, as well as Antec '96 and '97. The authors belong mostly but not exclusively to industry, specifically to some of the major players in the metallocene based polymer technology like Dow Chemical (with its INSITE™ process), Exxon Chemical, DuPont-Dow Elastomers, Phillips Petroleum, Montell Polyolefins, Fina Oil and Chemical Company, and Union Carbide Co. Academia, as well as other industrial companies, either already participating or contemplating to play a role in this field are also represented.

Based on the subjects treated, the papers can naturally be grouped into the four selected areas:

- *Materials*, dealing with the more recent advances in catalyst and polymer production technology,
- *Properties*, presenting various areas where metallocene based polymers differentiate themselves from their Ziegler-Natta cousins,
- *Processing*, concentrating on the effect of the different metallocene polymer molecular structures and properties on their processability, and lastly,
- *Markets*, encompassing both some general views of the directions where this technology is heading in the years ahead, as well as some specific attempts to replace other existing, well established, commodity polymers such as PVC.

It would probably be unfair to direct the reader to only a few of the almost 50 papers included. They have all been selected for inclusion here since each contributes to the grand picture, through theoretical or practical studies, small scale or large-scale applications, inventions or innovations. What is important to mention are the major issues, the solution of which will dictate the economics of the successful technology in the targeted market application.

In the *Materials* section the cost of the catalyst at various production scales is an important issue. The cost itself (in $/lb) is misleading though, if one neglects the productivity of the catalyst and this could vary widely depending on the ligand/metal and process selection in order to obtain a tailored composition. The engineer's desire to develop a "drop-in" system into the existing commercial installations is on one hand a natural drive, but on the other it is bound to require some more or less extensive modifications of the existing production lines. Due to the variety of target polymer compositions (and where applicable microtacticity and sequence distribution), from polyethylene to polypropylene and polycyclic olefins with their copolymers, as well as from elastomers to thermoplastics, not one single process is expected to dominate: solution, slurry, liquid bulk, gas phase, combined liquid-gas phase. Low pressure as well as high pressure and high temperature "drop-in" processes are described. In a few instances, some compositions are only accessible via metallocene based chemistry, such as syndiotactic polystyrene or syndiotactic polypropylene. The production systems have to be selected and optimized for these compositions. The use of a mixed metallocene catalyst targeted at obtaining a bimodal molecular weight distribution for improved processability is also discussed.

In the *Properties* section the effect of: the metallocene polymers' narrower molecular weight distributions with low molecular weight extractables, together with their controlled content of long chain branching, as well as their ultralow densities, on the morphology, crystallization kinetics, melt rheology and relaxation is discussed. Thermal stability with and without antioxidants, the toughening of clear polypropylene by a metallocene poly(ethylene-co-butene or octene) plastomer, frictional behavior, adhesion (including self-adhesion) and paintability are also important property issues presented. A special mention needs to be made to the development of reactive silane grafted elastomeric metallocene polymers allowing for the use of a secondary moisture cure in foam and wire and cable applications, optimized via modeling of the corresponding architecture.

In the *Processing* section the practical issues of metallocene polymer injection molding, film extrusion, glass fiber reinforcement, fiber spinning, blending and alloying are presented comparatively, when appropriate, with the conventional Ziegler-Natta analogs.

In the *Markets* section the various homo and copolymers available via metallocene chemistry, and which cover elastomers as well as plastomers are discussed in terms of their present and future in specific applications. Food packaging films, electrical insulation applications, impact modifiers in blends, molded goods for automotive, medical and other durables, as well as foams and adhesives are just the first commercial target markets. Obviously, attempts to overtake PVC markets, especially in medical applications, are to be expected and, indeed, there is research directed at replacing this commodity material if the price and performance could be improved.

George M. Benedikt and Brian L. Goodall
BFGoodrich Co.
9921 Brecksville Rd.
Brecksville, Ohio 44141-3289
USA

Economic Factors for the Production of Metallocene and Perfluorinated Boron Cocatalysts

Jeffrey M. Sullivan

Boulder Scientific Company, P.O. Box 548 Mead, Colorado 80542, USA

ABSTRACT

Metallocene production costs are evaluated for a wide range of molecules currently under development for polyolefin catalyst systems, including bridged and unbridged ligands prepared from cyclopentadienes, indenes, and fluorenes. Production costs are evaluated on a life cycle basis and on a functional basis. Lifecycle costs are useful for budgeting and planning, and functional costs are useful for comparisons between competing alternatives. Examples from Boulder Scientific's production experience are given for each category of metallocenes and for perfluorinated boron cocatalysts.

INTRODUCTION

Metallocenes were developed in the 1950's, beginning with the discovery of ferrocene[4] followed shortly thereafter by the synthesis of titanocene dichloride and zirconocene dichloride.[9] By 1960, hundreds of metallocene molecules had been synthesized and characterized. Extensive reviews[3,9] described the state of the art at that time. Interest in metallocenes continued to grow gradually during the past 30 years until their recent application to polyolefin catalysis.

Within the past 3 years, commercial interest in metallocene production for catalyst applications has increased dramatically. While some of the more common metallocenes (ferrocene and zirconocene, for example) have been commercially available in large quantities for a number of years, most of the metallocenes and cocatalysts currently under development for polyolefin production were either unknown or produced only for laboratory research. The sudden surge in demand for kilogram and megagram quantities of complex metallocene molecules has exerted pressure to get into production rapidly and drive costs down as soon as possible.

The Boulder Scientific Company, located in Mead, Colorado, has been producing metallocenes commercially for over 15 years. Recent interest in substituted cyclopentadienyl metallocenes, indenyl metallocenes, silyl-bridged metallocenes, and boron-based cocatalysts has spawned extensive laboratory development and commercial-scale production at the Mead facility. In response to the rapidly changing polyolefin market, Boulder Scientific now offers over 30 metallocene products, many of which are produced in kilogram to 100-kilogram quantities.

Boulder Scientific has scaled approximately 20 metallocene production processes from 10 grams to batch sizes as large as 100 kg. As these production processes have matured, many lessons have been learned and trends in production costs have been identified. This paper summarizes production economics for a variety of metallocenes and boron cocatalysts, so that polyolefin producers may have a guide to the relative cost of candidate catalyst molecules. Just as new drug molecules are screened for effectiveness and production costs, new metallocene molecules can be evaluated on a similar basis. The costs in this paper represent production costs of bulk metallocenes and boron cocatalysts but do not include the costs of producing the finished supported catalyst.

COST METHODOLOGY

Metallocene production costs can be evaluated on a product life-cycle basis and on a functional basis. Product life-cycle evaluation is chronologically oriented and includes development, initial scale-up (or pilot plant), early commercial production, and mature commercial production. Production costs are (in general) inversely related to quantity. It is not uncommon to experience a 4 to 10-fold reduction in cost from early kilogram-scale production to mature commercial production. Functional evaluation is dependent on the molecular structure and synthetic building blocks and can be conducted at any point in the product life cycle. Factors such as raw-material cost, air and moisture sensitivity, number of synthetic steps, purification requirements (in process and final), and yield are all factored into a functional cost analysis. Table 1 summarizes various cost contributions and whether they contribute to a functional analysis, life-cycle analysis, or both.

Table 1. Functional and Life-Cycle Cost Contributions

Economic Factor	Functional	Life Cycle
Raw materials	1	3
Labor	3	1
Number of steps	1	3
Air or moisture sensitivity	1	3
Yield	1	3
Production scale	2	1
Purification requirements	1	2
Final delivery form	2	2

Key: 1-significant impact; 2-moderate impact; 3-minimal impact

The two cost approaches in Table 1 serve different purposes. Functional analysis provides relative cost between two or more alternatives, but life-cycle cost analysis provides actual production costs for budgets, planning, or accounting.

Functional analysis is very useful early in the development cycle, particularly for comparing between competing alternatives. If, for example, 5 metallocenes were under evaluation for use in polyolefin production, a functional cost analysis could rank the products by production cost. Functional cost analysis is also useful in comparing competing metallocenes under development by different polyolefin producers. Although the catalytic activity (which is not always available in the public domain) is required for a complete analysis, the relative metallocene cost can be easily determined using a functional cost analysis. Life-cycle cost analysis is more useful once a production commitment has been made, and ultimate cost projections are required for long-term planning. Since the cost of the metallocene is often a significant portion of the overall catalyst cost, economic viability of polyolefin production may depend on the ultimate price of the bulk metallocene.

PRODUCT LIFE-CYCLE COST ANALYSIS

Detailed cost models for multi-step metallocene production can be extremely complex and time consuming. At Boulder Scientific, cost projections follow these four steps.

1. Development cost. A substantial effort is invested to develop a "plant-ready" process on the laboratory bench top. Typically, a process is "ready" for the pilot plant when a large-scale (12 L glassware at Boulder Scientific) process runs successfully and repeatably on the bench. The development cost associated with this phase varies widely, depending on the number of steps and the starting point for development. Syntheses reported in the literature are sometimes difficult to reproduce and scale up. Costs are usually directly proportional to the number of chemist hours required (raw materials are rarely a significant contribution to development cost at the bench scale). Development times at Boulder Scientific range from one week (40 chemist hours) to six weeks (240 chemist hours) per synthetic step. An average of 100 chemist hours per synthetic step is useful for planning purposes, but should be replaced by actual hours or hours from a comparable process if these data are available.

   The cost of a chemist-hour should include direct costs, as well as R&D overhead (analytical support, raw material purchasing, and any other administrative support). General supplies and chemicals can be accounted for by adding a weekly "use charge" to cover the cost of one chemist working for one week. Depending on the nature of the process and on the definition of R&D overhead, use charges between $500.00 and $2000.00 per week may be appropriate. Using the average values of 100 chemist-hours per step, $1000.00 per week use fee, and a $100.00/hr fully burdened labor rate, the development cost for a four-step process would be $50,000.00 and would require 10 weeks. Any special equipment or unusually expensive raw materials should be added to this estimate.

2. Pilot-plant scale. Once a process is developed at the bench scale, production begins in the pilot plant. A pilot plant may consist of 25, 50, and 75-L glasslined reactors, or may involve conventional reactors in sizes from a few liters up to 1000 L. The size of the pilot plant is generally dependent on the size of the company's full-scale manufacturing operations. Costs in the pilot-plant phase are highly influenced by labor costs, since the number of labor hours per kg of material produced is quite large compared to full-scale manufacturing. It is not unusual to require as many or more hours of labor for a batch in the pilot plant as in the full-scale manufacturing plant, but the quantity produced is a fraction of full-scale production. For these reasons, product pricing is often very flat in the 3 to 10 kg range. Pilot-plant costs include set-up, raw material usage, production labor, support labor (R&D chemists, analytical chemists, and QA/QC), and waste treatment or disposal. At Boulder Scientific, these costs are modeled on a simple spreadsheet program with sections for each cost category. Production labor is a function of elapsed time and number of workers required for each step. For example, a multi-step process might require 110 hours of continuous clock time and an average of 1.25 workers, resulting a total of 137.5 production hours. In general, it is difficult to accomplish more than one synthetic step per day in the pilot plant, and 2(24-hour) days is a better general estimate. For complex processes, one full week (120 hours) is not unusual for each synthetic step. Raw material costs may be as low as 1/10 the labor cost or as

high as the pilot-plant labor costs. The impact of raw material cost is discussed in more detail in other sections.

3. Early full-scale manufacturing. In a traditional product development cycle, substantial study at the laboratory and pilot-plant level provides ample design detail for full-scale manufacturing. In the rapidly changing metallocenes field, however, projects are frequently rushed from laboratory to pilot scale to full scale manufacturing. Consequently, the first few production batches serve to debug the process and work out final scale-up details. Production campaigns at full scale are usually short, because metallocene demand for polyolefins has not yet reached maximum levels. Production costs in this phase are noticeably reduced because of economy of scale. Raw material costs are relatively constant (on a per kg of metallocene basis), but labor per kg produced declines substantially. Actual costs from this phase form the first reliable basis for projecting long-term manufacturing cost. Batch turnaround time is a good indicator of labor costs, and batch yields can be combined with turnaround time to produce a reliable production cost. Raw material to labor cost ratios vary from 3 to 1 down to 0.25 to 1, depending on the process.

4. Mature full-scale manufacturing. Within the next 3 years, as polyolefin technology matures and penetrates the market, several metallocenes and cocatalysts will enter this phase. This phase is marked by relatively large volumes, stable demand, and regular delivery schedules. For competitive reasons, cost pressure continues to be exerted on suppliers and the delivered price continues to fall until a plateau is reached. The level of the plateau is determined by raw material costs, scale of manufacturing, production technology, and labor cost. Production costs may include capital recovery for new equipment installed to meet the ongoing demand for metallocene products.

FUNCTIONAL COST ANALYSIS

Functional cost analysis provides insight into the relative cost of molecules based on their structure, properties, and synthesis. It can be useful for comparing different competing molecules, or different processes for producing the same molecule. The following factors go into a functional cost analysis.

1. Raw material cost, weighted at each step for yield
2. Overall yield (the mass yield is more useful than a molar yield)
3. Number of synthetic steps
4. Number of purification steps (distillation, recrystallization, etc.)
5. Special handling requirements (e.g. air or moisture sensitivity)

   At Boulder Scientific, spreadsheet programs calculate raw material requirements for each step based on stoichiometry, reaction yield, and purification losses. In indene-based ligands, for example, it is not uncommon to require 5 to 10 of indene for every kg of finished metallocene. The raw material cost, in total dollars per kg of finished metallocene, can then be readily calculated. The spreadsheets also reveal the most costly raw materials, which might suggest alternate routes, substitute raw materials, or more competitive sources to reduce costs. Finally, the raw material cost per kg of finished metallocene may be a useful screening tool between competing molecules or synthetic routes.

   Yields can be compared on a molar basis or a mass basis. Molar yields (those typically reported in the literature) can be converted to mass yields using the appropriate molecular weights. For example, a metallation reaction might proceed in 47% molar yield, but the increase in molecular weight might produce a mass yield of nearly 100% (1 kg of metallocene per kg of ligand). Multi-step processes typically have low overall molar yields because of yield compounding. A 90% molar yield over 5 steps produces an overall yield of 59%; an 80% molar yield over 5 steps produces a 33% overall yield. Even when converted to mass yields, these low overall yields result in substantial losses in raw material conversion. When calculating raw material costs, mass yields are required, but when comparing processes or products, molar yields are a better indicator of overall effectiveness.

EXAMPLES OF SPECIFIC COMPOUNDS

Examples from several different classes of metallocenes provide different insights into production costs and issues. In order to span the wide range of metallocenes under consideration today, the following molecules have been included: Brintzingers catalyst (an ethylene-bridged indenyl zirconocene), a substituted cyclopentadienyl zirconocene, a mono-cyclopentadienyl catalyst (a silyl bridged molecule), and a perfluorinated boron cocatalyst molecule.

BIS(INDENYL)ZIRCONIUM DICHLORIDE AND RELATED DERIVATIVES

In the early 1980's Brintzingers and coworkers synthesized racemic ethylene-bridged bis(indenyl)zirconium dichloride, $Et(Ind)_2ZrCl_2$, and evaluated its properties as a polymerization catalyst.[7,8] Since then, Brintzinger's catalyst and various derivatives have been evaluated by companies worldwide for activity and performance. Commercial preparation of the metallocene (or its substituted forms) can be divided into 4 parts:
1. Preparation and purification of the starting indene (substituted or not).
2. Bridging the indene molecule to form the bis-indenyl ethane ligand.
3. Reacting the ligand with zirconium tetrachloride to form the metallocene.
4. Isolating and purifying the racemic isomer of the metallocene.

Step 1 can require a simple distillation or drying (commercial indene is typically 90% pure and contains some water), or can be more involved if a functional group must be attached to the indene molecule. Step 2 is usually conducted using ethylene dibromide, and the resulting bis-indenyl ethane can be isolated as a solid. There are a variety of reaction schemes for step 3, and the crude material may be isolated as a solid or passed as a solution to step 4. The finished metallocene is/is not sensitive to air and moisture and must be handled accordingly.

At Boulder Scientific, one indenyl process required low-temperature operations for steps 2 and 3. The overall yield, from indene to finished metallocene varied from 10% to 25%, which made the resulting metallocene quite expensive. The process was run at the 10-kg scale, and the production costs were in excess of $10,000 per kg. Clearly, this process was not economically viable in practice, even though paper analysis suggested that the cost should be substantially lower. An improved process with better yields and costs is currently under development for production of this metallocene.

SUBSTITUTED BIS(CYCLOPENTADIENYL)ZIRCONIUM DICHLORIDES

The preparation of bis(cyclopentadienyl)zirconium dichloride (zirconocene dichloride) was developed in the 1950's, and applications of zirconocene and titanocene to polymerization occurred in 1957.[2,5] In addition to the unsubstituted forms of titanocene and zirconocene dichloride, many substituted cyclopentadienyl (Cp) metallocenes of zirconium and titanium are actively under development as polyolefin catalysts. Typically, these molecules consist of alkyl-substituted cyclopentadienes reacted with zirconium tetrachloride to form the corresponding metallocene. The degree of substitution ranges from mono-substituted to penta-substituted cyclopentadiene. The production process for these metallocenes can be broadly described as follows:

1. Prepare substituted, purified cyclopentadiene
2. React cyclopentadiene with zirconium tetrachloride to form the metallocene
3. Isolate and purify the correct isomer of the metallocene.

Step 1, of course, can be the most difficult, since it may involve multiple chemical steps and purifications. Synthetic chemists have developed numerous methods of preparing substituted cyclopentadienes. Boulder Scientific developed a proprietary 4-step process for producing a di-substituted cyclopentadiene, and scaled the process from laboratory to approximately 20-kg scale. Production cost data as a function of scale are given in Table 2.

Table 2. Production Costs for Substituted Cp Metallocene.

Run#	Scale	Cost, $/kg	Relative Cost
1	2 kg	9,356	1.00
2	5 kg	6,973	0.75
3	10 kg	3,623	0.39
4	20 kg	1,469	0.16

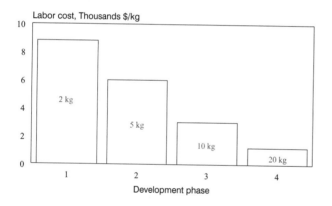

Figure 1. Labor Costs During Development of a Substituted Cyclopentadienyl Zirconocene.

This process was not highly sensitive to raw material cost (at the 20 kg scale, raw materials were approximately 20% of the total cost) and demonstrated excellent cost decreases with increasing scale and production experience. Unlike the Brintzingers process, yields in the plant continued to improve with time and experience. The overall economics were highly dependent on the production of a key intermediate (a substituted cyclopentadiene). Labor costs (in U.S. dollars) during development are plotted in Figure 1.

CONSTRAINED-GEOMETRY SILYL-BRIDGED MOLECULES

In addition to "traditional" metallocene catalysts, constrained-geometry (CG) mono-cyclopentadienyl catalysts have been developed by Dow Plastics.[6] These titanium-based catalysts can provide unique polyolefin properties, and follow a generic synthetic route similar to the previous examples. In the first step, a multi-step process is required to prepare purified ligand. In the second step, the ligand is attached to the metal (titanium, in this case). In the third step, the finished molecule is separated and purified.

Evaluating all metallocenes on a similar basis (ligand prep, metallocene prep, and final purification) facilitates cost comparisons for different processes and molecules. For example, consider the hypothetical comparison between 3 metallocenes in Table 3.

Table 3. Hypothetical Metallocene Comparison

Type	RM Index	Ligand Steps	Purifications	Ligand Yield	Metallocene Yield	Overall Yield
Substituted-bridged indenyl	1.0	3	2	40%	45%	18%
Substituted cyclopentadienyl	0.8	4	1	50%	45%	22.5%
Silyl-bridged CG	0.6	3	3	60%	35%	21%

The relative raw-material cost for each metallocene in Table 3 is the quotient of the RM Index (actual raw material cost per unit excluding yield) and the overall yield. Carrying out this multiplication provides the following rankings for raw material cost:

Silyl-bridged CG   2.85
Cp-based   3.56
Indenyl   5.56

After accounting for raw material costs, labor costs can be incorporated by evaluating the number of steps and number of purifications. In this example, assume that all steps require equal amounts of labor. If the labor cost per step is 1/2 the raw material cost of the indenyl metallocene, then the overall relative costs can be obtained as follows:

Labor cost per step = $(0.5)(5.56) = 2.78$
Total labor cost = (number of steps)(2.78)

Table 4. Relative Cost Contributions for Hypothetical Metallocenes

Type	R.M.	Labor	Total
Silyl-bridged CG	2.85	16.68	19.53
Cp-based	3.56	13.90	17.46
Indenyl	5.56	13.90	19.46

The relative cost contributions are given in the Table 4. In Table 4, the compound with the lowest unit raw material cost wound up having the highest overall cost because of the labor contribution. This is not unusual for metallocene production at the 10 to 100 kg production scale. Labor costs become less significant as production increases in capacity, as illustrated in the final example.

TRIS(PENTAFLUOROPHENYL) BORON COCATALYST

All metallocene catalysts require cocatalysts for activation. Historically, organo-aluminum compounds, and specifically, methylalumoxanes, MAO, have filled this role. Some catalysts, however, have been activated by perfluorinated phenyl boron compounds, such as tris(pentafluorophenyl) boron, TPFPB. Boulder Scientific produces TPFPB in large quantities, and its cost evaluation completes the examples in this article.

Unlike many metallocene molecules, the cost of TPFPB is highly driven by raw materials. The production process requires several chemical steps, but the cost of labor is far outweighed by the cost of key raw materials. Consequently, the improvement in production costs with scale are primarily related to discounted purchases of raw materials in larger quantities, not improvements in labor. Table 5 summarizes labor and raw material costs as a function of project history (which is directly related to scale).

Production costs decreased by approximately 80% from the first pilot run to large-scale manufacturing, even though labor costs decreased by more than 90%. If production were dominated by labor cost, even greater cost reductions would have been achieved. The reduction in raw material cost reflects cost savings from buying in bulk. Since most economy of scale is achieved by reducing the labor cost per unit of production, raw material costs only decrease with scale if the raw-material supplier can also take advantage of economy of scale. In this example, considerable cost reductions were achieved in raw material purchases, but a metallocene process dependent on a costly material like butyl lithium would probably not achieve similar reductions.

Table 5. Relative Production Costs for TPBFB

Phase	Scale	RM Cost	Labor Cost	RM/Labor Ratio	Total Cost
1	3 kg	1.0	1.0	0.34	1.0
2	100 kg	0.94	0.12	2.8	0.32
3	400+ kg	0.51	0.07	2.4	0.18

Production cost data from a series of cocatalyst production runs are plotted in Figure 2. The initial pilot run was normalized to 1.0, and the other costs were calculated from that basis. Note the step decrease in raw material cost associated with larger-scale manufacturing, and the steady decrease in labor cost as production rates increased and the process efficiency improved. The "developmental" label refers to the first runs at large scale, before the process was sufficiently opti-

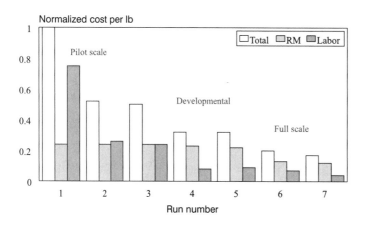

Figure 2. Cocatalyst production costs (normalized) during these three phases of process development. Note: First pilot-run costs were normalized to 1.0.

Table 6. Estimated Metallocene Prices

Compound	$/10 kg	$/500 kg
bis(Cp)TiCl$_2$	350	285
bis(Cp)ZrCl$_2$	500	325
Et(Ind)$_2$ZrCl$_2$	9,000	6,500
bis(1,2-dimethyl Cp)ZrCl$_2$	12,000	7,500
bis(1,3-dimethyl Cp)ZrCl$_2$	3,000	2,000
bis(n-butyl Cp)ZrCl$_2$	2,000	1,500
bis(Me$_4$Cp)ZrCl$_2$	2,000	1,500
bis(Me$_5$Cp)ZrCl$_2$	2,500	2,000
bis(Ind)ZrCl$_2$	1,000	750
bis(2-MeInd)ZrCl$_2$	3,000	2,000
(Me)$_2$Si-bis(Cp)ZrCl$_2$	2,000	1,500
(Me)$_2$Si-bis(Ind)ZrCl$_2$	5,000	3,000
(Me)$_2$Si-bis(9-Flu)ZrCl$_2$	3,000	2,000
Ipl(Cp-9-Flu)TiCl$_2$	9,000	6,000
Ipl(Cp-9-Flu)ZrCl$_2$	7,000	5,000
Ipl(3-MeCp-9-Flu)ZrCl$_2$	8,000	6,000

Me=methyl;Ind=indenyl;   Cp=cyclopentadienyl;   Flu=fluorenyl; Ipl=isopropylidene

Table 7. Estimate Cocatalyst Prices

Compound	$/10 kg	$/500 kg
tris(pentafluorophenyl)boron	1,600	1,200
Methylalumoxanes, MAO	500	380

mized. In the costs for Run #6 and #7, labor constitutes such a small part of the total cost that pricing would be dictated almost completely by raw material costs, not by further scale-up or process improvements.

PRICE PROJECTIONS

Based on the costing strategies outlined above, Boulder Scientific has estimated prices for metallocenes and cocatalysts in quantities of 10 kg and 500 kg. Table 6 summarizes estimated prices for a wide variety of metallocenes, and Table 7 lists estimated prices for a boron cocatalyst and an aluminum cocatalyst. All estimated prices are in U.S. dollars.

CONCLUSIONS

The final chapter on metallocene production costs has not yet been written, because the production scale of these molecules is still changing rapidly. Since large-scale metallocene production is in its infancy, current prices reflect high development costs and minimal economy of scale. In most cases, production costs will improve considerably as demand stabilizes at higher levels and production processes are optimized.

Costs can be evaluated from a functional basis, which includes production of ligand, metallocene, and final purified product, or costs can be evaluated from a life-cycle approach. As the examples illustrated, costs are influenced by the number of production steps, overall yield, and raw material cost. Due to the complexity of metallocenes, there are no hard and fast cost "rules" that can be applied for comparing approaches or molecules. Since most metallocenes are labor intensive, considerable improvement in production methods, scale, and ultimately, costs should be realized in the next 5 years.

REFERENCES

1.   J. Birmingham, **Synthesis of Cyclopentadienyl Metal Compounds**, Advances in Organometallic Chemistry, Eds. F. G. A. Stone and R. West, *Academic Press*, New York, 1964.
2.   D. S. Breslow and N. R. Newburg, *J. Am. Chem. Soc.,* **79**, 5072 (1957).
3.   E. O. Fischer and W. Semmlinger, **Darstellungsmethoden fur Cyclopentadienyl-Metall-Verbindungen**, *University of Munich*, August, 1961.
4.   T. J. Kealy and P. L. Pauson, *Nature*, **168**, 1039 (1951).
5.   G. Natta, R. Pino, G. Mazzanti, and R. Lanzo, *Chim. Ind. (Milan)*, **39**, 1032 (1957).
6.   D. Schwank, *Modern Plastics*, August, 49 (1993).
7.   F. R. W. P. Wild, M. Wasincionek, G. Huttner, and H. H. Brintzinger, *J. Organomet. Chem.*, **288**, 63 (1985).
8.   F. R. W. P. Wild, L. Zsolnai, G. Huttner, and H. H. Brintzinger, *J. Organomet. Chem.*, **232**, 233 (1982).
9.   G. Wilkinson and J. M. Birmingham, *J. Am. Chem. Soc.*, **76**, 4281 (1954).

Polypropylene Reinvented - Cost of Using Metallocene Catalysts

Norman F. Brockmeier
Argonne National Laboratory

ABSTRACT

This study develops scoping estimates of the required capital investment and manufacturing costs to make a zirconocene catalyst/cocatalyst system $[(F_6\text{-acen})Zr(CH_2C\text{-}Me_3)(NMe_2Ph)][B(C_6F_5)_4]$ immobilized on a specialized silica support. The costs for this fluorine-based system are compared with corresponding estimates for two other metallocene catalysts using methylaluminoxane (MAO)-based cocatalysts. Including the weight of the support and cocatalyst, each of the production facilities for making the three zirconocene catalyst systems is sized in the range of 364 to 484 tonnes per year. The cost to make the F-based catalyst system is estimated to be $10,780/kg, assuming a 20% return on capital invested. By comparison, the costs for the two MAO-based catalyst systems fall into the range of $10,950/kg to $12,160/kg, assuming the same return. Within the ±50% accuracy of these estimates, these differences are not significant. Given a catalyst productivity of 250 kg of resin per gram of zirconocene, the cost contribution in the finished ethylene-propylene copolymer resin is 4.4 cents/kg, excluding costs for selling, administration, and research.

BACKGROUND

There has been a great deal of excitement about the potential for polyolefin demand and sales growth to shift into the range of 10% to 20% annually. The impetus behind this dramatic shift is the promise afforded by the physical properties that can be achieved by using metallocene or Single-Site-Catalysts, SSCs, to manufacture new polyolefins. Examples of improved properties include higher flexural modulus, higher heat distortion temperature, improved impact strength and toughness even at low temperature,[1] and better film clarity. Announcements in various trade journals and conferences indicate that a few metallocene-made polyolefins are already available in the commercial marketplace; most notably polyethylenes such as LLDPE and LDPE, and to a lesser extent, isotactic and even syndiotactic polypropylene, PP. New developments are announced almost weekly.

The purpose of this study is to provide some scoping estimates of the capital investments and manufacturing costs for facilities to make the new polyolefin catalyst systems. Previous studies have presented estimates for metallocene catalyst systems based on the use of methylaluminoxane, MAO, cocatalyst. The focus of this study is the cost picture for metallocene systems based on the use of fluorine, in particular using the perfluorophenyl borate anion as the active cocatalyst. The complete catalyst/cocatalyst is the zirconocene system $[(F_6\text{-acen})Zr(CH_2CMe_3)(NMe_2Ph)][B(C_6F_5)_4]$ immobilized on a specialized silica support. This complete system will be abbreviated hereinafter as F-SSC. The assumption is that this heterogeneous form of F-SSC will be essentially a drop-in substitute for the Ziegler-Natta, Z-N, supported high-activity catalysts in common use today.

The polyolefin selected as an example product for this study is a specialty ethylene-propylene, EP, copolymer having excellent toughness, high flex mod, and high impact strength at low temperature. While this F-SSC is not necessarily the optimum choice to make this EP copolymer, the assumption is that the costs involved will be reasonable estimate of reality.

DISCUSSION

For ease of comparison with the results of other cost studies,[2-4] this study uses the same estimating bases wherever possible. Plant locations are assumed to be the U.S. Gulf Coast with a year-end 1994 completion date. Capital investments include all facilities needed for a complete and operable plant (process, laboratories, utilities, rail access, and roads). The catalyst productivity for polymer is assumed to be the same as that achieved with MAO-based SSCs in a previous study:[3] 250 kg polymer/g zirconocene. The polyolefin EP copolymer is assumed to be made in a gas-phase process. The annual capacity of plant A to make F-SSC is set at 6,800 kg (15,000 lb), based on the zirconocene alone. This output is enough to supply 8-12 world-scale polyolefin plants with catalyst. The capital needed for making the F-SSC catalyst and immobilizing it is estimated to be 9.0 million dollars (MM$). The companion plant B to make the F-based cocatalyst is sized for 25,000 kg (55,000 lb) per year, in a batchwise operation (six days per batch, 300 days/yr). The required capital investment for plant B is 3.7 MM$. This cocatalyst is a major raw material for the catalyst plant — a facility that also immobilizes the F-SSC at a weight ratio of 50:1 support to zirconocene. Thus, the annual production by weight of the supported catalyst system is: 341,000 kg silica support, 6,800 kg zirconocene, and 17,000 kg of tris(pentafluorophenyl)boron for a total solid product of 364,000 kg (~803,000 lb). The total capital investment for the two combined (A and B) facilities is 12.7 MM$.

The F-based cocatalyst made in plant B is tris(pentafluorophenyl)boron, with a formula of $B(C_6F_5)_3$. Hereinafter this will be called TPFPB. The synthesis of TPFPB has been documented in the literature, starting from pentafluorophenyl bromide.[5] Jordan and coworkers have described the subsequent synthesis of the zirconocene-based catalyst that includes TPFPB as the tetrakis(pentafluoro-phenyl)borate anion.[6]

Table 1 shows scoping estimates of the costs to manufacture three different cocatalysts for metallocene SSCs. When the cost of capital is included, the netback is that price needed for a 20% Return on Investment, ROI. The right-hand column shows the netback estimated for TPFPB: $1,480/kg ($673/lb). The other two columns in Table 1 show netback estimates of $220/kg for MAO and $164/kg for modified MAO with capital investments, respectively, of 6 MM$ and 7 MM$.[2] At first glance, there does not seem to be any comparison between the cost of TPFPB and the other cocatalysts. However, it is better to compare all three of these systems in the completed, immobilized final form, to be shown later. Note that the ratio of variable to fixed cost for TPFPB is 20 because the major raw material is nearly $500/kg. Assume that variable cost does not change with plant scale. If the facilities were scaled to half the output, the fixed costs per kg would roughly double, but the netback would not be increased much. By contrast, the cost ratio for MAO is 13; scaling down this process would increase the netback to a much greater extent.

Table 2 provides a much more meaningful basis on which to compare the costs for using any of these metallocene systems. The zirconocene is immobilized on a suitable support, and treated with a cocatalyst to produce a complete system that can successfully be metered into a polymerization

Table 1. Metallocene cocatalyst manufacturing cost capacity: 25,000 kg/yr

Required Capital Investment, million $	6.0	7.0	3.7
Variable costs	MAO, $/kg	MMAO, $/kg	TPFPB, $/kg
Key raw materials	128	57	1,175
Other chemicals	7	18	156
Utilities	35	35	13
Total Variable Cost	170	110	1,344
Fixed costs (ex. depreciation)	13	15	64
Total MFG costs (ex. depreciation)	183	125	1,408
Depreciation (10% of total)	4	4	15
Add 20% ROI + work cap.	33	35	57
Netback reqd. (ex. SAR)	220	164	1,480

MAO: methylaluminoxane; TPFPB: tris(pentafluorophenyl) boron

Table 2. Metallocene catalyst manufacturing cost. Capital investment: 9mm$, capacity 6800 kg/yr

	CAT B.1, $/kg	CAT B.2, $/kg	F-SSC, $/kg
Recipe, mass ratio (Zr:Cocat:t-buAl:Silica)	1:18:0:50	1:16:55:50	1:2.4:0:50
Variable costs			
Zirconocene Compound	5,060	5,060	4,830
Cocatalyst (MAO-based or TPFPB)	4,840	2,640	3,773
Tri-isobutyl Al Alkyl	-	1,016	-
Silica support	1,100	1,100	1,100
Other chemicals	184	184	184
Utilities	27	27	27
Total variable cost	11,211	10,027	9,902
Fixed costs (ex. depreciation)	372	372	372
Total MFG cost (ex. depreciation)	11,583	10,399	10,274
Depreciation (10% of total)	132	132	132
Add 20 % ROI	442	442	442
Netback reqd. (ex. SAR)	12,157	10,973	10,778

Table 3. Catalyst systems & price ranges 20% return on investment

Catalyst	Cat B.1	Cat B.2	F-SSC
Catalyst capacity, tonne/yr	6.8	6.8	6.8
Capital investment, MM$	15	16	12.7
Productivity based on active catalyst, kg/g	250	250	250
Netback range, $/kg active catalyst	10,800-14,300	9,600-12,600	9,400-12,200
Productivity based on total solids delivered, kg/g	3.6	3.4	4.7
Netback range, $/kg total solids	156-207	136-173	176-228
Catalyst cost in finished resin, ¢/kg (ex. SAR)	4.3-5.7	4.0-5.1	3.7-4.9

Table 4. Catalyst systems & price ranges 20% return on investment

Catalyst	TiCl$_3$	Z-N HAHS	Cat B.1	Cat B.2	F-SSC
Catalyst capacity, tonne/yr	455	68	6.8	6.8	6.8
Capital investment, MM$	18	30	15	16	12.7
Productivity based on active catalyst, kg/g		1,000	250	250	250
Netback range, $/kg active catalyst		14,400-17,600	10,800-14,300	9,600-12,600	9,400-12,200
Productivity based on total solids delivered, kg/g	1	2.5	3.6	2.0	4.7
Netback range, $/kg total solids	19-26	360-440	156-207	79-103	176-228
Catalyst cost in finished resin, ¢/kg (ex. SAR)	1.9-2.6	1.4-1.8	4.3-5.7	4.0-5.1	3.7-4.9

reactor. The systems based on MAO or MMAO both use the same zirconocene:[1] Me$_2$Si(2-Me-benzyl indenyl)$_2$ZrCl$_2$. They are both estimated to have the same capital investment of 9 MM$.[3] The combined facilities for the MAO-based systems thus cost 15 MM$ versus 12.7 MM$ for F-SSC. They each require a recipe with a mass ratio of 1:18:50 of zirconocene:co-catalyst:silica. On the other hand, the recipe for F-SSC is 1:2.4:50 of zirconocene:cocatalyst:silica (Zr:TPFPB:Si). The right-hand column in Table 2 shows that the netback for F-SSC is estimated to be $10,778/kg ($4,899/lb). The

Table A-1. Metallocene catalyst manufacturing process. Scoping economics
Process: Batchwise, Bridged & Heterogenized on SiO_2 support;[1] polymer: EP copolymer; capacity: 6800 kg/yr; Cap. investment: 9 MM$ (6 ISBL+3 OSBL)

Variable costs, ¢/kg	Price, ¢/kg	Usage, g/g	Metallo., $/kg
Aromatic solvent	132	14	18
Zirconocene compound[1]	460,000	1.05	4,830
Specialized solvent	880	10	88
TPFPB[2]	148,000	2.55	3,773
Alkane solvent	264	10	26
Si compound	1,320	2	26
Silica support	2,200	50	1,100
Aromatic ester	264	1.5	4
Organic waste disposal	66	30	10
Utilities			
Electricity	6.0		9
Steam	1.1		9
Cooling water			2
Inert gas	124		7
Subtotal variable cost			9902
Fixed cost (ex. depreciation)			
Operating wages (W)	$23/hr	45,600hr/yr	
		1049	154
Direct main, matl & labor (M)	(7% ISBL)	420	62
Direct overhead	45% (W+M)	661	97
Allocated fixed cost	(T+I+M)	408	59
Subtotal cost		2538	372
Total MFG costs (ex dep.)			10,274
Depreciation, 10% of total			132
Add 20% ROI + work cap			442
Netback reqd. (ex. SAR)			10,778

[1]catalyst $[(Fe_6\text{-acen})Zr(CH_2CMe_3)(NMe_2Ph)][B(C_6F_5)_4]$
[2]cocatalystTris(pentafluorophenyl)boron

system based on MMAO has only 2% higher netback. The MAO-based catalyst has a netback that is 13% higher than F-SSC. Within the accuracy of making scoping estimates, these differences are not believed to be significant. A polyolefin manufacturer will probably make a selection of catalyst system based more heavily on the physical properties achieved in the product grades that can be made. Metal

Table A-2. TPFPB catalyst manufacturing process. Scoping economics
Process: Batchwise; capacity: 25,000 kg/yr; Cap. investment: 3.7 MM$ (2.8 ISBL+0.9 OSBL)

Itemized costs	Price, ¢/kg	Usage, tonne	$
Annual revenues			
Cocatalyst sales	148,000	25.0	3,701,000
Operating expenses			
Operating labor	$23/hr	31,300hr	720,000
Maintenance, material & labor	7% ISBL		193,000
Direct overhead	45% (W+M)		411,000
Allocated fixed cost	T+I+M		293,000
Utilities			
Electricity	6¢kWh	3000M KWh	180,000
Steam	1.1	5454.5	60,000
Cooling Water			15,000
N_2 Inert gas	124	36.3	45,000
Organic waste disp.	66	545.5	360,000
PFPB[1]	49,740	59.0	29,348,000
Butyl lithium	4,620	6.3	289,000
Boron trichloride	1,670	11.5	192,000
Si Compound	1,320		
Tetrahydrofuran	299		
Pentane	264	727.3	1,920,000
Diethyl ether	128	375	479,000
Misc. other	44	113.6	50,000
Subtotal			35,195,000
Gross profit			1,820,000
Depreciation (10% of total inv.)			-370,000
Interest (9%)			-333,000
Net realized, $			1,117,000

[1]pentafluorophenyl bromide

Table A-3. TPFPB catalyst manufacturing process. Scoping economics
Process: Batchwise; capacity: 25,000 kg/yr; Cap. investment: 3.7 MM$ (2.8 ISBL+0.9 OSBL)

Variable costs, ¢/kg	Price, ¢/kg	Usage, g/g	TPFPB, $/kg
PFPB[1]	49,740	2.36	1173.9
Butyl lithium	4,620	0.25	11.7
Boron trichloride	1,670	0.46	7.7
Tetrahydrofuran	299	8.55	25.5
Pentane	264	29.0	76.8
Diethyl ether	128	15.0	19.1
Misc. other	44	4.55	2.0
Organic waste disposal	66	21.8	14.3
Utilities			
Electricity	6.0kWh	120 kWh/kg	7.2
Steam	1.1		2.4
Cooling water			0.7
Inert gas	124		1.8
Subtotal variable cost			1,343.1
Fixed cost (ex. depreciation)			
Operating wages (W)	$23/hr	2,608 hr/mo	
		720	18.8
Direct main, matl & labor (M)	(7% ISBL)	193	7.7
Direct overhead	45% (W+M)	411	16.5
Allocated fixed cost	(T+I+M)	293	11.7
Subtotal cost		1,617	64.7
Total MFG costs (ex dep.)			1,407.8
Depreciation, 10% of total			14.7
Add 20% ROI + work cap			57.9
Netback reqd. (ex. SAR)			1,480.4

[1]pentafluorophenyl bromide

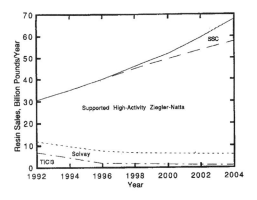

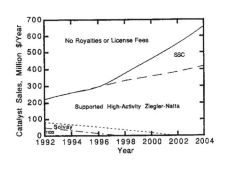

Figure 1. Catalyst systems used for global PP & EP production projections.   Figure 2. Catalyst revenue projections for global PP & EP.

residues left in the product will be somewhat different because the catalyst compositions are different, even at the same Zr productivities assumed here.

A catalyst vendor will most likely sell these metallocene systems priced on the basis of the solids delivered, which will include the silica support. Table 3 compares netback prices and productivities for all three systems based on solids. The MAO-based catalyst B.1 would produce 3,600 g EP copolymer/g solids costing $163/kg. The MMAO-based catalyst B.2 would produce 2,000 g/g solids costing $90/kg. For comparison, the F-SSC system would produce 4,700 g copolymer/g of solids that would have a netback of $202/kg. The bottom line, however, is the catalyst cost contribution in the finished polymer: Just 4 to 5 cents/kg for either the F-SSC or for the MMAO-based systems; Cat B.1 is slightly higher. The cost for a Z-N system is 1.4 - 1.8 cents/kg of resin.

SUMMARY

The cost estimates for three different metallocene catalyst systems shown in Tables 1-3 indicate that the F-SSC system (and the MMAO-based system) may have a slight cost advantage over the MAO-based system. However, the 13% advantage is probably not significant relative to the uncertainties in a scoping estimate. A polymer producer is more likely to choose between catalyst systems based on the physical properties that can be achieved in the finished product. A comparison of properties is beyond the scope of this study. The expectation is that the enhanced properties (relative to products made with Z-N catalysts) will create sufficient demand to more than overcome the higher price of metallocene-made resins — prices higher by $50/tonne excluding the SAR costs.

ACKNOWLEDGMENT

Valuable information has been contributed from conversations with R. F. Jordan and G. G. Arzoumanidis, Argonne National Laboratory has given the author the freedom to participate in this conference.

REFERENCES

1.   K. D. Hungenberg, J. Kerth, F. Langhauser, and P. Muller, *Proc. Polyolefines'94*, Kanazawa, Japan, March 10, 1994.
2.   N. F. Brockmeier, *Proc. MetCon'94*, Houston, Texas, May 26, 1994.
3.   N. F. Brockmeier, *Proc. MetCon'95*, Houston, Texas, May 17, 1995.
4.   J. M. Birmingham, G. J. Hanna, and J. M. Sullivan, *Proc. Metallocenes'95*, Brussels, Belgium, April, 1995.
5.   A. G. Massey and A. J. Park, *J. Organometallic Chem.*, **2**, 245 (1964).
6.   E. B. Tjaden, D. C. Swenson, R. F. Jordan, and J. L. Petersen, *Organometallics*, **14**, 371 (1995).

New Approaches for Ziegler-Natta Catalysts for Polypropylene

Edward S. Shamshoum and David Rauscher

Fina Oil and Chemical Co., Research and Technology Center, Deer Park, TX, USA

INTRODUCTION

Interest in olefin polymerization using cationic metallocene complexes continues to increase very rapidly. This phenomenon is reflected in the number of patents, publications, and conferences dealing with this vast growing field. Work in this area, especially for zirconocene/methylaluminoxane (or other ion pair cocatalyst systems) catalysts is beginning to see commercial development and application. Metallocenes development at Fina, and especially the syndiospecific metallocene catalyst system, led to a commercial test run in Fina's LaPorte polypropylene plant in April of 1993. This experimental run was historical for being the first successful commercial production of syndiotactic polypropylene and also was the first reported commercial test run using a stereospecific metallocene catalyst for the polymerization of propylene in a liquid, full loop reactor system conventional for Ziegler-Natta catalysts. In the polyethylene area, Exxon, Dow, and others began reporting the commercialization of metallocene produced LLDPE in 1991 and 1992. Due to the more complex nature of polypropylene (or other α-olefins) structure, the polymerization technology and the commercialization of metallocenes in this area has been lagging three to five years behind that of polyethylene. This paper will bring out some of the differences; between the two metallocene systems.

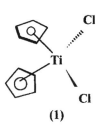

(1)

Metallocene systems such as bis-cyclopentadienyl-titanium or zirconium complexes[1] have been known since the early development of Ziegler-Natta catalysis. Examples of these metallocenes are those reported by Breslow and Chien in 1957 and 1958.[1] The study of such complexes led to the initial proposal that cationic Ti centers were the active sites for polymerization. For propylene polymerization, these metallocenes are non-stereoselective and produce atactic polymer, but such metallocenes are used for the polymerization of ethylene.

The successful application of metallocenes to propylene polymerization began with two breakthroughs: the use of methylaluminoxane, MAO, as a cocatalyst and the use of bridged zirconocenes (*ansa* complexes). In 1976, Kaminsky *et al.*[2] reported the use of MAO as a cocatalyst with Cp_2TiCl_2 in the polymerization of ethylene to obtain an increase in catalyst activity by several orders of magnitude. Pursuing the proposal that MAO readily forms an active cationic form of the metallocene, other ionizing reagents were later shown to generate active catalyst systems. Examples include compounds such as $AgBPh_4$, $[Cp_2Fe]BPh_4^f$, $Bu_3NHBPh_4^f$,

Scheme I

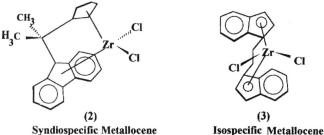

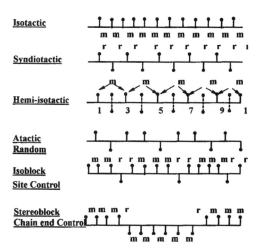

(2)
Syndiospecific Metallocene

(3)
Isospecific Metallocene

Isotactic

Syndiotactic

Hemi-isotactic

Atactic
Random

Isoblock
Site Control

Stereoblock
Chain end Control

Figure 1: Microstructures of Polypropylene

and Ph₃CBPh₄ᶠ. The chemical reaction occurring between the metallocene and the co-catalyst is shown in scheme I.

The second break-through for propylene was the discovery that bridged or *ansa*-zirconocene catalysts could produce isotactic (iPP)[3] and syndiotactic polypropylene (sPP)[4] (3 and 2, respectively). The addition of a bridging substituent made the complex stereorigid, imparting Cs symmetry to the syndiospecific complex (complex 2), and C_2 symmetry to the isospecific complex (i.e., the *rac* isomer, complex 3). In 1984, Ewen demonstrated that *rac*-ethylene-*bis*-indenyl titanium could effect the isospecific polymerization of propylene.[5] In 1988, Fina reported the preparation of the first syndiospecific metallocene catalyst, iPr(CpFlu)ZrCl₂, which afforded high yields and high tacticity (>70%) syndiotactic polypropylene.[4]

The discovery of such complexes initiated the enormous interest in metallocenes today. The activity of these catalysts can be high, and they have proven to be very versatile in the ready production of atactic, isotactic, hemiisotactic, syndiotactic, and isoblock polymers (Figure 1). These catalysts are also useful for the copolymerization of monomers (e.g., C₄ + α-olefins, cyclic olefins) which show poor activity with Ziegler-Natta catalysts. Moreover, the "single site" nature of metallocenes lends these catalysts towards detailed investigation of their behavior. The synthetic variability of the ligand environment for metallocene catalysts holds promise for the rational design and tailoring of metallocene catalyst systems.

This work describes some of the key features in stereospecific polymerization of propylene using metallocene catalysts, with an emphasis on syndiospecific polypropylene, sPP.

ADVANTAGES OF METALLOCENE FOR sPP

Prior to the development of the iPr(CpFlu)ZrCl$_2$/MAO catalyst system, sPP was known only as an extractable byproduct or impurity in iPP polymerization using first generation Ziegler-Natta catalysts. Later, soluble vanadium systems were discovered which produced sPP but only at low temperature and with poor microtacticity. Such limitations led to the belief that, like atactic polypropylene, the commercial applications of sPP were limited.

The iPr(CpFlu)ZrCl$_2$ catalyst system opened the possibility of producing a highly stereoregular sPP polymer and culminated in Fina's plant trial to produce sPP resin in a commercial facility.

ORIGIN OF STEREOSELECTIVITY

The origin of stereospecificity for metallocene catalysts[6] such as *rac*-ethylene-*bis*-indenyl and iPr(CpFlu)ZrCl$_2$ stems from the π-face enantioselectivity of the active complexes together (in the most general case) with a continuous chain migration mechanism. Following the activation with MAO or another ionizing reagent (see above), the two chloride ligands in the metallocene are replaced by a monomer coordination site and a nascent polymer chain. Propagation by *cis*-migratory insertion of the polymer chain to a coordinated monomer (that is, a chain migration) exchanges the sites of monomer coordination and the polymer chain (now extended by one monomer unit). In this way, the two possible active sites ideally alternate with each insertion. In the "enantiomorphic, site control mechanism" each active site stereochemically controls the π-coordination of the incoming prochiral propylene monomer (Figure 2) or the conformation of the growing polymer chain which in turn controls the p-coordination of the monomer. This determines the stereochemistry of insertion of each monomer unit into the growing polymer chain. Whether the stereochemistry of the polymer chain or the monomer is first controlled by the metallocene is irrelevant. The important point is that the enantioface selectivity is ultimately controlled by the stereochemistry of the metallocene.

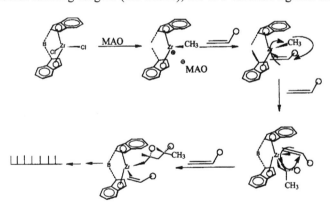

Figure **2** Propagation Mechanism for Isospecific Metallocene

Isospecific metallocenes such as *rac*-(C$_2$H$_4$)(Ind)$_2$ZrCl$_2$ are stereorigid due to the bridging group (that is, free rotation of the η^5-indenyl ligand is prevented) and belong to the C$_2$ symmetry group. Being dissymmetric, these complexes are chiral, but the two active sites within each complex are stereochemically equivalent due to C$_2$ symmetry which maintains the same enantioface selectivity for propylene at both possible sites of coordination. In Figure 2, both monomer coordination steps depict the propylene methyl group directed away from the backbone (i.e., for 1,2 insertion) and away from the "cis" C$_6$ indenyl ring. These successive coordinations both take place at the propylene **si** enantioface as depicted. In this example, it is important to note that only the chiral *rac* isomer produces iPP, the achiral *meso* isomer only produces atactic polypropylene.

In contrast, the syndiospecific metallocene iPr(CpFlu)ZrCl$_2$ belongs to the Cs symmetry group where the bridge carbon atom and the zirconium atom lie in the plane of symmetry which bisects the Cl-Zr-Cl. In the active complex, e.g., iPr(CpFlu)ZrP$^+$ (P=polymer chain), the symmetry is lost and the molecule is chiral but the two possible enantiomers are related by the mirror plane. Therefore, the two possible active sites (one on each side of the mirror plane) and the enantioface selectivity of each site are stereochemically inverted. Given a *cis* ligand migratory insertion mechanism, these sites will alternate when there is a monomer insertion and the site of monomer coordination moves across the mirror plane. In this way, the preferred enantioface coordination of propylene alternates and in turn, alternates the stereochemistry of each incorporation into the polymer

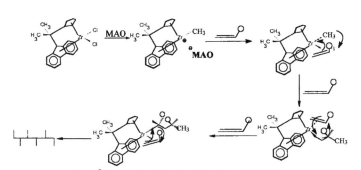

Figure 3: Propagation Mechanism for Syndiospecific Metallocene

chain to produce sPP.[7] In the sPP example, both monomer coordination steps again depict the propylene methyl group directed away from the backbone and away from the "cis" C$_6$ fluorenyl ring, but one coordination takes place at the **si** enantioface and the next at the **re** enantioface as depicted (Figure 3).

However, the important detail to note here is not the symmetry of the complex. In the most general case, the tacticity of the polymer chain produced depends only on the *sequence of enantioface selectivity*. It is important to note that if propagation occurred from only one side in the syndiospecific metallocene (i.e., no site alternation by chain migration) isotactic polypropylene would be produced. Discussed below is an example where syndiospecific iPr(CpFlu)ZrCl$_2$ can be modified to produce hemi-isotactic and isotactic polymers where the complex has no symmetry.

METALLOCENE VARIATIONS

For iPr(CpFlu)ZrCl$_2$ (complex 2), substituents on the Cp ligand position distal to the bridge will clearly have some contact with incoming propylene monomer or the growing polymer chain. The degree of enantioface selectivity can be affected. Beginning with iPr(CpFlu)ZrCl$_2$ the addition of a methyl group (complex 4)[11] causes the catalyst system to produce hemi-isotactic polypropylene. If the methyl group

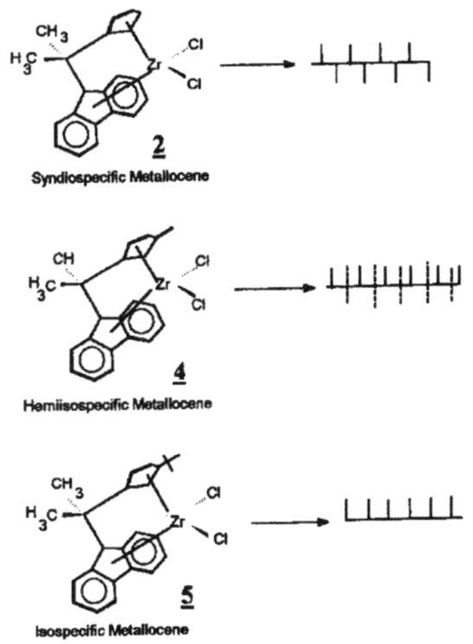

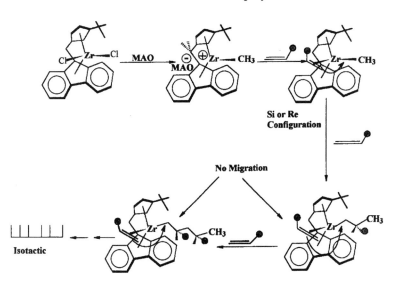

2
Syndiospecific Metallocene

4
Hemiisospecific Metallocene

5
Isospecific Metallocene

Figure 4: Methyl (4) and t-butyl (5) substitutions
on the Cp distal position for iPr[Cp-Flu]ZrCl$_2$ (2)

is in turn substituted with a bulky t-butyl group (complex 5),[12] an isotactic polymer is obtained (Figure 4).

With both substitutions (Me[11] and t-butyl[12]), the syndiospecific metallocene changes from C$_s$ to C$_1$ symmetry (this means 4 and 5 no longer contain mirror plane symmetry). But in each case, the tacticity of the polymer produced can be empirically explained by considering the sequence of enantioface selectivity. For the methyl complex, one coordination site will be little influenced by the methyl group and should have good monomer enantioface selectivity as for the analogous site in iPr(CpFlu)ZrCl$_2$. However, following migratory insertion, the next coordination site (on the other side of the metal atom) will necessarily be influenced by the methyl group. The simplest explanation is that steric contact associated with the methyl group decreases enantioface selectivity at the adjacent coordination site. Therefore, every other monomer is inserted into the polymer chain stereoselectivity at the site not influenced by methyl creating the alternant isotactic character of the hemi-isotactic structure. The alternant atactic character comes from the loss of enantioselectivity for coordination adjacent to the methyl group.

The situation may be clearer in the case of t-butyl complex (see Figure 5). The authors as well as Razavi and Ewen (both of whom reported a similar mechanism at the

**Si or Re
Configuration**

No Migration

Isotactic

Figure 5: t-butyl complexes.

Table 1. Effect of Cp substitution in iPr(CpFlu)ZrCl$_2$

Metallocene	%rrrr	%mmmm
iPr(CpFlu)ZrCl$_2$ (2)	84	-
iPR(Me-CpFlu)ZrCl$_2$ (4)	18	24
iPr(t-Bu-CpFlu)ZrCl$_2$ (5)	-	78

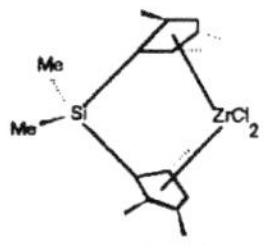

2,3,5-Me$_3$, 2',4',5'-Me$_3$
6

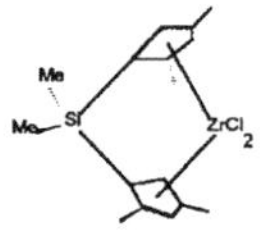

2,4-Me$_2$ 3',5'-Me$_2$
7

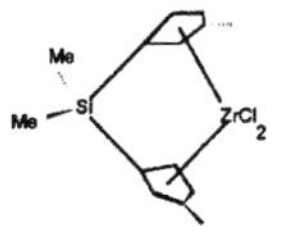

3-Me, 4'-Me
8

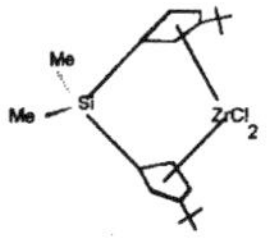

3-t-But, 4'-t-But
9

Table 2

M	Rac/Meso	R$_n$	R$_m$	Activity^a	M$_w$/1000	mp °C	%mmmm
Zr	85/15	2,3,5-Me$_3$	2',4',5'-Me$_3$	1.59	33.9	162.0	97.7
	94/16	2,4-Me$_2$	3',5'-Me$_2$	11.1	86.5	160.4	97.1
	88/12	3-Me	4'-Me	16.3	13.7	147.8	93.4
		2,3,5-Me$_3$	H	7.35	15.5	85.1	49.9
		2,4-Me$_2$	H	5.23	10.6	68.9	41.0
		3-Me	H	6.69	6.6	79.5	58.4
	73/27	3-t-But	4'-t-But	0.31	9.6	149.9	93.4
		3-t-But	H	7.94	4.3	125.7	77.9
Hf	93/7	2,3,5-Me$_3$	2',4',5'-Me$_3$	0.30	256.1	162.8	98.7
		2,4-Me$_2$	3',5'-Me$_2$	0.14	139.2	162.4	98.5
		3-Me	4'-Me	1.61	66.8	148.2	
		3-Me	H	0.06	52.8	103.8	
	26/74	3-t-But	4'-t-But	0.03	17.1	157.4	

^{a} 9 kg PP/mmole catalyst xh

1994 Ziegler conference) believe that the steric bulk of the t-butyl group prevents the migration of the polymer chain thus causing the incorporation of monomer to occur from only one side of the complex.

Therefore monomer is coordinated at a single enantioselective site. Each monomer unit is incorporated into the polymer chain with the same stereochemistry and isotactic polypropylene is produced.

There are also examples of Cp substitution in C_2 symmetric isospecific metallocenes. Miya, *et al.*[8] have reported a series of isospecific catalysts of the formula $[Me_2Si(R_n-C_5H_{4-n})-(R_m-C_5H_{4-m})MCl_2$ (M=Zr,Hf, complexes 6-9). Ligands with methyl groups at the 2- or 5- positions on both Cp groups (proximal to the Si-bridge) afforded the highest melting point polymers but in comparison ligands with methyl groups only at the 3- and 4- positions (distal to the Si bridge) significantly lower the melting point and molecular weight. Substitution of only one Cp group (e.g., R_n=Me, R_m=H), destroys the C_2 symmetry of the complex to yield significantly lower melting points.

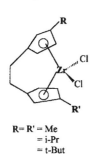

R= R' = Me
    = i-Pr
    = t-But

10

Increasing the steric bulk of the Cp substituents also tends to lower the polymer molecular weight.

The influence of the distal substituents is, however, largely dependent upon the nature of the bridge. In sharp contrast to Miya's work, for a similar series of ethylene-bridged metallocenes, $Et(R-Cp)_2ZrCl_2$ (R=Me, iPr, t-But), substituted distal to the bridge (Table 3 - complex 10). Collins, *et al.*[9] have shown that isotacticity increases with the steric bulk of the R group. However, *both* distal positions (also called β-positions) cannot be substituted since Fina has shown that the metallocene *rac*-Et(3-Me-Ind)$_2$ZrCl$_2$ produces aspecific polypropylene.

Spaleck *et al.*[10] have demonstrated that substituted bis-indenyl complexes, $Me_2Si(R,R'-Ind)_2ZrCl_2$ (complex 11) afford a high molecular weight

Table 3

M	rac/meso	R_n	R_m	$M_w/1000$	mp °C	%mmmm
Zr	55/45	3-Me	4'-Me	19.6	133	92.2
	46/54	3-iPr	4'-iPr	19.4	136	94.6
	42/58	3-t-But	4'-t-But	17.4	141	97.6

Table 4

M	2-	4	activity	$M_w/1000$	mp °C	%mmmm
Zr	H	H	190	36	137	81.7
	Me	H	99	195	145	88.5
	Me	iPr	245	213	150	88.6

polymer when the indenyl groups contain an alkyl group proximal to the bridge (indenyl 2-position) and isopropyl substitution at the 4-position.

For the isospecific metallocenes discussed thus far, only Miya and Fina's iPr(t-But-Cp,Flu)ZrCl$_2$ (4) have given examples of complexes which cannot be C$_2$-symmetric. In Miya's complexes the two η^5-Cp moieties; are substituted differently. However, for iPr(CpFlu)ZrCl$_2$ one η^5-ligand moiety is fluorenyl which creates new opportunities for substitution and variation of the ligand. In particular, the fluorenyl 2- and 7-positions appear to be well removed from any direct influence on the active coordination sites for polymerization; therefore, any substitution at these positions will be demonstrative of the influence of an electronic effect (it is difficult to be sure that steric influence is not playing a role).

Substitution at the fluorenyl 2- and 7-positions (Table 5) with more electron-donating ligands (e.g., t-butyl)[13] is consistent with an increase in the syndiotacticity of the polymer as evidenced by an increase in the melting point. But electron-donating groups (e.g., alkylamino and alkoxy) which can coordinate with acidic species such as MAO have the opposite effect. In contrast, at the fluorenyl 4-position, this coordination is believed to be sterically disfavored and the electron-donating effect increases the melting point. However, weakly electron donating substituents (e.g. t-butyl) have little effect in the 4-position.

Table 5

Fluorenyl Position	R=	Activity/1000	mp °C	M$_w$/1000
2,7	H	320	138	100
	F	98	132	83
	Me		135	
	t-But	116	141	75
4	Me		112	42
2	OMe		117	69
4	OMe		136	189
4,5	Phenanthryl		113	

The influence of the ligand environment adjacent to the active coordination sites is demonstrated by the addition of an annelated 6-membered ring at the fluorenyl 4- and 5-positions in iPr(Cp-Flu)ZrCl$_2$ (i.e., to give a phenanthrenyl structure). This significantly lowers the melting point of the resin from 138 to 113°C.

Though removed from any direct influences on the metal center's active coordination sites for polymerization, the nature of the bridge in these *ansa*-metallocenes is important. The dimethylcarbyl (isopropylidene) bridge in (e.g.) Ipr(CpFlu)ZrCl$_2$ is the shortest bridge. Replacement with a dimethylsilyl or ethylene (-CH$_2$-CH$_2$-) group lengthens the bridge.[5] Having less "flexibility", the one atom silyl bridge[5] is believed to impose greater rigidity to the ligand over the two atom ethyl bridge. A direct comparison of isospecific metallocenes differing only in the bridging moiety shows that the silyl-bridged analogues afford higher melting points (Table 6).

Table 6

Complex	Activity/1000	mp °C	%mmmm
Et[Ind]$_2$ZrCl$_2$	188	132	78.5
Me$_2$Si[Ind]$_2$ZrCl$_2$	190	137	81.7

Table 7

Complex	Activity/1000	mp °C	M$_w$/1000
iPr[Cp-Flu]ZrCl$_2$	320	138	100
Et$_2$Si[Cp-Flu]ZrCl$_2$	8.3	107	286
Me$_2$C[Cp-Phen]ZrCl$_2$	23	119	155
Me$_2$Si[Cp-Phen]ZrCl$_2$	2.2	none	65

The syndiospecific iPr(CpFlu)ZrCl$_2$ behaves differently. In this case, increasing the bridge length with a silyl bridge leads to a decrease in melting points (Table 7).

Clearly, in these examples, the nature of the influence of the bridge differs between isospecific and syndiospecific metallocenes.

Table 8[10]

Complex	Activity/1000	mp °C	M$_w$/1000
iPr[Cp-Flu]ZrCl$_2$	320	138	100
Ph$_2$C[Cp-Flu]ZrCl$_2$	108	137	507

In addition to the bridging atom, the bridge substituents also play a role, but in this case they influence the final molecular weight of the resin. The larger substituents, as shown in Table 8, allow the metallocene in producing higher molecular weight resins.

Table 9

Complex/Metal	Yield	M_w/1000	% rrrr
iPr[Cp-Flu]ZrCl$_2$	177	69	81
iPr[Cp-Flu]HfCl$_2$	27	777	74

In the first report on syndiospecific polymerization[4] using iPr(CpFlu)ZrCl$_2$, it was also reported that the hafnium analogue afforded a lower polymer yield, lower syndiotacticity, but significantly higher molecular weights (Table 9).

Table 10

Metal	R_n	R_m	Activ/1000	M_w/1000	mp °C	% 4m
Zr	2,4-Me$_2$	3',5'-Me$_2$	11.1	86.5	160.4	97.1
Hf	2,4-Me$_2$	3',5'-Me$_2$	0.14	139.2	162.4	98.5

The Miya et al.[8] isospecific complexes (above) also afforded lower yields and higher molecular weights when hafnium was used, but with similar or slightly higher tacticity. Since these hafnium complex-produced polymers were more stereoregular, higher melting points were exhibited (Table 10).

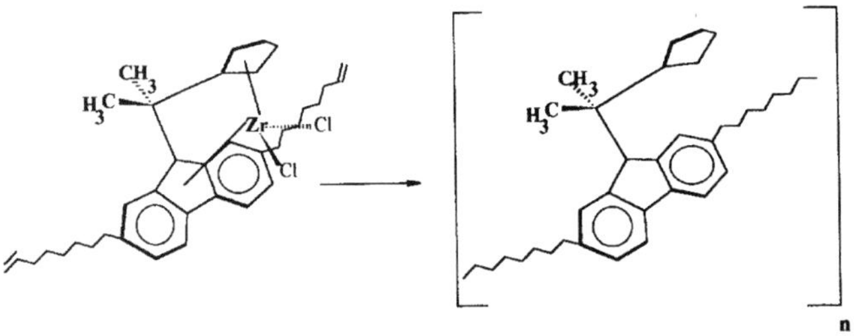

CATALYST SUPPORTING

The polymer fluff or flake produced by metallocene catalysts typically has a low bulk density (i.e., high specific volume) and would be difficult to handle using existing production equipment. Moreover, metallocenes can lead to "fouling" during the polymerization where a polymer deposit builds up in the reactor. Fluff morphology is, to a large extent, determined by the "replicant effect" and the catalyst morphology. In order to produce a more readily handled polymer fluff (i.e. different morphology) and eliminate fouling, some means of changing the catalyst morphology is required. Since the "morphology" of the metallocene or the MAO cannot be changed, efforts have been made to place the catalyst on a support. Conventional supports and approaches can be used, but another possibility is to construct a "self-supporting" metallocene. In another example of the synthetic versatility of the metallocene ligand, work was carried out at Fina to substitute polymerizable α-olefin groups such as octenyl at 2- and 7-positions in the fluorenyl group of iPr(CpFlu)ZrCl₂. Polymerization of these pendant olefins into polymer chains during a prepolymerization of the catalyst will result in the *in situ* formation of a polymeric catalyst system.[14] This approach was consistent with an improvement of the polymer bulk density with some reduction in the polymer melting point and catalyst activity (Table 11).

Table 11

Metallocene	Activity/1000	mp °C	bd
iPr[Cp-Flu]ZrCl₂	150	138	0.16
iPr[Cp-(2,7(CH₂)₆CHCH₂)₂Flu]ZrCl₂	45	128	0.27

COCATALYST

While the focus of the paper, thus far, has been on the structural modifications of the metallocenes and their influence on polymer properties, the nature of the cocatalyst can also play a role. Table 12 compares the effect of three cocatalysts among which a new one that was recently developed by Fina oil and chemical.[15]

Table 12

Cocatalyst	Metallocene	Pol. temp. °C	mp °C
[Ph₃C] [B(Phf)₄]	Et[Ind]ZrCl₂	60	135
MAO	same	60	135
[Ph₃C] [Al(Phf)₄]	same	60	141
[Ph₃C] [B)(Phf)₄]	iPr[Cp-Flu]ZrCl₂	50	131
MAO	same	50	140
[Ph₃C] [Al(Phf)₄]	same	50	142

As shown in Table 12, Fina developed cocatalyst system is superior to those reported in the literature.

COPOLYMERIZATIONS

Table 13

Catalyst	% C$_2$ Gaseous Feed	% C$_2$ Polymer	Randomness Factor	T$_m$ °C	M$_w$/1000
Ziegler-Natta	4	3	2.6	144	-
	6	4.5	1.7	138	493
	8	5.9	1.5	134	-
Et[Ind]$_2$ZrCl$_2$	2	1.2	very high	121	-
	4	2.4	120	113	31.9
	6	3.3	11.5	103	-
iPr[Cp-Flu]ZrCl$_2$	2	1.5	very high	121	-
	4	2.9	7.5	102	372
	6	4.3	4.6	92	-
Ph$_2$C[Cp-Flu]ZrCl$_2$	2	3.5	1.1	115	-
	4	3.6	3.7	99	372
	6	5.4	3.4	-	-
Supported-Metcn.	2	4.3	1.0	113	-
	4	6	1.2	100	-
	6	5.2	1.6	100	-

The versatility of metallocenes in homopolymerization raises the possibility of novel or improved copolymerization. Metallocenes which differ in the stereoregularity of the propylene homopolymer produced could have markedly different selectivities for the incorporation of comonomer (Table 13).

Random copolymers produced using conventional isospecific Ziegler-Natta catalysts are important commercial products. Ethylene incorporation disrupts the stereoregularity of the polymer to disrupt crystallization. Some consequences are a lower melting point for the resin, lower moduli, and greater impact resistance. The nature of the ethylene incorporation is important, and conventional Ziegler-Natta catalysts can give "blocky" incorporation at high ethylene levels. In contrast, metallocene catalysts afford significantly more random incorporation as demonstrated by an increase in a randomness, or R-factor for a given level of ethylene incorporation.[10]

While Ziegler-Natta catalysts are inefficient for the incorporation of higher α-olefin, metallocenes can be very efficient. However, the nature of the metallocene changes when it is supported. Compared to the unsupported metallocenes, the supported example above exhibits a high level of ethylene incorporation with randomness factors lower than Ziegler-Natta catalyst. Moreover, the single site nature of metallocenes will produce a more uniform composition of copolymer, unlike Ziegler-Natta catalysts which will produce a mixture of compositions due to the heterogeneity of active sites.

CONCLUSION

Metallocene catalysts have opened a new era for olefin polymerization with the promise of rational catalyst design to optimize new systems or to afford previously unknown polymer structures. The examples given in this paper already demonstrate that there is a great deal of flexibility and much is being learned about the relationship between the metallocene structure and polymerization performance.

REFERENCES

1.   a. D. S. Breslow and N. R. NewGuy, *JACS*, **79**, 5072 (1957).
      b. D. S. Breslow, U.S. **Patent 2 827 446** (1958).
2.   A. Anderson, H. G. Cordes, J. Herwig, W. Kaminsky, A. Mark, R. Mottweiles, J. H. Sum, and H. J. Vollmes, *Angew. Chem. Int. Ed. Engl.*, **15**, 630 (1976).
3.   Brintzinger *et al.*, *Angew. Chem. Int. Ed. Engl.*, **24**, 507 (1985).
4.   a. Fina Technology, **U.S. 4 892 851**.
      b. Fina Technology, **U.S. 5 334 677**.
      c. J. Ewen *et al.*, *JACS*, **110**, 6255 (1988).
5.   a. J. A. Ewen, *JACS*, **106**, 6355 (1984).
      b. J. Ewen *et al.*, *Makromol. Chem. Macromol. Symp.*, **48/49**, 253 (1991).
      c. Fina Technology, **E.P. Appl. 0284 708**.
6.   P. Corradini *et al*, *Macromolecules*, **24**, 1784 (1991).
7.   J. Ewen *et al.*, *Proc. Int'l Symposium on Recent Developments in Olefin Polymerization Catalysts*, Tokyo, **56**, 439 (1989).
8.   S. Miya *et al.*, **Catalytic Olefin Polymerization**, Vol. S6, p. 531, Eds. T. Keii and K. Soga, *Elsevier*, 1990.
9.   S. Collins *et al.*, *Organometallics*, **10**, 2349 (1991).
10.   a. Spaleck *et al.*, *Angew. Chem. Int. Ed. Engl.*, **31(11)** 1347 (1992).
      b. Spaleck *et al.*, *Organometallics*, **13**, 954 (1994).
11.   Fina Technology, **U.S. 5 036 034**.
12.   Fina Technology, **E.P. Aappl. 0537 130**.
13.   a. E. S. Shamshoum and D. Rauscher, *MetCon 93*, Houston, TX, May, 1993.
      b. Reddy *et al.*, Fina Technology, **EP-0577 581-A2**.
14.   B. R. Reddy and E. S. Shamshoum, Fina Technology, **U.S. 5 308 818**.
15.   Fina Technology Japanese Patent L-O-P Public, # 220119/1994, Aug. 9, 1994.

UNIPOL® Gas Phase Copolymerization with SSC Metallocene Technology

Frederick J. Karol

Union Carbide Corporation, Polymers Research and Development, Bound Brook, NJ 08805

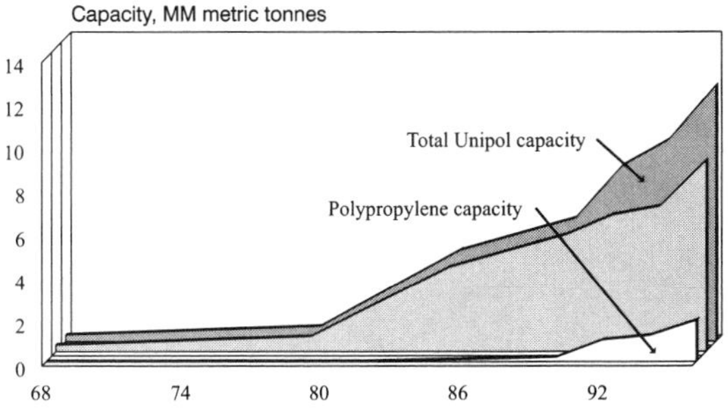

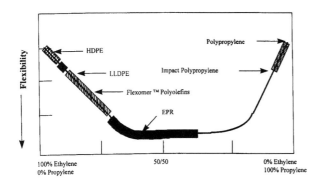

Figure 1. Total worldwide UNIPOL capacity.

Figure 2. The versatility of the gas-phase process.

INTRODUCTION

Catalysis in gas phase reactions for the production of olefin polymers is well known in the polyolefin industry. The growth of the UNIPOL® process, initially commercialized in 1968, has truly been awesome (Figure 1).[1,2] Developments in the UNIPOL® Process for polyethylene, EPR/EPDM, and polypropylene continue to demonstrate the broad versatility of the process (Figure 2). The emergence of metallocene catalysis in many laboratories around the world in the last decade has added yet another significant catalytic tool for manipulating and controlling the molecular framework of polyolefin polymers. Selection of the appropriate metallocene catalyst, with control of ligand environment at the active site, continues to provide the basis of improved process operations and unique product opportunities. Recent attention has focused on the UNIPOL® gas phase process using proprietary metallocene technology.[3]

With over 900 reactor years of UNIPOL® operations around the world, a broad base of proprietary technology and know-how has accumulated to enhance process operability and reactor continuity. These learnings form the basis for developments with new catalyst compositions in the UNIPOL® Process.

Catalysts used in the UNIPOL® Process must satisfy numerous requirements for efficient process operations and as a means to provide unique product opportunities. The continuous demand for improved product characteristics has required constant, new developments in catalyst technology. This paper describes the polymerization characteristics and certain polymer property indicators for a number of "single-site" metallocene catalysts, SSC, examined in a gas phase process for ethylene, α-olefin copolymerization.[4]

The ability to manipulate long chain branching in the polyethylene chain adds another dimension to the scope and versatility of the UNIPOL® Process.[5]

CATALYSTS MAKE IT POSSIBLE

$$\frac{d[M_1]}{d[M_2]} = \frac{[M_1]}{[M_2]} \frac{r_1[M_1]+[M_2]}{[M_1]+r_2[M_2]}$$

$$when\ r_1 = 1/r$$

$$\frac{d[M_1]}{d[M_2]} = r_1 \frac{[M_1]}{[M_2]}$$

Figure 3. Copolymerization equation.

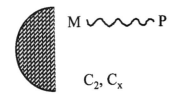

M $\sim\!\!\sim\!\!\sim$ P

C_2, C_x

Figure 4. Probe for nature of active site.

Table 1. Reactive ratios, C_2/C_6, for metallocene catalysts

Metallocene	r_1	r_2	$r_1\bullet r_2$
[Ind]$_2$ZrCl$_2$(InZ)	88	0.005	0.44
Cp$_2$ZrCl$_2$	62	0.003	0.19
(CH$_3$)$_2$SiCp$_2$ZrCl$_2$	33	0.027	0.89
C$_2$H$_4$[Ind]$_2$ZrCl$_2$(EBIZ)	31	0.013	0.40
Me$_2$Si[Ind]$_2$ZrCl$_2$	(10.3)	(0.022)	-
Me$_2$C[FluCp]ZrCl$_2$	(3.2)[1]	(0.065)	-

Activities and overall productivities with metallocene catalysts can be extremely high in the gas phase leading to a residual transition metal content of ppm or less in the polymer. In many cases, the catalyst sustains a high level of activity for many hours, with little evidence of catalyst decay. However, changes in catalyst compositions and process conditions can alter such a kinetic profile. With methylalumoxane as an activator, alumoxane/zirconium molar ratios may be 300/1 or lower depending upon the specific metallocene used and reaction conditions under which polymerization takes place. Selection of the final catalyst composition reflects a consideration of both polymerization rates in the reactor and the costs associated with a specific catalyst.

Many members of the metallocene family display an outstanding capacity to incorporate comonomer in ethylene, α-olefin copolymerization (Figures 3, 4, and Tables 1-3). These catalysts are among the best of all coordination catalysts for copolymerization.[6] One way to describe such a capacity

Table 2. Copolymerization observations with metallocene catalysts

- Copolymerization depends upon ligand structure of catalyst
- Bridged catalysts are usually better at incorporating comonomer
- α-olefins generally show an accelerating rate effect in ethylene copolymerization

Table 3. Key copolymerization results

- Copolymerization results consistent with ion-pair model
- Each zirconocene catalyst displays its own characteristics
- Copolymer unsaturation and chain branching useful probes for changes in the nature of the active site
- Reactivity ratios and copolymerization kinetics consistent with first-order Markov statistics

for copolymerization of ethylene with 1-hexene is to examine the C_6/C_2 molar ratio necessary to be fed to the reactor to achieve a copolymer of 0.92 g/cc density. Such ratios for catalysts with outstanding copolymerization capabilities would be in the range of C_6/C_2=0.004-0.005. Comparable ratios in ethylene, 1-butene copolymerizations were C_4/C_2=0.02-0.03. With a high activity Ziegler-Natta titanium catalyst such ratios would be C_6/C_2=0.12 and C_4/C_2=0.35. For a chromium oxide catalyst a typical value would be C_4/C_2=0.083.

Polymer molecular weights produced with metallocene catalysts depend upon the specific family of metallocene catalysts (e.g., bridged or unbridged) as well as the specific ligand composition. Many of the unbridged metallocenes have a tendency to produce ethylene copolymers of lower molecular weight (melt index > 1). Bridged metallocene catalysts typically provide a means to reach higher molecular weights. Precise control of ligand environment at the metal center in unbridged metallocenes can provide a route to higher molecular weight. Bridged metallocene catalysts typically provide a means to reach higher molecular weights. In copolymerization with α-olefins, chain transfer to comonomer can be a significant kinetic process leading to lower molecular weights for the resultant copolymers (Scheme A).[6]

Metallocene catalysts have been identified as single-site catalysts (SSC) capable of providing polyethylenes of narrow molecular weight distribution (polydispersity ~ 2). Such catalysts run in a gas phase process also produce polyethylenes of narrow molecular weight distribution. It is not necessary for these single-site catalysts to remain in solution in order to provide polyethylenes of narrow molecular weight distributions. However, there are other metallocene catalysts that behave differently as they produce polyethylenes with molecular weight distributions as high as 4-5 (Table 4). In-situ routes to more than one type of active site during the polymerization process may be one explanation for such a broadening of molecular weight distribution.

Scheme A. Chain end structures by β-hydride elimination

Table 4. Comparison of MWD and CCD for four metallocene catalysts in gas phase

Catalyst	M_w/M_n	Lw/Ln
A	2.1-2.3	3.0-3.1
B	2.3-2.6	3.6-4.5
C	2.8-3.7	1.5-2.2
D	4.0-5.3	1.4-2.4

Metallocene catalysts can produce ethylene, α-olefin copolymers of narrow comonomer compositional distribution. Such a capability of these catalysts have been used to differentiate metallocene catalysts from traditional titanium-based Ziegler-Natta catalysts. In gas phase copolymerizations with metallocene catalysts, the degree of uniformity in compositional distribution may be controlled over a wide range depending upon the catalyst and process conditions under investigation. Such control of compositional and molecular weight distributions by selection of the appropriate metallocene catalyst provides the basis of expanding the range of product opportunities.

Temperature Rising Elution Fractionation, TREF, was used to provide information on comonomer compositional distribution. From the TREF temperature data, the branch frequency may be obtained for a given comonomer. Consequently, the main chain lengths between branches, expressed as Lw and Ln may be calculated. Lw is the weight average chain length between branches:

$$Lw = \sum iWiLi$$

and Ln is the number average chain length between branches:

$$Ln = 1/\sum i(Wi/Li)$$

wherein Wi is the weight fraction of the component i having an average backbone spacing Li between two adjacent branch points. A Crystallizable Chain Length Distribution, CCLD, Lw/Ln, of less than about 3, preferably less than about 2, is indicative of narrow comonomer compositional distribution.

LONG CHAIN BRANCHING

Polyethylenes containing long chain branches possess high melt strength and exhibit low viscosity under high shear conditions, permitting high processing rates. Early work to introduce and take advantage of long chain branching in polyethylene was carried out using low pressure, chromium oxide catalysis.[7] Introduction of long chain branches into polyethylene can occur when vinyl-terminated long chain polymer molecules copolymerize with ethylene monomer during polymerization. Such unsaturated long chain polymer molecules are not expected to be in the vapor state under the reaction conditions typically employed in gas phase processes, but rather are expected to be solids or high boiling liquids having lower mobility. Accordingly, it was believed that serious difficulties might be encountered in attempting to introduce long chain branches into polyethylene in a gas phase polymerization, and that it might not be possible to accomplish this under industrially acceptable conditions. It has been shown that long chain branches can be introduced into polyethylene in a gas phase process using certain bridged metallocene catalysts and appropriate reaction conditions Table 5.[5,8]

Table 5. introduction of long chain branching in C_2/C_6 copolymers bridged metallocene catalysts

Cat E/Cat F	1:1	5:1	3:1	0.1
Reaction Temp., °C	85	85	65	85
Density, g/cc	0.933	0.940	0.908	0.920
LCB/1000 Carbon atoms	0.3	2.0	0.0	0.8

In gas phase polymerizations the polymer particle may be viewed as a microreactor within which polymer chains are created. In the process of polymer chain transfer, some polymer chains will be terminated and generate terminal vinyl groups. Polymer chains with these vinyl groups remain in the vicinity of the active sites. Such a presence creates an opportunity for the vinyl-terminated polymer chain to incorporate as a comonomer into a growing polymer chain within the polymer particle (Figure 5). Such incorporation provides a means of introducing long chain branching. Proximity effects of this type are particularly significant in gas phase polymerizations where polymer chains and active sites are contained within the growing polymer particle.

PROPERTY INDICATORS AND CATALYST KINETICS

Use of metallocene catalysis in the UNIPOL® Process provides the opportunity to expand the product opportunities available from the process. Metallocene polyethylene products available from the

Gas Phase Polymerization

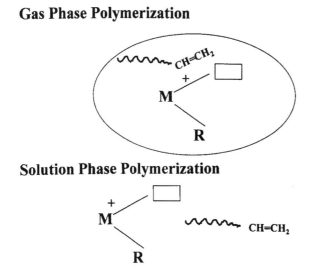

Solution Phase Polymerization

Figure 5. Proximity effect for long chain branching in particle microreactor.

UNIPOL® Process will include high performance LLDPEs for blown film applications.[3] Catalyst compositions from different catalyst families can provide a significant route to polymer compositions showing different end-use properties. An arsenal of characterization methods is currently used to define the molecular and rheological features of the polyolefins produced with metallocene catalysts. Such an arsenal includes size exclusion chromatography, SEC, TREF, rheological relaxation spectrum measurements, etc. Various measurements to determine the amount of long chain branching in polyethylenes have developed as a result of the introduction of metallocene catalysts.

The breadth of polymer molecular weight distribution is usually associated with an active sites (>2) present in the polymerizing system. A broad spectrum of active sites species would be expected to produce a polymer of broad molecular weight distribution. This effect is caused by differences in rates of chain propagation and chain transfer for the individual active centers. The dominant chain transfer process with metallocene catalysts is β-hydride elimination (Scheme A).

Comonomer compositional distribution between polymer chains relates to relative monomer/comonomer incorporation rates during chain propagation. Hence it is possible by selection of the appropriate catalyst to manipulate, independently, the breadth of polymer molecular weight distribution and comonomer compositional distribution.

The effect of catalyst parameters on ethylene, α-olefin copolymerization can best be interpreted in terms of an ion-pair structure for numerous metallocenes studied (Scheme B). In slurry and solution polymerizations, ligand environment at the active ion-pair site is believed to influence the degree of ion-pairing from intimate contact ion-pairs to ones that behave like solvent-separated ion-pairs.[6]

In gas phase polymerizations, the ion-pair active sites become immobilized. Such an immobilization minimizes various possible equilibria of the active sites with excess alumoxane. Such an effect with some metallocene catalysts leads to gas phase catalysts that are somewhat more homogeneous, tighter ion-pairs than for the centers in the corresponding soluble catalysts (Table 6).

COMMERCIAL IMPACT AND OUTLOOK FOR THE FUTURE

The importance of ligand modification in metallocene catalysis has already made the 1990s the "decade of the ligand" in polyolefins. The unprecedented ability of these catalysts to regulate and control the molecular architecture of individual polymer molecules provides the foundation for rational control of polymer properties and for synthesis of entirely new polyolefin compositions. The combination of these new catalysts and their use in the UNIPOL® Process creates exciting new opportunities in product design and process improvements (Table 7).

$$Cp_2MX_2 + nMAO \rightleftharpoons [Cp_2M\text{-}CH_3]^+ \bullet X \bullet [X\text{-}MAO]^- \bullet [MAO]_{n-1}$$

Contact Ion Pair

$$[Cp_2M\text{-}CH_3]^+ \bullet X \bullet [X\text{-}MAO]^- \bullet [MAO]_{n-1} \rightleftharpoons$$

$$[Cp_2Zr\text{-}CH_3]^+ \| \| [X_2\text{-}MAO]^- \bullet [MAO]_{n-1}$$

Solvent Separated Ion Pair

Scheme B. Ion-pair model.

Table 6. Effect of heterogenization on activity, M_w, PDI, BBF, and TCE of an unbridged metallocene catalyzed/1-hexane copolymerization

Al/Zr	Activity[1]	M_w	PDI	BBF	TCE
250	26,737	110,292	2.2	6.6	0.0261
500	60,858	96,262	2.1	8.0	0.0256
1000	78,105	98,770	2.1	8.2	0.0269
500	75,053	125,768	2.3	-	0.0262

Table 7. Key points

- Outstanding catalyst productivities and comonomer incorporation demonstrated in gas phase for metallocene catalysts
- Homogeneous ethylene copolymers can be produced in UNIPOL process
- Selection of metallocene catalyst composition provides route to control molecular and comonomer distribution

REFERENCES

1.   F. J. Karol, **History of Polyolefins**, Eds. Seymour & Cheng, *Reidel Publ.*, 1996.

2.   F. J. Karol, *Macromol. Symp.*, **89**, 563 (1995).

3.   W. A. Fraser, *Metallocenes'96,* Dusseldorf, Germany, March 6-7, 1996.

4.   **U.S. Patents 5 317 036** (1994**)** and **5 527 752** (1996) to Union Carbide Corporation.

5.   **European Patent Application 0659 773** (Priority Date 12/21/93) to Union Carbide Corporation.

6.   F. J. Karol, S.-C. Kao, E. P. Wasserman, and R. C. Brady, *New J. Chem..*

7.   J. P. Hogan, C. T. Levett, and R. T. Werkman, *SPE Journal*, **23**, 87 (1967).

8.   M. Ohgizawa, M. Takahashi, and N. Kashiwa, *Metallocenes'95*, Houston, Texas, May 17-19, 1995.

High-Molecular-Weight Atactic Polypropylene from Metallocene Catalysts. Influence of Ligand Structure and Polymerization Conditions on Molecular Weight

Luigi Resconi, Fabrizio Piemontesi, and Robert L. Jones

Montell Polyolefins, G. Natta Research Center, P. Le Donegani 12, 44100 Ferrara, Italy

INTRODUCTION

The substitution pattern of the π-Cp ligands of metallocene polymerization catalysts is the major factor influencing both propagation rate and type and extent of chain transfer reactions, that in turn determine the molecular weight of the poly(α-olefin) produced.[1] Other factors that heavily affect the molecular weight are the polymerization temperature and monomer concentration, while the Al/Zr ratio (for MAO activated metallocene catalysts) has in most cases only a minor influence.[2]

We have previously reported the synthesis[3] and characterization[4] of high molecular weight atactic polypropylene (HMW-aPP), a material having unique properties. The catalyst we designed for the synthesis of this polymer is the C_{2v}-symmetric dimethylsilanediylbis(9-fluorenyl) zirconium dichloride (**1**),[3] which is able to produce HMW-aPP with molecular weights in the M_w range from 100,000 to 500,000 at practical (50 to 70°C) polymerization temperatures.

Disadvantages associated with the preparation and use of **1** included insolubility of the complex in common solvents. This feature led to difficulty in obtaining purified catalyst samples. To overcome this problem, the solubility of the complex was increased by 1) substituting methyl groups for chlorine atoms on the zirconium metal center or 2) substituting longer chain alkyl groups for methyl groups on the silicon atom bridging the two fluorenyl ring systems.

The synthesis of $Me_2SiFlu_2ZrMe_2$ was accomplished by treatment of the $Me_2SiFlu_2ZrCl_2$ complex **1** with 2 equivalents of methyl magnesium bromide in an ethereal solution followed by standard work-up.[3] This modification allowed greater purification of the complex which afforded a crystalline material suitable for structural determination.[3] The catalytic properties of the complex were improved by the higher purity, however, the solubility of the complex was not substantially higher compared to the parent **1**.

To further increase the solubility of the catalytic system, the aliphatic character of the complex was increased by substituting butyl groups for methyl groups on the bridging silicon atom. The synthesis of di-n-butylbis(9-fluorenyl)silane and the corresponding (n-Bu)$_2$Si(9-fluorenyl) zirconium dichloride (**2**),[5] its characterization, and polymerization performance are described in this paper.

Alt and coworkers reported the synthesis and ethylene polymerization behavior of two analogous systems, the C_{2v}-symmetric ethylenebis(9-fluorenyl) zirconium dichloride (**3**/MAO) and its hafnium analogue, but did not report the characterization of aPP obtained from them.[6] Rausch and Chien discussed the polymerization of propylene catalyzed by **3** and claimed that **3** produced aPP with very high molecular weights even at relatively high T_p with $\overline{M}_w$ being almost insensitive of this latter parameter.[7]

We have previously discussed the influence of monomer concentration on aPP molecular weight with **1**/MAO.[3] Here we report our results from studies on the influence of polymerization parameters, such as the polymerization temperature and the catalyst/cocatalyst ratio, on the polymerization behavior of **1**/MAO, **2**/MAO, and **3**/MAO, and the non bridged bis(2-methyl-indenyl) zirconium dichloride (**4**)[8] (see Chart 1).

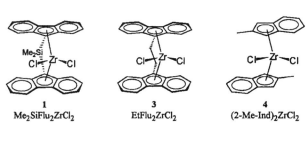

| **1** | **3** | **4** |
| Me$_2$SiFlu$_2$ZrCl$_2$ | EtFlu$_2$ZrCl$_2$ | (2-Me-Ind)$_2$ZrCl$_2$ |

Chart 1

We also discuss the results of our investigation on the influence of hydrogen and ethylene addition on catalyst activity and aPP molecular weight.

RESULTS AND DISCUSSION

INFLUENCE OF POLYMERIZATION TEMPERATURE

We have already shown that **1**/MAO, at 50°C and among a large number of zirconocene/MAO catalysts including **3**, produces aPP with the highest molecular weights.[3] An updated list of polymerization results from aspecific metallocene/MAO catalysts is shown in Table 1.

Polymerization results show an increased activity for the **2**/MAO over the **1**/MAO system.[5] As mentioned above, the insolubility of the Me$_2$SiFlu$_2$ZrCl$_2$ precatalyst does not readily permit the isolation of purified samples. **1** was obtained in 85% purity. Inevitably, LiCl(Et$_2$O)$_x$ which is produced as a byproduct of the reaction between the ligand dianion and the metal tetrachloride, is isolated with the zirconocene complex. Lithium chloride complexed with diethyl ether can lower catalyst activity by reacting with active catalyst sites.

The synthesis and characterization of the more soluble n-Bu$_2$SiFlu$_2$ZrCl$_2$ pre-catalyst **2**,[5] which was recrystallized from CH$_2$Cl$_2$, is reported in the experimental section. The use of diethyl ether as a solvent system instead of tetrahydrofuran is a modification to the previously reported method. The use of a single solvent system, as opposed to the use of both THF and diethylether is preferred.

As the molecular weight is a function of both the rate of chain transfer and the rate of propagation, the influence of polymerization temperature, T_p, was investigated to obtain the energy of activation for propagation versus transfer. Polymerization results for propylene polymerizations at different Al/Zr ratios in liquid monomer with **1**/MAO, **2**/MAO, and **3**/MAO are compared in Table 2.

Ethylenebis(9-fluorenyl) zirconium dichloride (**3**/MAO), even at low Al/Zr ratios, always gives lower molecular weight polypropylenes than **1**/MAO. The lower molecular weights modify the physi-

Table 1. Propylene polymerization with aspecific metallocene/MAO catalysts[a]

Metallocene	Amount µmol	Al/Zr	$g_{pp}/mmol_M h$	$[\eta]^b$	$\overline{M}_w^c$	$\overline{P}_n$
Cp_2ZrCl_2	5.8	1,500	5,700			19^d
Cp_2HfCl_2	5.8	1,400	4,600			137^d
$(MeCp)_2ZrCl_2$	5.9	1,400	24,500			35^d
$Cp^*_2ZrCl_2$	5.8	1,500	3,600			4.5^d
$Cp^*_2HfCl_2$	5.8	1,500	12,600			3.4^d
$Me_2SiCp_2ZrCl_2$	5.7	1,500	9,100			17^d
$Me_2Si(Me_4Cp)_2ZrCl_2$	5.9	1,500	3,700	0.23		316^d
Ind_2ZrCl_2	5.8	1,500	18,000			110^d
$(2\text{-Me-Ind})_2ZrCl_2$	8.8	1,000	10,000	0.29	21,600	220^d
$(2\text{-Me-H}_4Ind)_2ZrCl_2$	9.3	3,000	9,200	0.23	15,800	188^e
$4,7\text{-Me}_2\text{-Ind})_2ZrCl_2$	8.9	1,000	200			376^d
$(4,6\text{-Me}_2\text{-Ind})_2ZrCl_2$	8.9	1,000	400	0.19	12,200	145^e
$(2,4,6,\text{-Me}_3\text{-Ind})_2ZrCl_2$	8.4	1,000	3,300	0.15	8,800	105^e
$EtFlu_2ZrCl_2$	7.7	1,000	17,700	1.15	140,400	1,670
$EtFlu_2HfCl_2$	6.6	1,900	8,800	1.14	138,700	$1,650^e$
$Me_2SiFlu_2ZrCl_2$	6.2	2,300	16,500	2.58	420,900	$5,000^e$
$Me_2SiFlu_2ZrMe_2$	2.7	2,800	29,000	2.10	317,800	$3,780^e$
$Bu_2SiFlu_2ZrCl_2$	1.6	2,000	32,700	2.76	460,400	$5,470^e$
$Bu_2SiFlu_2ZrMe_2$	1.7	2,000	25,300	2.79	320,500	$3,800^e$

[a]general polymerization conditions: 1-L or 2-L stainless-steel autoclave, liquid propylene, 50°C, 1h. Metallocene/MAO solutions aged in toluene (10-20 ml) for 5-10 min at room temperature
[b]intrinsic viscosity measured in THF at 135°C
[c]calculated from the $[\eta]=1.85 \times 10^{-4} M_w^{0.737}$, see ref. 9
[d]number-average polymerization degree from ^1H- NMR, assuming one double bond per polymer chain
[e]number-average polymerization degree from viscosity, assuming $\overline{M}_w/\overline{M}_n=2$

Table 2. Influence of T_p on molecular weight for catalysts 1-3[a]

Catalyst	T_p °C	Al/Zr molar	Visc[b] $[\eta]$ dL/g	Visc[b] $\overline{M}_w^c$	SEC[d] $\overline{M}_w$	SEC[d] $\overline{M}_n$	SEC[d] $\overline{M}_w/\overline{M}_n$
1/M-MAO	10	1,000	5.46	1,162,000	1,253,000	699,000	1.79
1/M-MAO	20	1,000	4.63	929,000	1,070,000	584,000	1.83
1/M-MAO	30	1,000	3.70	685,000	852,000	460,000	1.85
1/M-MAO	40	1,000	3.19	560,000	625,000	271,000	2.31
1/MAO	50	150	2.94	502,000			
1/MAO	50	300	3.26	577,000			
1/MAO	50	600	2.83	476,000			
1/MAO	50	1,000	2.62	429,000			
1/M-MAO	50	1,000	2.56	415,700	595,000	290,000	2.05
1/MAO	50	2,300	2.58	420,900			
1/M-MAO	60	1,000	1.99	295,000	419,000	171,000	2.45
1/MAO	70	2,000	1.09	130,000			
2/M-MAO	30	2,000	3.30	586,700			
2/MAO	50	2,000	2.76	460,400			
2/MAO	70	1,000	1.24	139,900			
2/MAO	70	2,000	1.15	126,300			
3/M-MAO	30	1,000	1.78	253,900	257,600	124,800	2.06
3/M-MAO	40	1,000	1.32	169,200	171,300	77,900	2.20
3/MAO	50	1,000	1.15	140,300			
3/M-MAO	50	1,000	0.87	96,100	115,800	47,300	2.45
3/M-MAO	60	1,000	0.75	78,600	92,200	36,600	2.52
3/MAO	70	1,000	0.47	41,700			
3/M-MAO	70	1,000	0.39	32,400			

[a]T_p=polymerization temperature, typically ± 2 °C. Polymerization conditions: 1-L or 2-L stainless-steel autoclave, 0.5 or 1 L liquid propylene, 1-2 h, metallocene/MAO solutions aged in toluene (10-20 ml) for 10 min at room temperature
[b]intrinsic viscosity measured in THF at 135°C, averages of different samples
[c]estimated from the $[\eta]=1.85 \times 10^{-4} M_w^{0.737}$, see ref. 10

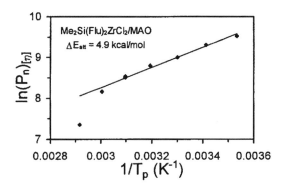

Figure 1. $\ln(\overline{P}_n)$ vs. $1/T_p$ for 1/MAO.

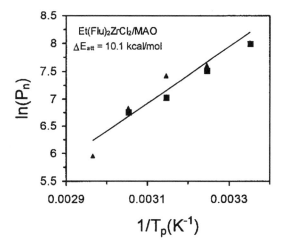

Figure 2. $\ln(\overline{P}_n)$ vs $1/T_p$ for 3/MAO. ■ - $\overline{P}_n$ from SEC; ▲ - $\overline{P}_n$ from [η].

cal properties of the material: while aPP from 1 and 2 are solid, non-sticky, elastomeric materials, samples from 3 show enhanced cold-flow or at T_p=70°C are viscous liquids. The Arrhenius plots ($\ln P_n$ vs $1/T_p$) for 1/MAO (Figure 1) and 3/MAO (Figure 2) give $\Delta E^{\ddagger}$ values of 4.9 (T_p=10-60°C, R=0.994) and 10.1 kcal/mol (T_p=30-70°C, R=0.956) respectively. At 70°C, we observe a lowering in catalyst activity for all three catalysts. As for the molecular weights, we observe both a deviation from, the linear behavior in the Arrhenius plot (molecular weights lower than expected) and a broadening of the molecular weight distributions for both 1/MAO and 3/MAO. These observations indicate some modification in the structure of the active centers induced by the temperature.

Finally, we have found that the non-bridged aspecific catalyst, (2-methyl-indenyl)$_2$ZrCl$_2$ (4/MAO) shows a much stronger dependence of molecular weight on T_p. The polymerization results for propylene polymerizations in liquid monomer with 4/MAO are compared to the known (Ind)$_2$ZrCl$_2$/MAO in Table 3. The ln P_n vs $1/T_p$ correlation is not linear, leveling off at high $1/T_p$ values (viscosity values above 11 dL/g). This result implies that either the M_w/viscosity correlation is not respected at the very high molecular weight values, or that a mechanism change occurs at polymerization temperatures below 0°C.

NMR ANALYSIS

^{13}C NMR analysis confirms that, at any polymerization temperatures, both the C_{2v}-symmetric metallocenes and the freely rotating (2-Me-Ind)$_2$ZrCl$_2$ produce atactic or nearly atactic polypropylenes, and the polymerization mechanism is in all cases Bernoullian (at the dyad level). All samples are highly regioregular, as no head-to-head or tail-to-tail propylene units were detected. The methyl triad analysis of selected samples from 3/MAO and 4/MAO is reported in Table 4. The Arrhenius plots (ln(m/r) vs $1/T_p$) are compared in Figure 3. It can be observed that aPP from 3, like 1 and 2, is slightly syndiotactic,

Table 3. Propylene polymerization with Ind$_2$ZrCl$_2$/MAO and 4/MAO catalysts^a

Metallocene	Amount ηmol	T$_p$ °C	Al/Zr molar	t min	Yield g$_{pp}$/mmol $_M$h	[η]^b dL/g	$\overline{M}_w^c$	$\overline{P}_n^d$
Ind$_2$ZrCl$_2$	5.8	50	1,500	60	18,000			110
Ind$_2$ZrCl$_2$	2.5	20	3,000	60	4,700	0.63	62,000	
Ind$_2$ZrCl$_2$	2.5	0	3,000	60	1,800	1.82	261,700	
4	8.8	50	1,000	120	10,000	0.29	21,700	220
4	2.4	30	3,000	60	4,000	1.45	192,000	
4	2.4	20	3,000	15	15,000	4.18	808,600	
4	2.4	10	3,000	60	7,300	6.46	1,460,000	
4	2.4	0	3,000	60	8,000	11.43	3,166,000	
4	2.4	-10	1,000	20	9,100	11.55	3,211,000	
4	2.4	-20	2,000	25	7,900	13.24	3,865,000	

^{a}general polymerization conditions: 1-L stainless-steel autoclave, liquid propylene. Metallocene/MAO solutions aged in toluene (10-20 ml) for 5-10 min at room temperature
^{b}intrinsic viscosity measured in THF at 135°C
^{c}calculated from the [η]=1.85x10^{-4} M$_w^{0.737}$, see ref. 9
^{d}number-average polymerization degree from ^{1}H- NMR, assuming one double bond per polymer chain

while **4** is slightly isospecific by chain end control. In any case, the degree of chain end control is very low and is the same for both catalysts ($\Delta E^{\ddagger} = 0.17$ kcal/mol).

EFFECT OF HYDROGEN

Most metallocenes are highly sensitive towards the addition of hydrogen: in general, an activating effect is observed at low levels of hydrogen addition for isospecific zirconocenes.[9]

In the case of **1**/MAO, we observed a negative effect of hydrogen on catalyst activity: at 50°C in liquid monomer neither the activity nor the molecular weight are affected by the addition of 50 mL H$_2$ (Activity=21,400 g$_{pp}$/mmol$_M$h, [η]=2.09, $\overline{M}_w$=315,700 corresponding to $\overline{P}_n$ ca. 3,700); in the presence of 200 mL H$_2$, however, the activity drops to one fourth and the resulting aPP has a lower molecular weight (Activity=5,400 g$_{pp}$/mmol$_M$h, [η]=1.52, $\overline{M}_w$=204,900 corresponding to $\overline{P}_n$ ca. 2,400). The negative effect of hydrogen on catalyst activity has been confirmed also in the case of ethylene polymerization: addition of hydrogen to a suspension polymerization of ethylene (hexane, 4 bar, 50°C totally

Table 4. ^{13}C NMR triad analysis of aPP obtained with 3/MAO and 4/MAO

Catalyst	T_p	mm	mr	rr	B=4 mmrr/(mr)^2
3	30	20.93	52.83	26.24	0.79
3	40	20.51	52.87	26.61	0.78
3	50	21.00	51.96	27.04	0.84
3	60	19.72	52.47	27.81	0.80
3	70	17.80	52.32	29.88	0.78
4	-20	39.07	46.61	14.32	1.03
4	-10	37.59	46.27	16.15	1.13
4	0	35.95	47.11	16.93	1.10
4	10	37.48	46.72	15.80	1.09
4	20	36.56	47.03	16.41	1.08
4	30	34.85	47.68	17.47	1.07
4	30	35.30	47.75	16.95	1.05

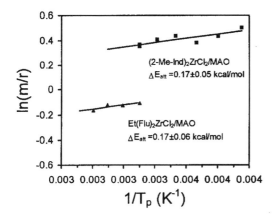

Figure 3. ln(m/r) versus 1/T_p for **3**/MAO and **4**/MAO.

suppresses polymer formation. This effect is reversible: venting the hydrogen/ethylene gas mixture and replacing with fresh ethylene restarts the polymerization activity.

SYNTHESIS OF ATACTIC PROPYLENE/ETHYLENE COPOLYMERS, APEC

Copolymers were prepared in slurry using **2**/M-MAO and **4**/M-MAO catalysts. The experimental conditions and polymerization results are reported in Table 5 and copolymer characterizations are shown in Table 6. Ethylene increases polymerization activity for both catalytic systems while t[η] values comparable to those obtained with

2. In the latter system, ethylene is a very effective chain transfer agent: with 2/MAO ethylene can be used as a molecular weight regulator in place of hydrogen.

The addition of ethylene gives rise to a positive effect on the glass transition temperature, T_g, of atactic polypropylene: the inclusion of ethylene in the polymer chain produces a material as APECs with a lower T_g than aPP. This effect is linear with the ethylene content.

Table 5. Synthesis of atactic propylene-ethylene copolymers, aPEC, with 4/M-MAO and 4M/MAO

Run[a]	Catalyst	mmol	Al/Zr molar	Auto-clave L	C_3H_6[b] g	C_2H_4[b] g	T_p °C	t min	H_2 mL	Yield g/m $mol_M h$
C1	2/M-MAO	3.2	500	2.6	904	31	20	60	0	41,000
C2	2/M-MAO	3.2	500	2.6	835	84	25	72	0	70,100
C3	2/M-MAO	1.6	500	1.4	463	1	30	29	0	39,200
C4	2/M-MAO	1.6	500	1.4	413	41	30	48	0	97,300
C5	2/M-MAO	3.2	500	4.3	1454	8	50	120	0	30,400
C6	2/M-MAO	3.2	500	4.3	1453	21	50	120	0	47,500
C7	2/M-MAO	1.6	2,000	4.3	1453	21	50	120	0	91,100
C8[c]	2/M-MAO	1.6	500	4.3	988	265	50	60	0	167,700
C9	4/M-MAO	5.9	2,000	1.35	407	11	30	120	0	4,600
C10	4/M-MAO	4.8	2,000	1.35	407	11	30	120	7	4,900
C11	4/M-MAO	3.6	2,000	1.35	397	18	30	120	8	5,700
C12	4M/-MAO	2.4	2,000	1.35	378	34	30	120	9	10,200
C13	4/M-MAO	2.4	2,000	1.35	333	71	30	120	13	14,000

[a]metallocene/M-MAO solutions aged in ispar C/hexane (5-10 mL) for 5-10 min at room temperature
[b]monomer amounts charged in the autoclave before injection the catalyst
[c]hexane (21 mL) added in the autoclave before injecting the catalyst

CONCLUSIONS

The influence of the polymerization temperature and the catalyst/cocatalyst ratio has been investigated in the polymerization of propylene to atactic polypropylene, aPP, with four aspecific zirconocenes: the bridged C_{2v}-symmetric [dimethylsilanediylbis(9-fluorenyl)] zirconium dichloride **1**, [di(n-butyl)silanediylbis(9-fluorenyl)]zirconium dichloride **2**, and [ethylenebis(9-fluorenyl)] zirconium dichloride **3**, and the non bridged [bis(2-methylindenyl)]zirconium dichloride **4**.

Our previous findings demonstrated that there is a close similarity between the ethylenebisfluorenyl and dimethylsilanediylbisfluorenyl ligands in the selection of chain transfer modes.[3] Here we have shown that, at polymerization temperatures above ambient, 1/MAO and 2/MAO produce aPP with the highest molecular weight, in contrast to the results reported by Rausch and Chien for 3/MAO.[7]

The necessity of performing polymerization experiments under constant monomer concentration (preferably in liquid monomer) and carefully controlled polymerization temperature when evaluating metallocene catalysts for propylene polymerization, and (equally important) a reliable measurement of the molecular weight, is again confirmed.

Table 6. Characterization of aPEC with 2/M-MAO and 4/M-MAO

Run	$[\eta]^a$ dL/g	$\overline{M}_w^b$	Ethylene content[c] %mol	T_g K
C1	0.67	67,400	40.8	223
C2	0.65	64,700	45.1	218
C3	1.31	167,500	8.9	
C4	0.6	58,000	48.6	216
C5	1.32	169,200	5.5	260
C6	1.36	176,200	13.6	255
C7	1.22	152,000	12.3	256
C8	0.65	64,700	51.0	
C9	1.38	179,800	22.6	247
C10	0.96	109,800	23.7	247
C11	1.19	147,000	29.5	238
C12	1.43	188,600	42.3	227
C13	2.34	368,000	59.2	216

[a]intrinsic viscosity measured in THN at 135°C
[b]calculated from $[\eta]=1.85 \times 10^{-4} M_w^{0.737}$, see ref. 9
[c]from ^{13}C NMR

The influence of the addition of hydrogen or ethylene on the polymerization of propylene with 1/MAO and the more active 2/MAO catalyst has also been investigated. While hydrogen lowers the catalyst activity without being a good chain transfer agent, ethylene proved very effective in lowering the molecular weight of aPP even at very low ethylene/propylene ratios. At the same time, ethylene addition increases the catalyst activity and produces amorphous propylene/ethylene copolymers, APEC, with lower T_g.

ACKNOWLEDGMENT

We thank G. Baruzzi and M. Colonnesi for the polymerization experiments, I. Mingozzi, A. Marzo, and C. Baraldi for the molecular weight measurements, H. Rychlicki, O. Sudmejier and I. Camurati for the NMR measurements, M. Cappati and A. Celli for the DSC analysis, Prof I. E. Nifant'ev for a sample of (tBuCp)2ZrCl2 and V. A. Dang for the synthesis of 2-methylindene.

EXPERIMENTAL SECTION
GENERAL PROCEDURES

All catalyst manipulations were performed under nitrogen and using conventional Schlenk-line techniques. Toluene was distilled over $AliBu_3$ and stored under nitrogen. Hexane was purified over alumina. Typical residual water content was 2 ppm. Polymerization grade propylene was received directly from the Montell Ferrara plant. Commercial MAO (Witco, 30% w/w in toluene) was brought to dryness and further treated under vacuum (2 hours, 60°C) in order to remove most of the unreacted TMA. Commercial M-MAO (Albemarle, 2.3 molar solution in isopar C) was used as received. $Me_2Si(Flu)_2ZrCl_2$ **1** and $Et(Flu)_2ZrCl_2$ **2** were prepared as previously described.[3]

The chemical purities of all compounds were confirmed by ^1H NMR (200 MHz, $CDCl_3$).

POLYMERIZATION

Polymerizations were carried out in liquid propylene in either a 1-L (0.5 L propylene) or 2-L (1 L propylene) stainless-steel autoclave at constant pressure for 1 hour. Temperature control was within ±2°C. Stirring was kept at 800 RPM by means of a three-blade propeller. Metallocene and MAO were precontacted for 10 min in toluene solution (10 mL), then added to the monomer/solvent mixture at the polymerization temperature. The polymerizations were quenched with CO; the polymers were isolated by distilling off the solvents under reduced pressure, then dried at 50°C in vacuo.

PROPYLENE/ETHYLENE COPOLYMERIZATION

A stainless-steel autoclave (volume reported in Table 5) equipped with magnetically driven stirrer, 35-100 mL, stainless steel vial, thermoresistance connected to a thermostat for temperature control was dried at 70°C in a propylene stream.

In the autoclave, thermostated at the chosen polymerization temperature (see Table 5), were charged the initial amounts of monomers.

The metallocene was then dissolved in the alumoxane solution with some hexane added. After stirring for 10 minutes the solution was injected in the autoclave by means of ethylene pressure through the stainless-steel vial. The polymerization was carried out at constant temperature, continuously supplying a mixture of the two comonomers with a composition near to that required by the copolymer, during the time reported in Table 5. The reaction was stopped by injecting 600 mL of carbon monoxide. After degassing the residual monomers the polymer was collected and dried at 70°C for 2 hours.

POLYMER ANALYSIS

Intrinsic viscosities were measured in tetrahydronaphthalene, THN, at 135°C.

The weight average molecular weights of aPP were obtained from its intrinsic viscosity and the Mark-Houwink-Sakurada parameters derived by Pearson and Fetters,[29] and the $\overline{M}_n$ values were derived assuming $\overline{M}_w/\overline{M}_n=2$ for all samples (experimental values for selected samples are in the 2-2.2 range).

GPC measurements were carried out on a Waters 150-C GPC equipped with TSK columns (model GM-HXL-HT) at 135°C with 1,2-dichlorobenzene as solvent. Monodisperse fractions of polystyrene were used as standard. Solution ^{13}C NMR spectra were run at 75.4 MHz on a Varian UNITY-

300 NMR spectrometer. Samples were run as 15 % solutions in $C_2D_2Cl_4$ at 130°C. Chemical shifts are referenced to TMS using as a secondary reference the mmmm methyl peak of polypropylene at 21.8 ppm. 6,000 transients were accumulated for each spectrum with a 12 second delay between pulses.

SYNTHESIS AND ANALYSIS OF THE DI-N-BUTYLBIS(9-FLUORENYL)SILANE LIGAND

Synthetic methods similar to those developed for the dimethylbis(9-fluorenyl)silane ligand were followed. Fluorene (23.27 g, 140 mmol) was dissolved in 100 mL diethyl ether and the solution was cooled to -78°C. A solution containing 140 mmol methyllithium (1.4 M in diethylether, 100 mL) was added dropwise to the stirred solution while maintaining the temperature at -78°C. After the addition was complete, the solution was allowed to warm to room temperature. Stirring was continued overnight.

In a separate flask, di-n-butyldichlorosilane (14.9 g, 70 mmol) was dissolved in 50 mL diethylether. The temperature was reduced to -78°C, and the solution containing the fluorene anion (prepared above) was added dropwise. After the addition was complete, the reaction was allowed to warm slowly to room temperature and stirred overnight. The reaction was then treated with a saturated aqueous solution of ammonium chloride, the organic layer was collected and dried over magnesium sulfate, then the solvents were removed in vacuo. The material was further purified by washing with methanol and drying in vacuo. Yield: 23.39 g (70.7% @ 97% purity by GCMS). ^{1}H-NMR (CD_2Cl_2), d, ppm: 7.81 (d, 4H), 7.30 (m, 12H), 3.95 (s, 2H), 1.05 (m, 4H), 0.85 (m, 10H), 0.5 (m, 4H).

SYNTHESIS AND ANALYSIS OF THE CATALYTIC COMPLEXES

Di-n-butylsilanediylbis(9-fluorenyl) zirconium dichloride, 2

Di-n-butylbis(9-fluorenyl)silane (8.83 g, 18.7 mmol) was dissolved in 150 mL diethyl ether and the temperature was lowered to -78°C. A solution containing 37.4 mmol methyllithium (1.4 M in diethylether, 26.7 mL) was added dropwise to the stirred solution. After the addition was complete, the reaction was allowed to warm to room temperature and stirring was continued overnight.

In a separate flask, zirconium tetrachloride (4.36g, 18.7 mmol) was slurried in 100 mL pentane and the temperature reduced to -78°C. The dianion prepared above was added in a dropwise fashion. The reaction was allowed to warm to room temperature and stirring was continued overnight. The next morning, the solids in the flask were collected by filtration and washed with fresh diethylether. Purification of the pre-catalytic zirconocene complex from the lithium chloride byproduct was accomplished by removing the ether/pentane solvents in vacuo, suspending the dried solids in methylene chloride, then filtering to collect the soluble product. Methylene chloride was removed in vacuo, leaving a bright red free flowing powder. Yield: 9.64 g (@ 81.5% purity by ^{1}H-NMR, final yield of complex was 75.7%). ^{1}H-NMR (CD_2Cl_2), d, ppm: 7.85 (d, 8H), 7.35 (t, 4H), 7.10 (t, 4H), 2.3 (m, 4H), 2.10 (m, 4H), 1.9 (m, 4H), 1.05 (t, 6H).

Synthesis of di-n-butylsilyl bis(fluorenyl) zirconium dimethyl

To a stirred solution containing di-n-butyl bis fluorene zirconium dichloride (1.58 g, 2.5 mmol) in 75 mLs diethylether at -78°C, was added 5 mmol methyllithium (1.4 M solution in diethylether, 3.57 mL). The solution was stirred overnight, allowing the temperature to warm to ambient slowly overnight. The next morning, the solvents were removed in vacuo, the dark brown solids were taken up in methylene chloride and filtered. The methylene chloride was then removed in vacuo, leaving 0.96 g of a dark brownish red free flowing solid. Yield: 64%; ^{1}H-NMR indicates ~90% purity, the remainder being the monomethyl derivative. ^{1}H-NMR (CD_2Cl_2, δ, ppm): 7.9 (d, 4H), 7.65 (d, 4H), 7.3 (t, 4H), 7.05 (t, 4H), 2.0 (m, 8H), 1.7 (m, 4H), 1.05 (t, 6H), -2.5 (s, 6M). (The ratio of monomethyl to the dimethyl complexes is calculated from the integral of the singlet at -2.1 ppm).

(2-methyl-indenyl)$_2$ZrCl$_2$

2-methylindan-2-ol: a mixture of indan-2-one (freshly distilled before use, 36 g, 272 mmol) in 400 mL of anhydrous Et_2O was slowly added to a mixture of methylmagnesium bromide (3M solution in hexane, 100 mL, 300 mmol) in 200 mL of Et_2O at 0°C. The reaction mixture was stirred at room temperature for three hours, and then quenched with 350 g of ice and a NH_4Cl solution (30 g in 500 mL of water). The organic layer was separated, washed with saturated $NaHCO_3$ (500 mL) and then water (500 mL), dried over Na_2SO_4 and concentrated to give a slightly yellow solid (37.8 g). This solid was identified as 2-methyl-indan-2-ol by NMR and GC/MS. ^{1}H NMR ($CDCl_3$), δ (ppm): 7.08 (broad s, 4H), 2.84 (s, 4H), 2.78 (s, 1H), 1.40 (s, 3H).

2-methylindene: without any further purification, 2-methyl-indan-2-ol (25 g, 169 mmol), toluenesulfonic acid monohydrate (1 g) in 100 mL of toluene was refluxed for 2 hours. GC analysis of the reaction crude indicated that 96% of 2-methylindene was formed (NMP was used as an internal standard). The reaction mixture was concentrated and then distilled (a small amount of 4-t-butylcatecol and two NaOH pellets were added before the distillation) to yield 2-methylindene, b.p. 58-60°C @ 2 mm Hg (16.7 g, 76%). ^{1}H NMR ($CDCl_3$, δ, ppm): 7.4-7.0 (m, 4M), 6.11 (s, 1H), 3.21 (s, 2H), 2.10 (s, 3H).

(2-methyl-indenyl)$_2$ZrCl$_2$: to a solution of 10.9 mmol (1.42 g) of 2-methylindene in 30 mL THF at 0°C were added 4.4 mL of a 2.5 M solution of n-butyllithium (11 mmol). After addition, the reaction was allowed to warm to room temperature and stirred an additional 4 hours. The solvents were then removed in vacuo and the remaining solids were washed with pentane. Zirconium tetrachloride (1.27 g, 5 mmol) was added as a dry powder to the above product, and a slurry was made in pentane. THF was added (1 mL) to facilitate the reaction, which was allowed to proceed at room temperature overnight with stirring. The solids were then collected by filtration, and washed with pentane. Further purification of the complex was done by dissolving the product in methylene chloride, filtering, and removing the solvent from the filtrate in vacuo. Yield 1.51 g (71.8 %). ^{1}H NMR ($CDCl_3$, δ, ppm): 7.75-7.55 (m, 4H, Ar), 7.35-7.15 (m, 4H, Ar), 5.81 (s, 4H, H1 e H3), 2.04 (s, 6H, Me).

REFERENCES

1.   H. H. Brintzinger, D. Fischer, R. Mulhaupt, B. Rieger, and R. M. Waymouth, *Angew. Chem. Int. Ed Engl.*, **34**, 1143 (1995).
2.   S. Jungling and R. J. Mulhaupt, *Organomet. Chem.*, **497**, 27 (1995).
3.   L. Resconi, R. L. Jones, E. Albizzati, I. Camurati, F. Piemontesi, F. Guglielmi, and G. Balbontin, *Polymer Preprints*, **35(1)** 366 (1994); L. Resconi, R. L. Jones, A. L. Rheingold, and G. P. A. Yap, *Organometallics*, **15**, 998 (1996); R. L. Jones, L. Resconi, and A. L. Rheingold, *Proc. 5th International Business Forum on Specialty Polyolefins*, September 20-22, 1995, Schotland Business Research Inc., Skillman, NJ, 1995.
4.   R. Silvestri, L. Resconi, and A. Pelliconi, *Proc. Metallocenes'95*, Brussels, April 26-27 (1995); E. Marchetti, P. Sgarzi, R. Silvestri, and L. Resconi, *Europhysics Conference on Macromolecular Physics: Morphology of Polymers*, Prague, July 17-20 (1995); A. Pelliconi, R. Silvestri, V. Braga, and L. Resconi, **Eur. Pat. Appl.**, **95101217.8** to Spherilene, 1994; R. Silvestri, L. Resconi, and A. Pelliconi, **Eur. Pat. Appl. 95112230.8** to Spherilene, 1994.

5.   L. Resconi and R. L. Jones, **Eur. Pat. Appl., 96103118.4** to Montell, 1996.

6.   H. Alt, W. Milius, and S. J. Palackal, *Organomet. Chem.*, **472**, 113 (1994); H. Alt, S. Palackal, K. Patsidis, M. Welch, R. Geerts, E. Hsieh, M. McDaniel, G. Hawley, P. Smith, **Eur. Pat. Appl., 524 624** to Phillips Pet., 1993.

7.   Y.-X. Chen, M. D. Rausch, and I. C. W. Chien, *Macromolecules*, **28**, 5399 (1995).

8.   J. Van Beek, J. De Vries, R. Persad, and G. Van Doremaele, **Int. Appl. WO 94/11406** to DSM, 1994; L. Resconi, F. Piemontesi, and D. Balboni, **Eur. Pat. Appl., 693,506** to Spherilene, 1996.

9.   T. Tsutsui, N. Kashiwa, and A. Mizuno, *Makromol. Chem., Rapid Commun.*, **11**, 565 (1990); N. Kashiwa and M. Kioka, *Polym. Mat. Sci. Eng.*, **64**, 43 (1991); M. Kioka, A. Mizuno, T. Tsutsui, and N. Kashiwa, **Catalysis in Polymer Synthesis**, Eds. E. J. Vandenberg and J. C. Salamone, *ACS Symposium Series 496*; ACS, Washington, DC, 1992; V. Busico, R. Cipullo, and P. Corradini, *Makromol. Chem., Rapid Commun.*, **14**, 97 (1993); S. Jungling, R. Mulhaupt, U. Stehling, H. H. Brintzinger, D. Fischer, and F. Langhauser, *J. Polym. Sci., Polym. Chem.*, **33A**, 1305 (1995).

Homo- and Co-polymers Derived from Multicyclic Olefin Monomers: The Quest for Higher T_g Materials

George M. Benedikt, Brian L. Goodall, Lester H. McIntosh III,
Larry F. Rhodes, Louis M. Wojcinski II

The BFGoodrich Company, 9921 Brecksville Rd., Brecksville, OH 44141

INTRODUCTION

Developments in the area of metallocene-catalyzed olefin polymerizations have advanced rapidly since the seminal discovery of Kaminsky and Sinn.[1] Extremely high activity catalysts have been developed that yield narrow molecular weight distributions for polyethylene, polyethylene with long chain branches,[2] highly isotactic polypropylene[3] with high molecular weights,[4] syndiotactic polypropylene,[5]

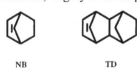

NB   TD

and syndiotactic polystyrene.[6] Metallocene technology also has been applied to copolymerization of unconventional monomers such as norbornene, NB, and tetracycloclodecene, TD, and with olefins such as ethylene. Copolymers containing high concentrations of multicyclic olefins such as NB or TD have attracted attention from the plastics industry because of their unique polymer properties. For example, at concentrations of norbornene higher than 12 mole %, a norbornene/ethylene copolymer becomes amorphous.[7] Moreover, increasing the content of the multicyclic monomer increases the glass transition temperature, T_g.[8]

Kaminsky has explored the relationship between norbornene content in norbornene/ethylene copolymers and glass transition temperature.[9] We have published data

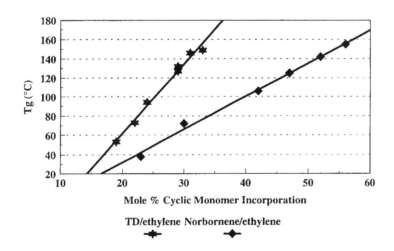

Figure 1. Effect of multicyclic monomer content on copolymer T_g.

on the influence of TD content on copolymer T_g.[10] The bulkier, more rigid TD monomer yields a higher T_g copolymer with ethylene than NB at a given incorporation (see Figure 1). From Figure 1, it is clear that polymers of norbornene and ethylene with glass transition temperatures higher than approximately 140°C would require increasing the norbornene content to at least 50 mole % and above. At 50 mole % norbornene an intriguing possibility exists: the material could be an alternating norbornene/ethylene copolymer. Above 50 mole %, of course, norbornene-norbornene enchainments would necessarily occur.

Based on the trend observed in Figure 1, a norbornene homopolymer would have an extremely high glass transition temperature. To date, however, zirconocene-based homopolynorbornene has been described as "insoluble"[11] and have "melting points [that] are surprisingly high ... over 600°C".[12]

In this contribution, we describe a method to prepare high T_g copolymers by increasing the norbornene content in the monomer feed stream in zirconocene-catalyzed polymerizations, NMR characterization of an apparent alternating copolymer of norbornene and ethylene, and the use of a cobalt catalyst to prepare a high T_g soluble homopolynorbornene.

RESULTS AND DISCUSSION

ZIRCONOCENE-CATALYZED POLYMERIZATION

I

Based on our previous studies of the copolymerization of TD and ethylene using zirconocene catalysts, it was clear that increasing the concentration of norbornene in the monomer feed stream would be essential in producing a high T_g copolymer.[13] Thus we chose conditions in which the polymerization was conducted in a very large excess of norbornene relative to ethylene (see Table 1 for details). We chose to investigate a well-known catalyst precursor: the ethylene-bridged bis(indenyl)zirconocene (I).

Catalyst I yielded poor activity (measured as g polymer/mmol Zr/hr). The molecular weight of the polymer is respectable (M_w=160,000) with a monomodal molecular weight distribution (M_w/M_n=2.35). The copolymer isolated using catalyst I is rather unique as judged by its thermal behavior (as measured by differential

Table 1. Copolymerization of norbornene and ethylene using zirconocene catalyst I*

Catalyst	Pressure psig	Activity g/mmol Zr/hr	M_w x10^{-3}	M_n x10^{-3}	T_g °C
I	20	450	160	68.2	142, 240

*polymerization was run essentially in neat norbornene (norbornene:toluene ratio=7.2;1 by weight) with a norbornene: Zr ratio=38,000:1, a Al (MAO):Zr ratio-300:1. Zr and MAO were prereacted for 15 min. The polymerization was run for 90 min at 65°C

scanning calorimetry). Note that the copolymer exhibits two T_g's: 142°C and 240°C. Evidently, two immiscible copolymers are produced using catalyst I.

Table 2. Characterization of copolymer fraction isolated from catalyst I

Fraction	M_w x10^{-3}	M_n x10^{-3}	T_g °C
toluene soluble	153	75.1	249
toluene insoluble	126	51.1	147

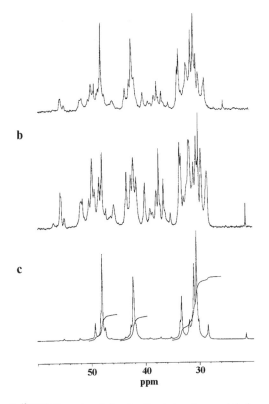

Figure 2. ^{13}C NMR spectra of a) unfractionated norbornene/ethylene copolymer, b) toluene soluble fraction, c) toluene insoluble fraction.

The immiscibility of the polymers from catalyst I suggested that the material may be fractionated. Indeed, two fractions can be isolated by stirring the polymer in toluene, filtering the insoluble portion (36% of total) and precipitating the soluble portion with methanol. The two fractions were then examined separately. The results are given in Table 2. The insoluble fraction exhibited a glass transition around 147°C, while the soluble fraction showed a T_g at about 249°C. These values match very well those found in the unseparated, bulk material. The molecular weight of the toluene insoluble fraction is somewhat lower than the soluble fraction; both copolymers have narrow polydispersities.

In Figure 2, the ^{13}C NMR spectra of the bulk material and the two fractions are presented. While the spectrum of the bulk material and the toluene soluble polymer are quite complicated, the toluene insoluble fraction is relatively simple. Note that there are essentially three regions of interest in the spectrum of the toluene insoluble fraction: 47-50 ppm, 41-44 ppm, and 28-34 ppm. Kaminsky previously has made ^{13}C NMR assignments for a norbornene/ethylene copolymer with a low incorporation of norbornene.[14] Based on this prior work, the most downfield resonance (48.0 ppm) is due to the former norbornene olefinic carbon 1, the bridgehead carbon 2 resonates at 42.2 ppm, and carbons 3 and 4 resonate at 31.0 and 33.3 ppm, respectively. The ethylene carbons come at 30.5 ppm. Integration of the three regions yield a 1:1:2.6 ratio, respectively, indicating that the ethylene incorporation in the toluene insoluble fraction is approximately 52 mole %. The T_g of this copolymer is in accord with the mole percent ethylene incorporation. Based on Figure 1, a norbornene: ethylene copolymer with a glass transition temperature of approximately 149°C would be expected to have an ethylene content of about 54 mole %.

With an ethylene incorporation of about 50 mole percent, the question remains whether the toluene insoluble material is an alternating or random copolymer. This question can be addressed with the aid of ^{13}C NMR spectra of hydrooligomers of norbornene recently published by Kaminsky and coworkers.[15] In this report, oligomers of norbornene (prepared from zirconocene catalysts under hydrogen pressure) were separated and then characterized by ^{13}C NMR spectroscopy. Below assignments for the dimers and trimers are given. A range of ^{13}C chemical shifts are given for each carbon since Kaminsky and coworkers report chemical shifts for all possible stereoisomers. (The structures drawn below do not specify a certain stereochemical sequence). From the hydrooligomer spectral data, one would expect a polymer containing a norbornene-norbornene-norbornene monomer sequence to exhibit resonances in the 53.2-55.0 ppm range (see the assignments for the hydrotrimers). Obviously, for

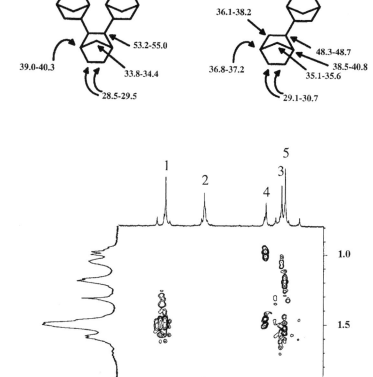

Figure 3. ^{1}H-^{13}C HETCOR NMR spectrum of alternating norbornene/ethylene copolymer.

the toluene insoluble copolymer (50:50 norbornene: ethylene) this is not the case (see Figure 2c), no significant peaks are observed above 50 ppm. Therefore, this material cannot be considered a random copolymer of norbornene and ethylene. The other alternative, an alternating copolymer seems to be a reasonable assumption based on the hydrooligomer data published thus far.[16]

The unique character of the alternating copolymer of norbornene and ethylene prompted us to investigate this material further using ^{2}D-NMR techniques. A ^{1}H-^{13}C heteronuclear correlation, HETCOR, NMR spectrum is presented in Figure 3. As we have found for TD/ethylene copolymers, the most downfield resonance in the ^{1}H NMR spectrum corresponds to the protons on carbon 2 (42.2 ppm). Carbon 1 (48.0 ppm) correlates with protons around 1.5 ppm. As expected, methine carbons 1 and 2 correlate with only one type of proton. The norbornene methylene carbons 3 (31.0 ppm) and 4 (33.3 ppm) as well as the ethylene carbon 5 (30.5 ppm) all correlate with two types of protons.

There are smaller peaks in the region of carbons 1 and 2. The smaller peaks in the carbon 2 (41-44 ppm) region, for example, all correlate with proton resonances in the same portion of the ^{1}H spectrum (1.9-2.2 ppm) as the large carbon 2 resonance. These smaller peaks are probably due to copolymer sequences of different tacticities. The presence of one dominant peak in the spectral region for both carbons 1 and 2 suggest that the alternating copolymer is highly tactic. The identification of the tacticity must await further investigation.

COBALT-CATALYZED POLYMERIZATION

Zirconocene complexes have been used to catalyze the homopolymerization of norbornene. However, the polymer isolated from these reactions has been described as insoluble and not exhibiting a melting transition below its decomposition temperature.[12] Other metals are known to polymerize norbornene. More than 10 years ago, Sen and coworkers described the synthesis of a THF-insoluble polymer using cationic Pd complexes.[17] However, these materials, upon reinvestigation, were found to be perfectly soluble in chlorinated hydrocarbons and to exhibit significant molecular weights.[18] We have found that, under the correct conditions, cobalt also will transform norbornene to high polymer.

If, for example, a heptane solution of norbornene monomer is treated with cobalt(II) neodecanoate in the presence of methaluminoxane, MAO, a viscous polymer cement is obtained. Significant amounts of polymer can be isolated (20%) after two hours. The polymer is soluble in hydrocarbons. NMR analysis of the polymer shows it to be 100% addition polymer; no ring-opening metathesis polymer is formed as evidenced by the lack of peaks in the olefinic region of the spectrum. The polymer exhibits a very high glass transition temperature of around 380°C.

The characteristics of this addition polymer synthesized using cobalt suggest that it is different than the polymer isolated from the previously reported zirconocene-based homopolymers. We have been concentrating our efforts on understanding the differences between the two polymers.

ACKNOWLEDGMENTS

The authors thank Dr. S. Huang for GPC measurements, P. Neal and J. Backsay for DSC measurements, and Professors J. Heppert and S. Collins for their comments. We thank Professor W. Kaminsky and Dr. M. Arndt for a preprint of their manuscript on norbornene hydrooligomers.

REFERENCES

1. H.-J. Sinn and W. Kaminsky, *Adv. Organometal. Chem.*, **18**, 99 (1980).
2. a) B. A. Story and G. W. Knight, *Proc. Metcon'93*, 111 (1993); b) J. C. Stevens, *Proc. Metcon'93*, 157 (1993).
3. W. Kaminsky, K. Kulper, H. H. Brintzinger, and F. R. W. P. Wild, *Angew. Chem., Int. Ed. Engl.*, **24**, 507 (1985).
4. W. Spaleck, M. Amberg, J. Rohrmann, A. Winter, B. Bachmann, P. Kiprof, J. Behm, and W. A. Herrmann, *Angew. Chem., Int. Ed. Engl.*, **31**, 1347 (1992).
5. J. A. Ewen, R. L. Jones, A. Razavi, and J. D. Ferrara, *J. Am. Chem. Soc.*, **110**, 6255 (1988).
6. T. H. Newman, R. E. Campbell, and M. T. Malanga, *Proc. Melcon'93*, 315 (1993).
7. W. Kaminsky, *Shokubai*, **33(8)**, 536 (1993).
8. H. Schnecko, R. Caspary, and G. Degler, *Angew. Makromol. Chem.*, **20**, 21 (1971).
9. W. Kaminsky and A. Noll, *Polym. Bull.*, **31**, 175 (1993).
10. G. M. Benedikt, B. L. Goodall, N. S. Marchant, and L. F. Rhodes, *Proc. Melcon'94*, (1994).
11. W. Kaminsky, M. Arndt, and A. Bark, *Polym. Prep.*, **32(1)**, 467 (1991).
12. W. Kaminsky, A. Bark, and I. Dake, *Stud. Surf. Sci. Catal.*, **56**, 425 (1990).
13. G. M. Benedikt, B. L. Goodall, N. S. Marchant, and L. F. Rhodes, *New. J. Chem.*, **18**, 105 (1994).
14. W. Kaminsky, A. Bark, and M. Arndt, *Makromol. Chem., Macromol. Symp.*, **47**, 83 (1991).
15. a) W. Kaminsky, M. Arndt, R. Engehausen, A. Noll, and K. Zoumis, *Poster Presentation at Forty Years of Ziegler Catalysis*, Freiburg, Germany, 1993; b) M. Arndt, **Ph. D. Dissertation**, *University of Hamburg*, 1993.
16. An alternating copolymer of the following sequence, norbornene-norbornene-ethylene-ethylene, cannot be ruled out. Note that the hydrodimers exhibits carbon 1 resonances between 48.3 and 48.7 ppm. The toluene insoluble, 50:50 norbornene:ethylene copolymer also exhibits resonances in this range. However, this possibility does not seem likely.
17. A. Sen and T.-W. Lai, *Organometallics*, **1**, 415 (1982).
18. a) C. Mehler and W. Risse, *Makromol. Chem., Rapid Commun.*, **12**, 255 (1991); b) N. Seehof, C. Mehler, S. Breunig, and W. Risse, *J. Mol. Catal.*, **76**, 219 (1992); c) C. Mehler and W. Risse, *Macromolecules*, **25**, 4226 (1992).

The Renaissance in Polyolefin Manufacturing Technology

Kenneth B. Sinclair

STA Research, 1042 Valley Forge Drive, Sunnyvale, CA 94087, USA

INTRODUCTION

Polyolefin manufacturing processes are mature technologies whose capabilities are well understood by the industry. The four primary process types – high pressure, solution phase, slurry phase, and gas phase processes – have been developed continuously over 30-40 years, improving their operating efficiencies, reducing investments, and expanding product range capabilities. By the end of the 1980s, the capabilities of each process type had been well defined, with all processes being closely cost competitive, but with each process having specific product range capabilities defining its optimum suitability for specific polymer markets.

The introduction of metallocene and other single site catalyst technologies has changed these well established competitive domains: new process/comonomer combinations are possible, the use of novel comonomers such as styrene, norbornene, and carbon monoxide is now commercially feasible, and seemingly impossible property combinations are now achievable. Higher catalyst activities, combined with remarkable process innovations, are writing the competitive order of technologies inside out invalidating previous conceptions of competitive ranking and process limitations. The product range capabilities of all processes are being broadened tremendously, and large new markets for high value products are becoming accessible to low cost "standard" polyolefin production processes.

This paper examines the impact of single site catalyst, SSC, technologies on the competitive domains of polyethylene manufacturing processes, and in particular:

- The product range capabilities of 1980s technologies compared to those of the current SSC-based technologies
- The new competitive position of slurry phase processes relative to gas phase processes in production of copolymers
- The extension of gas phase process capabilities to production of soft copolymers and elastomers
- The potential adaptation of polyolefin manufacturing technologies to production of engineering thermoplastics

HIGH PRESSURE POLYETHYLENE TECHNOLOGIES

High pressure processes are the "grandparents" of the polyolefin industry. Initially developed in the early 1940s, they typically involve the polymerization of ethylene at pressures of 1,000-3,000 atmospheres using free radical initiators to yield homopolymers containing both short and long chain branching. These conventional LDPE homopolymers have densities ranging from around 0.915 g/cc up to a maximum around 0.932 g/cc, and melt indices (I_2) ranging from about 0.1 dg/min to 100 dg/min. High pressure processes have very short reactor residence times of 30-120 seconds, and are therefore flexible with respect to grade changes. Wide variations in molecular structure are achievable, and the numerous commercial grades available have been closely tailored for specific application markets. However, it has never been possible with these free radical polymerization processes to produce stiff, linear, highly crystalline polymers: chain branching is inherent in the free radical polymerization mechanism and cannot be eliminated. Thus, these free radical technologies cannot make linear high density polyethylene, HDPE.

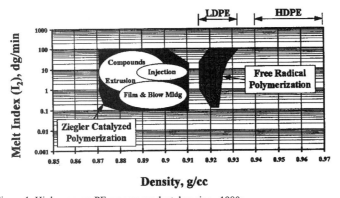

Figure 1. High pressure PE process product domains - 1980s.

Polyethylenes can have a very wide range of properties depending on molecular weight, molecular weight distribution, MWD, composition, and branching structure. If we look at the universe of property combinations that may be commercially useful, and define these combinations in terms of density and melt index, we could say broadly that melt indices ranging from virtually zero, to 1,000 dg/min are useful, combined with a density range from around 0.855 g/cc (completely amorphous elastomeric copolymers) up to about 0.97 g/cc (stiff highly crystalline homopolymers). Figure 1 plots the product domain of high pressure free radical processes within this useful melt index/density universe. It can be seen that, in spite of the very many available variations in LDPE products, the product domain of free radical polymerizations is really only a small patch in the middle of the useful universe.

Also shown in Figure 1 is a product domain for high pressure processes when used with Ziegler-Natta catalysts. High pressure Ziegler catalyzed processes are used in only a handful of plants around the world and account for a minute fraction of total high pressure PE production. However, they remain very interesting technologies because of their product range capabilities. They were developed in the late 1970s primarily by CdF/Ethylene Plastique (now Enichem), Elf Atochem, and Mitsubishi Petrochemical who used them commercially, as well as by BASF, Dow, and Rexene. These adaptations of the high pressure process greatly expanded its product range capabilities by providing control over chain linearity, short chain branching, and composition. As the figure shows, the Ziegler catalyzed polymerizations enabled the process to cover a significant proportion of the product universe, and in many respects made the high pressure process the most versatile technology available through the 1980s.

The process had limitations, however, related primarily to catalysis. The Ziegler catalysts employed were multi-site and produced low molecular weight oligomer and dimer byproducts which accumulated in recycle stream. These byproducts were reincorporated into the polymer chain as short chain branches, reducing the density of the product. Thus, the high densities typical of linear homopolymers were difficult to achieve. For this reason, among others, most commercial applications for the high pressure Ziegler catalyzed process have been in production of soft VLDPE resins.

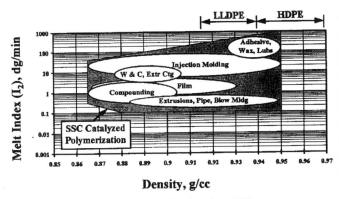

Figure 2. High pressure PE process product domains - 1990s.

Figure 2 shows a preliminary estimate of the new product range capabilities of the high pressure process when using single site catalysts. The very specific nature of SSCs provides close control over polymerization and eliminates low molecular weight byproduct fractions, thus pushing the upper density limits of the technology well into the domain of HDPE. At the same time, the good copolymerization capabilities of SSCs allow production of very soft polymers with closely controlled compositions and molecular weights, expanding accessible markets into new areas previously accessible only to high priced elastomers. Because the process is essentially solventless, it can also be used effectively for production of low molecular weight polymers such as crystalline waxes, base polymers for lube oil components, and for adhesives.

One significant limitation of current SSC-catalyzed high pressure technologies is their inability to copolymerize polar comonomers such as vinyl acetate and acrylates. These high pressure free radical copolymers are important elements in the global polyethylene industry, accounting for more than a million tons of PE production each year. They are used both for their flexibility and their chemical functionalities. While many of the currently produced EVA resins can be replaced by higher performing soft ethylene/α-olefin copolymers made with metallocenes, current SSC-based products cannot replace polar copolymers where they are used for their chemical functionality. New SSCs are being developed, however (such as those based on nickel and palladium), that have the required capabilities to incorporate polar comonomers. Thus, it seems that in the future, high pressure technologies combined with SSCs will continue to provide very attractive opportunities in production of specialties over an extremely broad product domain, including the polar copolymer domain.

The versatility of SSC technologies is also a good match for the inherent flexibility of high pressure processes: they can be used to economically produce a large number of specialty products (perhaps 30 to 40) in each production line. This compares with a practical limit of only 5-7 products from gas phase technologies.

High pressure SSC-based technologies have been used commercially by Exxon since 1991, and will be used by Mitsubishi Chemical starting in 1997. They are being developed also by BASF and Tosoh.

SOLUTION PHASE POLYETHYLENE PROCESSES

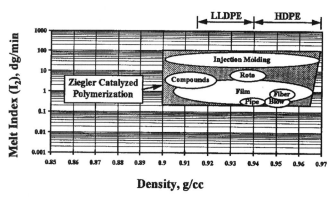

Figure 3. Solution phase PE process product domains - 1980s.

Solution processes were the first technologies to be used for production of linear polyolefins using coordination catalysts. They are used most notably by Dow, Nova, DSM, and Mitsui Petrochemical, and have been widely licensed over the past 15 years for production of LLDPE and HDPE. Primary attractions of the process include the ability to produce high melt index grades for injection molding as well as a broad range of copolymers, including soft copolymers with densities down to around 0.90 g/cc using conventional Ziegler-Natta catalysts. The overall product range is illustrated in Figure 3.

Comparison of this product domain with that illustrated earlier in Figure 1 for pre-SSC high pressure processes demonstrates the comparatively broad product domain of solution processes, particularly their coverage of the full LLDPE/HDPE density range from 0.915 g/cc to above 0.965 g/cc. The processes are not limited in the use of heavier comonomers such as octene-1. They also can be used with high comonomer concentrations in the reactor, enabling the production of lower density PE copolymers even with catalysts having relatively poor copolymerization capabilities.

The limitations of solution processes are related both to catalyst performance and to plant equipment capabilities. The poor copolymerization characteristics of Ziegler catalysts available in the 1980s restricted the practical lower density range to about 0.90 g/cc. At lower densities, the physical properties of the copolymers deteriorated: low molecular weight fractions of high comonomer content made the products sticky, while other properties such as clarity and melting point could not be improved further because of the presence of crystallizable homopolymer molecular chains.

The process limitation on product domain is the lower melt index limit of around 0.2 dg/min set by viscosity of the polymer solution as it exits the reactor. This limits the usefulness of the process in serving markets such as blow molding and thin film which typically employ higher molecular weight polymers with melt indices as low as 0.01 dg/min. Thus the broad compositional capabilities of solution processes are balanced by limited molecular weight capabilities. The process is therefore comple-

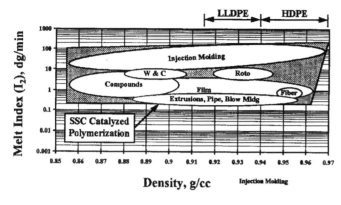

Figure 4. Solution phase PE process product domains - 1990s.

mentary to the slurry phase processes of the 1980s, and companies such as Dow use both slurry and solution processes in their production portfolio so as to cover the broadest product range.

The availability of SSCs for solution processes has broadened the product domains considerably as illustrated in Figure 4. Because of the good copolymerization characteristics of SSCs it is possible to produce virtually any ethylene copolymer composition ranging from completely amorphous copolymers to highly crystalline homopolymers. Operating conditions and production costs are similar across this entire product range. Thus, one of the leading low cost polyolefin processes can now be applied to production of a very broad range of elastomeric and soft specialties. However, the lower limit on melt index of around 0.2 dg/min remains.

Dow has used the solution process with SSCs since 1993 to produce plastomers and elastomers, and is expected to apply it also to production of volume LLDPE and HDPE products in 1997. Solution processes are also being used with SSCs by Dex Plastomers in Europe and by Mitsui Petrochemical in Japan. Their broad copolymer product range capabilities and inherent compatibility with homogeneous SSCs will place them among the most important polymer production technologies for the foreseeable future.

SLURRY PHASE POLYMERIZATION PROCESSES

Slurry processes involve the polymerization of ethylene at temperatures below the melting point of the polymer using a solid catalyst to form solid polymer particles suspended in an inert hydrocarbon diluent. The processes were developed in the late 1950s and early 1960s and quickly became the mainstay of the linear HDPE production industry. One of the best known examples of a slurry process, and the most widely licensed technology, is the Phillips continuous path pipe loop reactor "particle form" process; the reacting slurry is circulated through a wide bore cooled pipe loop by a circulating pump. This technology accounts for annual production of 4-5 million tons of linear PE, primarily HDPE.

The most attractive features of slurry processes include the simple and economical recovery of polymer from the reaction medium (by filtration, centrifugation, or flashing) and their capabilities to produce a very broad range of molecular weights, and particularly high molecular weight PEs. The chromium-oxide-on-silica catalyst systems developed by Phillips yield polymers with broad molecular weight distributions that are easy to process by blow molding and extrusion. The traditional process has significant limitations, however, including the inability to produce narrow MWD, high melt index grades using Phillips-type catalysts, and the inability generally to produce copolymers of density below about 0.937 g/cc.

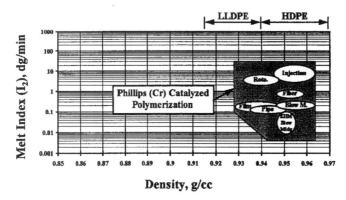

Figure 5. Loop slurry PE process product domains - 1980s.

Figure 5 shows the product domain for the loop reactor slurry PE process as it existed in the 1980s. The product domain was primarily in the HDPE region, covering polymers of melt index less than ~30 dg/min. When attempts were made with multi-site catalysts to produce copolymers of low density around 0.92 g/cc or below, plant output was seriously reduced due to swelling of the polymer particles, and the tendency of the polymer to become sticky and dissolve in the reaction diluent. In general, it was considered impractical, if not impossible, to produce PE copolymers of density below 0.920 g/cc using this technology.

The development of SSCs for the loop reactor slurry process has removed most of these process limitations, for the following reasons:

- SSCs yield polymers having very narrow MWD and compositional distribution
- The soluble low MW fraction of the polymers is virtually eliminated
- Polymer particles have a reduced tendency to swell and become sticky
- It is easier to incorporate comonomer into the polymer (higher reactivity relative to ethylene)

The effects on product domains for the process are quite dramatic as illustrated in Figure 6. It is now possible to efficiently and easily produce copolymers of density well below 0.915 g/cc (the nominal lower limit for LLDPE); the process has been used commercially since 1995 to produce ethylene/butene-1 copolymers of density 0.903 g/cc. Information in the patent literature indicates that densities as low as 0.88 g/cc may be achievable without encountering serious operational problems due to polymer stickiness or solubility. It may even be possible to produce EP copolymer elastomers of densities around 0.86 g/cc, although at significantly lower temperatures and reduced output.

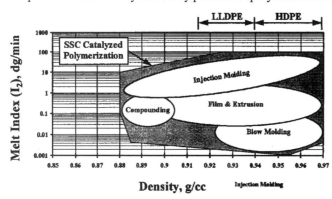

Figure 6. Loop slurry PE process product domains - 1990s.

The melt index range accessible to the process has also been expanded significantly, not only upward towards very high flow injection molding grades, but also down towards very high molecular weight polymers of almost immeasurable melt flow, and over a broad range of densities. Examples given in Phillips patents[1] include homopolymers having a high load melt index ($I_{21.6}$) of 0-0.06 dg/min, homopolymers of melt index (I_2) 1-2 dg/min with densities above 0.97 g/cc, and a hexene copolymer containing 30 wt % hexene, density 0.88 g/cc, high load melt index ($I_{21.6}$) 0.15 dg/min, and with a "super-random" comonomer distribution. The "super-random" comonomer distribution means that comonomer molecules are substantially all isolated between ethylene molecules, and the occurrence of comonomer dimers or longer sequences in the chain is significantly less than would be expected in a statistical random copolymerization.

Thus the loop reactor slurry process combined with SSCs is able to cover essentially the total product domain of currently commercial HDPE and LLDPE resins, as well as the developing domains of VLDPEs and plastomers. The process is currently (1997) being used commercially by BASF and Exxon to produce plastomers, and is being developed by Fina and others for production of high molecular weight medium and high density PEs, including bimodal MWD grades. Exxon has also indicated that it is considering the use of the technology to make LLDPE and HDPE in its existing production units at Baton Rouge, LA. Several other companies, including Borealis and Phillips, are developing applications for the technology over a very broad range of product densities and melt indices.

The new-found suitability of the loop reactor slurry process for production of LLDPE resins is perhaps the most significant change in the overall competitive position of the process relative to gas phase processes. In the 1970s, slurry processes were by far the most important technologies used for production of linear PEs. Gas phase processes, in particular those based on the use of Ziegler catalysts, were perceived to have significant product range limitations (particularly in production of broad MWD grades) and disadvantages in term of grades switching flexibility. In the 1980s, slurry processes lost this preferred position when gas phase processes were adapted (through better catalysis) to produce LLDPE. The total inability of slurry processes to economically make LLDPE precluded them from competing for this new and very large, high growth market for flexible lower density PEs. Perhaps if SSCs had been available in the 1970s, slurry processes would have retained their most favored position, and gas phase processes would never have developed to anything like their current dominant position in linear PE technology licensing.

What is the new competitive position of slurry processes relative to traditional gas phase LLDPE/HDPE technologies? It is beyond the scope of this paper to discuss these competitive factors in detail. Suffice it to say that the two technologies have been closely competitive in the past in production of HDPE resins, and the same is now likely to apply in production of LLDPE resins also. The gas phase process may have small advantage in terms of basic production economics, balanced by small advantages for the slurry process in term of great change flexibility. The slurry process has advantages with respect to the choice of comonomer for production of LLDPE: both technologies can use low molecular weight comonomers such as propylene, butene-1, and hexene-1, with essentially equal ease, but heavier comonomers such as octene-1, or even styrene or norbornene, could probably be used only in the slurry process. They would be difficult to use in the gas phase process even with the latest generation "super-condensing" technologies. It will be interesting to see whether these factors play an important role in future technology competitions.

GAS PHASE POLYETHYLENE PROCESSES

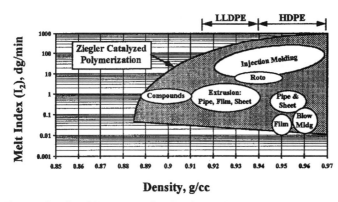

Figure 7. Gas phase PE process product domains - 1980s.

Gas phase processes have become the most widely licensed technologies in the polyolefins industry. Their primary advantages are simplicity, low cost, and a broad product range capability, particularly with respect to production of copolymers. This broad product range is illustrated in Figure 7 for 1980s technologies. Comparison of this figure with that for the slurry phase process in Figure 5 gives a clear indication of why gas phase processes became the preferred technologies: They could cover the complete range of interesting property combinations used in both LLDPE and HDPE markets.

Most developers of gas phase technologies, and all licensees, have been interested only in products of density higher than 0.915 g/cc. However, a few developers such as Union Carbide and Nippon Oil have focused on capabilities to produce soft VLDPE resins and plastomers. Both these companies have produced PEs of density between about 0.9 g/cc and 0.915 g/cc, used mainly in extrusions and compounds. However, with 1980s technologies, the lower density limit for gas phase processes has been about 0.885 g/cc, below which polymers become too sticky to fluidize in a gas phase reactor.

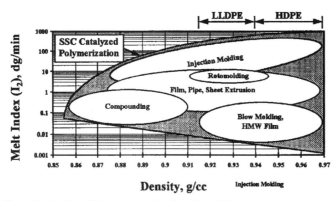

Figure 8. Gas phase PE process product domains - 1990s.

As with the slurry process, the introduction of SSCs removes many of the former product domain limitations of gas phase processes, and extends their capabilities well beyond past limits, as illustrated in Figure 8. For the same reasons as described for slurry processes, gas phase SSC-based polymers have less tendency to become sticky as density is decreased, permitting the safe production of polymers of density 0.885 g/cc; and below. At very low densities, however, down to the 0.855 g/cc of amorphous elastomers, even the lower stickiness of SSC-based products is not enough to ensure continuous plant operation. For these very soft polymers, Union Carbide has developed[2] techniques whereby a solid antistatic agent is added to the reactor to both prevent agglomeration due to static and to coat the polymer particles and prevent them sticking. Even EPDM can now be safely produced in a fluidized bed, yielding a free-flowing granular product. The preferred antistatic agent appears to be carbon black, meaning that all polymers produced with this technique are black, but colorless additives have also been developed by Union Carbide[3] for production of soft natural or colored polymers.

Union Carbide recently brought on stream a new gas phase reactor for production of EP rubber and EPDM. The Nippon Oil gas phase process has also been adapted to SSCs and will be used by Japan Polyolefins Company to produce plastomers and LLDPE in Japan.

SPECIALTY PLASTICS FROM STANDARD POLYOLEFIN TECHNOLOGIES

The very large global polyolefins industry of today is based on the use of only a handful of monomers: ethylene, propylene, butene-1, hexene-1, and octene-1. SSCs have unequaled capabilities to polymerize these same five basic monomers, but they also can effectively polymerize at least ten times this number of different vinyl monomers, including styrene, norbornene, vinyl chloride, and polar monomers such as carbon monoxide. Shell is already producing ethylene/carbon monoxide polymers using a palladium-based SSC, probably by a process route not unlike a slurry phase polyolefin process. BP and GE are working on similar technologies for production of E/CO polymers; perhaps we can expect to see a gas phase fluid bed process producing these very interesting engineering thermoplastics in the future. If this is achieved, and if the market for E/CO polyketones develops, then we could see the combination of the lowest cost polymerization technology with the lowest cost monomers available: carbon monoxide is a very cheap monomer!

Other interesting but speculative possibilities include the production of styrenic resins by solution, slurry, or gas phase polyolefin processes. These might range from LLDPE analogues based on the use of styrene comonomer rather than an α-olefin, through ESI elastomers, to ethylene modified glassy styrenic thermoplastics. Idemitsu is already producing sPS, and within a few years Dow will begin production of sPS also, possibly using a polyolefin-like gas phase process. Norbornene and other cyclic olefins are very interesting and potentially low cost monomers that could be polymerized in standard polyolefin production processes to make anything from clear elastomers to engineering thermoplastics with glass transition temperatures above 370°C. Mitsui Petrochemical and Hoechst are already producing ethylene/cyclic olefin copolymers for optical markets using SSCs. Both BF Goodrich and DuPont are developing nickel and palladium based SSCs with capabilities to produce anything from very high molecular weight liquids (M_w ~100,000) to engineering thermoplastics of M_w above 1 million. At the other end of the scale, Uniroyal is producing SSC-based low molecular weight functionalized EP copolymers, and Exxon appears to be developing SSC-based copolymers for the next generation of lube oil base stocks.

From this it is clear that SSCs provide a host of now opportunities both in production of large volume "standard" polyolefins as well as very high performance new specialties, across the whole range of polymer compositions, molecular weights, densities, melting points, and end use performance. Unlike past new technology introductions, this tremendous potential of SSCs can be made available to global markets very rapidly: they can be utilized in the existing manufacturing infrastructure, an infrastructure comprised of processes that have been developed and refined over decades to be what is now one of the most efficient infrastructures in the chemical industry. For materials manufacturers and their customers alike, the SSC/polyolefin process combination is a marriage made in heaven.

REFERENCES

1.   **U.S. Patent 5 571 880** to Phillips Petroleum, 1996.
2.   **U.S. Patent 4 994 534** to Union Carbide, 1991.
3.   **U.S. Patent 5 194 526** to Union Carbide, 1993.

Mixed Metallocenes for Designer Polymers

A. N. Speca and J. J. McAlpin
Exxon Chemical Company
Baytown Polymers Center, 5200 Bayway Drive, Baytown, TX 77522-6200, USA

INTRODUCTION

Metallocenes have inaugurated a new age of challenges and opportunities for the catalyst chemist and the polyolefin industry. Exxon Chemical is a leader in metallocenes and metallocene catalysts for use in olefin polymerization (EXXPOL® technology).[1] The catalytic ability of metallocenes to produce polymers in high yield and with narrow molecular weight distributions, NMWD, is well known.[2] That broader MWD polymers deliver improved melt processing in some polymer applications is also well known. Because of their single site nature, inter-compatibility and diverse molecular weight capabilities, metallocene mixtures in a single reactor can produce polymers with broader MWD. An attractive feature of the multiple metallocene BMWD polymers is the lack of low molecular weight waxes found in conventional BMWD polymer.

Another attractive feature of mixed metallocenes is the range of polymer properties accessible by carefully choosing the individual metallocenes and the process conditions. One now has a means of designing polymers with desired attributes. When the monomer is ethylene, the simplest (one dimensional) example of polymer design is obtained with control of MWD. With the addition of comonomer a second dimension is obtained by control of composition distribution. For propylene the one dimensional case occurs when both metallocenes have the same isotactic run length capability at different molecular weights. A second dimension is realized when the metallocenes are chosen to control not only molecular weight but stereoregularity (microtacticity). Another method for controlling microtacticity (crystallinity) is through the addition of comonomer. Further design dimensions for propylene polymers can be reached with process variables. These concepts will be discussed and illustrated with several EXXPOL® propylene polymer examples.

BACKGROUND

Theoretically the MWD of a polymer blend can be calculated from the equation in Figure 1 using the constituent number average molecular weights, M_{N_i}, weight average molecular weights, M_{W_i}, and weight fractions. As an example the equation is used to calculate the MWD shown by a 50:50 blend of polymer A with M_w of 85.0k and polymer B with M_w of 850.0k having a Schulz-Flory distribution of molecular weights (MWD=2). The resulting polymer has a MWD of 6 and a M_w of 467.5k. Unbalanced mixtures of the two metallocene polymers will have different M_ws and MWDs as shown in the graph of Figure 1. The maximum value of the MWD also depends on the individual polymer molecular weights.

Poly Blend Rule

Mw and MWD Range for Polymer Blend Of A and B

$$\mathbf{MWD} = \Sigma \ (f_i \times Mw_i) \times \Sigma \ (f_i \times 1/Mn_i)$$

For A $f_1 = 0.5$ $Mn_1 = 425k$ $Mw_1 = 850k$

For B $f_2 = 0.5$ $Mn_2 = 42.5k$ $Mw_2 = 85k$

$$\mathbf{MWD} = [(0.5 \times 850) + (0.5 \times 85)]$$
$$\times [0.5 \times 1/425) + (0.5 \times 1/42.5)]$$

$$\mathbf{MWD} = 6.0$$

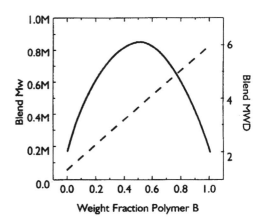

Figure 1. Polymer blends.

As these approach each other, the maximum MWD is reduced. For example, if polymer B is 425.0k then the broadest MWD has a weight to number average of 3.6.

The polymer blend rule cannot exactly predict the MWD of polymers produced with mixed metallocenes. In spite of a metallocene's ability to give NMWD polymer, in practice small deviations from the Schulz-Flory distribution are not unexpected. Factors such as mass transfer limitations, aluminoxane levels, metallocene purity, reactor impurities, metallocene and support interactions can impact the polymerization kinetics of either or both metallocenes and, hence, the individual and overall MWD. Furthermore, a one to one molar mixture of metallocene A (MCN A) and metallocene B (MCN B) can only produce a 50:50 polymer blend if the metallocenes exhibit equal activity in the reactor. These deviations afford challenges and opportunities for preparing new and novel polymers as will be seen further on.

DISCUSSION

INDIVIDUAL METALLOCENES

The relationship between metallocene ligand environment and metallocene polymerization capability[3] has fueled the growth and diversity of these compounds.[4] Thanks to our catalyst development work with Hoechst AG, a large number of metallocenes are available for making propylene polymers that cover a broad range of isotactic run lengths and molecular weights. The catalysts used in this study are racemic isomers of dimethylsilyl bridged indenyl metallocenes. They differ in the indenyl ring

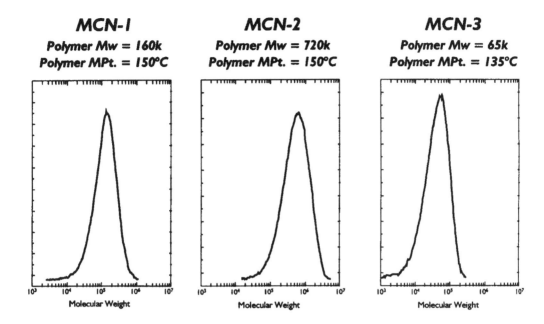

Figure 2. Polymers from example metallocenes.

substituents and have been named MCN-1, MCN-2 and MCN-3. The impact of the changing substituents is readily seen in the properties of the isotactic propylene homopolymers as shown in Figure 2. Two of the metallocenes give polymer with similar melting point but molecular weights that differ by about 4.5 fold (160.0k to 720.0k). The third gives polymer with both a lower melting point and molecular weight. With the formula of Figure 1 one calculates a 3.0 MWD for a 50:50 blend of MCN-1 and MCN-2 polymer components. How does this compare with the polymer made with a 1:1 molar mixed metallocene catalyst?

MIXED METALLOCENES

Figure 3 shows as solid and dashed lines the theoretical calculated response of MWD and M_w to the mole fraction change of MCN 2 in the mixed MCN catalyst. Overlaid on this graph are data points representing polymer MWD and M_w for produced polymer. The agreement between calculated and found is quite good and demonstrates the predictability and utility of polymer design in the base case.

Figure 4 shows the GPC traces for the individual metallocene polymers and the polymer from several mixed metallocene catalysts. Referring first to the 50:50 MCN-1 + MCN-2 traces, three important points are illustrated. First, the weight average molecular weights for the single metallocene polymers are matched by the peak positions of the mixed metallocene polymers within experimental error. This demonstrates the polymerization purity of the different metallocenes residing on the same support. Second, the mixed metallocene polymer GPC trace is not bimodal in the classical sense.

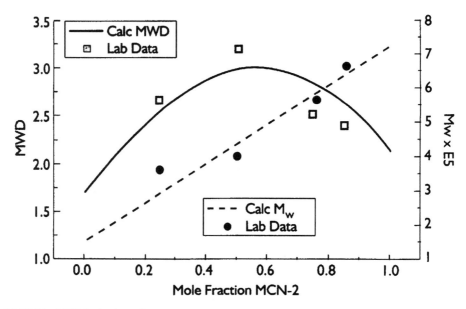

Figure 3. MCN-1 + MCN-2 mixed metallocenes.

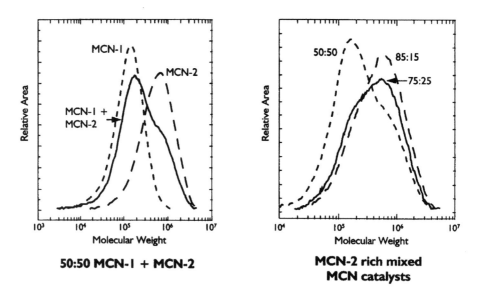

Figure 4. GPC traces for several molar ratio mixed MCN polymers.

The difference in the individual M_ws (about 4.5 fold) is not large enough for adequate separation of the individual peaks. Third, judging from the GPC shape it appears the MCN-1 polymer contribution is predominant since the MCN-2 polymer appears as the higher M_w shoulder. Individually the metallocene catalysts have about the same activity for propylene polymerization. Is MCN-2 less active in the presence of MCN-1? One alternate explanation lies in the slight asymmetry of the GPC curves evident for the individual metallocene polymers (dashed lines). Simply stated the shorter polymer chains from MCN-2 lie in the same M_w range as the weight average MW of MCN-1 polymer chains. This concentration enhancement unbalances the cumulative GPC trace. If one desires to suppress polymer chains in this molecular weight range, less MCN-1 should be used. This is demonstrated in the second graph where polymer from the 50:50 mixed metallocene catalyst is compared with those of polymers made with MCN-2 rich catalysts.

HYDROGEN AS A DESIGN TOOL

The addition of hydrogen to a polymerization process causes a boost in the rate of chain termination relative to chain propagation by a mechanism called chain transfer to hydrogen. This is one of the most fundamental means of reducing polymer M_w. If the metallocenes comprising the mixed MCN catalyst have identical hydrogen responses, i.e., the same degree of chain termination with hydrogen pressure, then the net result is a shift of the total polymer to lower M_ws without a change in the MWD broadness or shape. Such is not the case for our example metallocenes. The M_w vs. hydrogen graph in Figure 5 shows MCN-2 to be more sensitive to hydrogen since the polymer M_w drops faster with increasing hydrogen than the polymer of MCN-1. The impact on the mixed MCN polymer MWD is shown by the solid line GPC curve. There is no evidence of the high M_w shoulder seen previously. The presence of hydrogen has reduced the MCN-2 polymer M_w faster than the MCN-1 polymer M_w. MCN-2 polymer chains now overlap substantially the MCN-1 polymer chains. Evidence that MCN-2 polymer still contributes to broadening the overall distribution can be seen by comparing this 52 MFR mixed MCN + hydrogen polymer with a 57 MFR MCN-1 + hydrogen polymer (dashed line). The former has a broader MWD of 2.4 compared to 1.7 for the single MCN-1 polymer. An application where fine tuning of MWD could be useful is injection molding. Here one needs to optimize spiral flow without unduly sacrificing the inherent advantages brought to the molded part by the very narrow MWD polymer.

ETHYLENE AS A DESIGN TOOL

With single MCNs the addition of ethylene as comonomer can also reduce M_w because of a mechanism called chain transfer to monomer. The M_w-ethylene response, however, is small compared to hydrogen and many ethylene units become part of the growing chain without causing termination. The key role of ethylene is to reduce polymer crystallinity by disrupting the tacticity along the polymer backbone.[5] Surprisingly, with our mixed MCN catalyst example, the hydrogen effect on MCN-2, discussed above, appears to be reversed with the addition of ethylene as commoner. These products will be discussed in another SPE RETEC 97 paper.[6] This unanticipated impact of ethylene is illustrated in a series of GPC traces shown in Figure 6. One can see that the high M_w capability of MCN-2 reappears and gains strength as the amount of ethylene added to the reactor and hence, weight percent ethylene in the product increases. At about 1 wt % ethylene the mixed MCN polymer contains about equal amounts of MCN-1 and MCN-2 products and has a MWD of 3.6. This is a full 20% higher than the blending rule

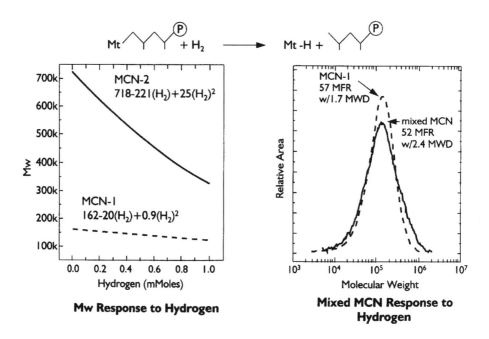

Figure 5. Molecular weight.

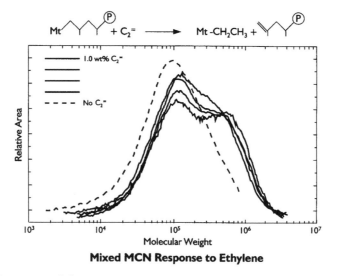

Figure 6. Mixed MCN response to ethylene.

predicts. Furthermore, neither polymer from MCN-1 nor MCN-2 are shifted to lower M_ws as predicted by the hydrogen-single metallocene experience.

MICROTACTICITY CONTROL

Another route to making lower crystallinity and hence lower melting propylene polymers is through choice of the metallocene compound. For example, while MCN-1 gives a 160.0k M_w, isotactic propylene polymer with a 150°C melting point, MCN-3 in Figure 2 provides a shorter isotactic run length polymer with 135°C melting point and 65.0k M_w.

A one to one molar ratio mixed MCN catalyst produced polymer with a measured 3.9 MWD. The MWD is much broader than predicted by the polymer blend rule and undoubtedly is due to the high concentration of shorter molecular weight chains as evident in the GPC trace shown in Figure 7. The impact of MCN-3 on broadening the melting range of the MCN-1 produced polymer can be seen in the Table. Data for the individual metallocenes show both to give narrow MWD polymers individually. Second melt DSC values show that MCN-3 polymer has a broader melting window because of lower microtacticity. The mixed MCN polymer has the same broad melting range and a broad MWD. Incidentally, this is similar to the result achieved with the addition of ethylene.

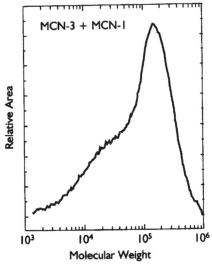

	MFR	MWD	M.Pt °C	°C Width at 1/4 Height	°C Width at 1/2 Height
MCN-1	35	1.8	149.6	9	7
MCN-3	1000	2.0	134.6	22	10
50:50 Mixed	40	3.9	144.3	23	11.5

GPC Trace for MCN Polymer

Figure 7. MCN-1 + MCN-3 polymer has broad MWD and melting range.

A broader melting range translates into a wider processing window and is highly desirable in certain applications such as Biaxially Oriented Propylene Polymers.

CONCLUSION

This paper has presented the idea and provided examples of a design hierarchy for producing polymers with desirable attributes using mixtures of metallocene compounds. The suitability of metallocenes resides in their compatibility within the catalyst while retaining their individual polymerization properties. The methodology employs the flexibility of choosing metallocenes with known individual responses to polymerization process variables as the basis for polymer design. Deviations from the simple melt blend rule present challenges but also opportunities for preparing new and novel polymers.

BIBLIOGRAPHY

1.   A. Montagna and J. C. Floyd, *MetCon '93*, Houston, May 26-28, 1993.
2.   H. Sinn *et al.*, *Makromol. Chem., Rapid Commun.*, **4**, 417 (1983).
3.   W. Spaleck *at al.*, *New J. Chem.*, **14**, 499-503 (1990); *Organometallics*, **13**, 954 (1994).
4.   J. H. Shut, *Plastics World*, April, 41 (1996).
5.   R. L. Hatterman, *Chem. Rev.*, **92(5)** 965 (1992); H. H. Brintzinger *et al.*, *Angew. Chem., Int. Ed, Engl.*, **34**, 1143 (1995).
6.   A. K. Mehta, M. C. Chen, and J. J. McAlpin, *SPE RETEC'97*, Houston, TX, February 23-26, 1997.

Advances in the Functionalized Polyolefins Synthesis and Applications

T. C. Chung

Department of Materials Science and Engineering
The Pennsylvania State University, University Park, PA 16802, USA

INTRODUCTION

Polyolefins, especially PE and PP, are used in a wide range of applications, since they incorporate an excellent combination of mechanical, chemical and electronic properties and processability. Nevertheless, deficiencies, such as the lack of reactive groups in the polymer structure, have limited some of their end uses, particularly those in which adhesion, dyeability, paintability, printability or compatibility with other functional polymers is paramount. Accordingly, the chemical modification of polyolefins has been an area of increasing interest as a route to higher value products and various methods of functionalization[1-4] have been employed to alter their chemical and physical properties.

In general, there are two routes to the functionalization of polymers, including direct copolymerization with functional monomers and the post modifications of preformed polymers. Unfortunately, Ziegler-Natta catalysts used in the preparation of polyolefins are normally incapable of incorporating functional group-containing monomers because of catalyst poisoning.[5] The Lewis acid components (Ti, V, Zr, and Al) of this catalyst will tend to complex with nonbonded electron pairs on N, 0, and X of functional monomers, in preference to complexation with the π-electrons of double bonds. The net result is the deactivation of the active sites, thus inhibiting polymerization. On the other hand, modification reactions of polyolefins are usually accompanied by some undesirable side reactions, such as crosslinking and degradation[6] of the polymer backbone. In addition, the control of functional group concentration and homogeneity in the polyolefin backbone is usually very difficult. Recently, there have been several successful examples of unsaturated polymers, with the double bonds at the chain ends[7] or pendant along the polymer backbone,[8] as the "reactive" prepolymers for subsequent modification reactions resulting in functionalized products.

RESULTS AND DISCUSSION

COPOLYMERIZATION OF α-OLEFINS AND BORANE MONOMERS[9-11]

This paper summarizes the new developments in the functionalization and graft reactions of polyolefin by Ziegler-Natta catalysts and borane monomers. One example involving copolymerization of ethylene and a borane monomer, such as 5-hexenyl-9-BBN, is shown in Equation 1. The copolymerization was carried out in a Parr reactor under N_2 atmosphere. Three catalyst systems, including two homoge-

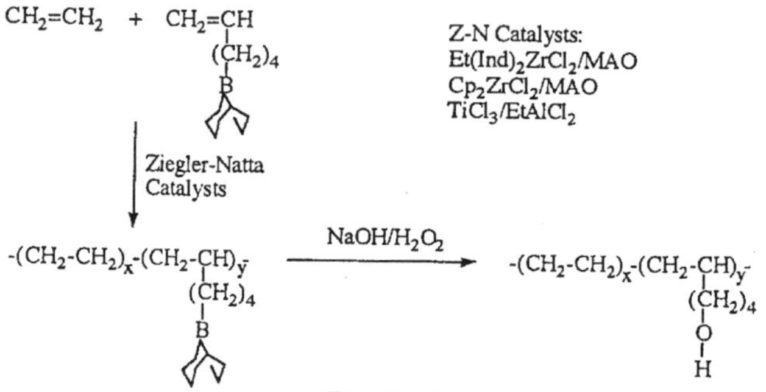

$CH_2=CH_2$ + $CH_2=CH$
$(CH_2)_4$
B

Z-N Catalysts:
Et(Ind)$_2$ZrCl$_2$/MAO
Cp$_2$ZrCl$_2$/MAO
TiCl$_3$/EtAlCl$_2$

Ziegler-Natta Catalysts

$-(CH_2-CH_2)_x-(CH_2-CH)_y$ $\xrightarrow{NaOH/H_2O_2}$ $-(CH_2-CH_2)_x-(CH_2-CH)_y$
$(CH_2)_4$ $(CH_2)_4$
B O
H

Equation 1

neous metallocene Et(Ind)$_2$ZrCl$_2$ and Cp$_2$ZrCl$_2$ with MAO catalysts and one heterogeneous TiCl$_3$AA/(Et)$_2$AlCl catalyst, were evaluated in the copolymerization reactions. Usually, the reaction was initiated by charging catalyst solution into the mixture of ethylene and 5-hexenyl-9-BBN, and a constant ethylene pressure was maintained throughout the polymerization process. Almost immediately white precipitate was observed in the beginning of reaction. The copolymerization was terminated by addition of IPA.

Table 1 summarizes the experimental results. Overall, the homogeneous zirconocene/MAO catalysts, especially, Et(Ind)$_2$ZrCl$_2$/MAO with stained ligand geometry and an opened active site, show satisfactory copolymerization results at ambient temperature.

Table 1. A summary of copolymerization reactions between ethylene, m_1, and 5-hexyl-9-BBN, m_2

Run no.	Cat. Type*	Comonomers m_1/m_2 psi/g	Reaction temp./time °C/min	Cat. activity kg/mol hr	Borane in copolymer mole%
A-1	I	45/0	30/60	644	0
A-2	I	45/0.56	30/30	1469	1.25
A-3	I	45/1.52	30/30	2020	2.15
A-4	I	45/2.10	30/30	2602	2.30
B-1	II	45/0	30/70	337	0
B-2	II	45/5	30/70	643	1.22
C-1	III	80/10	60/110	4.0	0

*catalysts: Et(Ind)$_2$ZrCl$_2$/MAO (I), Cp$_2$ZrCl$_2$/MAO (II) and TiCl$_3$AA/Et$_2$AlCl (III); solvent 100 ml toluene

Comparing runs A-1 to A-4, the concentration of borane groups in polyethylene is basically proportional to the concentration of borane monomer feed. About 50-60 % of borane monomers was incorporated into the PE copolymers after about half hour. It is very unexpected that the catalyst activity systematically increases with the concentration of borane monomer in Et(Ind)$_2$ZrCl$_2$/MAO catalyst system. Obviously, no retardation due to the borane groups is shown in these cases. The copolymerization of borane monomers in Cp$_2$ZrCl$_2$/MAO system (shown in run B-2) is significantly more difficult, only 1.22 mole % of borane monomer incorporated in PE copolymer even high concentration of borane monomer used. On the other hand, the heterogeneous TiCl$_3$AA/Et$_2$AlCl catalyst shows no detectable amount of borane group in the copolymer as shown in run C-1.

The borane containing copolymers are stable for long periods of time (6 months in dry-box) or at elevated temperatures (90°C during NMR measurement) as long as O$_2$ is excluded. By exposing a copolymer to air, the copolymer becomes insoluble at any temperature. The crosslinking reaction is due to free radical couplings, which are formed during the oxidation of borane groups by oxygen. In this study, the borane groups in polymers were reacted by ionic processes using NaOH/H$_2$O$_2$ reagents at 40°C for 3 hours. The borane groups were completely converted to the corresponding hydroxyl groups even in the heterogeneous conditions. The high surface area of borane groups in PE-B copolymer is apparently due to the semicrystalline microstructure. It is very interesting to note that the resulting functionalized polyethylene, LLDPE-OH, is structurally similar to that of linear low density polyethylene, LLDPE.

-(CH$_2$-CH$_2$)$_x$-(CH$_2$-CH)$_y$-
                        CH$_2$
                        CH$_2$
                        CH$_2$
LLDPE   CH$_3$

-(CH$_2$-CH$_2$)$_x$-(CH$_2$-CH)$_y$-
                        CH$_2$
                        CH$_2$
                        CH$_2$
LLDPE-OH   CH$_2$
                        O
                        H

The isospecific heterogeneous TiCl$_3$AA/Et$_2$AlCl catalyst was employed in the copolymerization reactions of 5-hexenyl-9-BBN and high α-olefins, such as 1-propene, 1-butene, and 1-octene. Typically, monomer pair was dissolved in toluene solution, then the polymerization was started by adding TiCl$_3$AA and Et$_2$AlCl which were premixed and aged for 1/2 hour in of toluene. Subsequently, borane groups in propylene copolymer were interconverted to various functional groups under mild conditions. One way to describe the copolymerization is to monitor the composition of copolymer during the reaction. An example using equal moles of borane monomer (5-hexenyl-9-BBN) and α-olefins, such as 1-propene, 1-butene, and 1-octene, was carried out at ambient temperature with TiCl$_3$AA and Al(Et)$_2$Cl catalyst. The borane groups in polymers were consequently oxidized to hydroxyl groups by NaOH/H$_2$O$_2$ reagents.

Figure 1 shows the plot of 5-hexenyl-9-BBN (mole %) in the copolymers versus the reaction time. In the copolymerization of 1-octene and 5-hexenyl-9-BBN, the copolymer compositions are quite flat and are close to the ideal 50 mole % mark as shown in Figure 1(a). Only small increase in borane content in the copolymer with the increasing conversion suggests a slightly higher reactivity of 1-octene as compared to 5-hexenyl-9-BBN. In the 1-butene case, the fluctuation of copolymer composition in Figure 1(b) is much more dramatic, especially in the beginning of copolymerization. The reactivity of 1-butene is significantly higher than that of 5-hexenyl-9-BBN. The initial incorporation of borane monomer in the polypropylene copolymer was very small as shown in Figure 1(c), only 1.6 mole % in the first 0.1 hour, and then increased to 3.5 and 6 mole % after 1 and 2 hours respectively. The reactivity of the monomers is obviously different, propylene >> 1-butene > 1-octene ~ 5-hexenyl-9-BBN. Basically, the reactivity of 5-hexenyl-9-BBN follows the same trend in heterogeneous

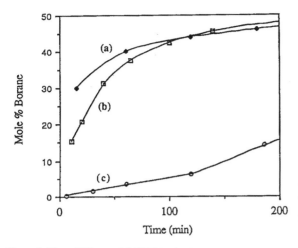

Figure 1. Plots of 5-hexenyl-9-BBN (mole%) in (a) poly(1-octene), (b) poly(1-butene) (c) polypropylene copolymers vs. time.

Ziegler-Natta polymerization, the smaller size the higher reactivity. The reactivity becomes less variable in high α-olefins.

It is feasible to obtain more uniform composition of copolymer by an engineering approach, such as the control of monomer feed ratio during the copolymerization. The more reactive α-olefin monomer was added gradually in order to keep its concentration constant relative to the borane monomer. This approach produced copolymer with a much narrower compositional distribution and at a higher yield of borane monomer than the corresponding one shot monomer addition in batch reactor. Table 2 compares the fractionation results of PP-OH copolymers obtained by both processes using 10/1 1-propene/borane monomer ratio and near complete monomer conversion. Both PP-OH-A (from batch process) and PP-OH-B (from continuous process) samples were subjected to fractionation by sequential Soxhlet extraction using various solvents, e.g., methanol, 2-butanone, MEK, heptane, and xylene all under N_2. The solvents were chosen so as to separate by polarity (OH content) and crystallinity (isotacticity and PP sequence length). The PP-OH copolymers with above 60% alcohol content are soluble in MeOH. The MEK fraction was rubbery tacky material indicative of low isotacticity. Due to the low boiling point of heptane, its fraction represents polymer with intermediate tacticity or with more hexenol units in it which reduce crystal formation. Xylene of course should dissolve all the remaining highly isotactic polymer. It is clear that the continuous reaction offers much narrower composition distribution. Most of PP-OH-B has isotactic microstructure and high crystallinity.

Table 2. Fractionation results, wt%, of PP-OH

Sample	Methanol	MEK	Heptane	Xylene	Insoluble
PP-OH-A	42.6	18.9	7.6	23.1	7.7
PP-OH-B	none	8.5	14.2	77.7	none

POLYOLEFIN GRAFT COPOLYMERS[12-15]

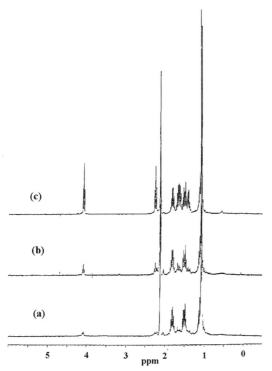

Equation 2

Figure 2. The ^{1}H NMR spectra of PP-g-PCL graft copolymers, containing (a) 17, (b) 33, and (c) 57 wt% PCL.

It is very interesting to expand the scope of borane-containing polymers, especially, to graft and block copolymers which contain both polyolefin and functional polymer segments. Both segments in the polymer are chemically bonded, but completely phase separated into different domains. A combination of physical properties can be achieved with the control of polymer morphology. In the other words, a high concentration of hydrophilic functional groups exist in polyolefin without disturbing polyolefin hydrophobic properties. The use of graft and block copolymers as emulsifiers and interfacial compatibilizers is an established technique to improve polymer interaction and morphology in polymer blends and composites. In chemistry, it is certainly very desirable to increase the efficiency of borane groups. Instead of converting one borane to one functional group, it is very important to use borane as initiator to create a functional polymer segment containing hundreds of functional groups. The general scheme in the preparation of polyolefin graft copolymers by using borane approach is shown in Equation 2.

By slowly exposing the copolymer (I) to air, the borane group becomes the reactive site for free radical polymerization. Many free radical polymerizable monomers, such as methyl methacrylate, can be polymerized to high molecular weight. In other reactions, the free radical polymerized functional polymers are chemically bonded to the side chains of polyolefin. Some interesting polymers, such as PP-g-PMMA, PE-g-PVA, and EP-g-PMMA, have been synthesized with controllable compositions and molecular microstructures. Most of them would be otherwise very difficult to prepare by the other existing methods.

Another type of PP graft copolymer, containing PP backbone and the condensation polymers on the side chains, was prepared by the hydroxylated polypropylene (PP-OH) (II) and the anionic ring opening reaction of ε-caprolactone (ε-CL). The ring opening reaction has taken place at room temperature by the insertion and ring opening of ε-CL into the Al-O bond. In all the reactions the polymer was separated into acetone soluble and insoluble fractions in a Soxhlet extractor using hot acetone. This method was proven an effective separation technique by isolating the PCL from a solution cast blend of i-PP and PCL homopolymers. The acetone-insoluble fraction is soluble in xylene at elevated temperatures. Figure 2 shows the ^{1}H NMR spectra of three acetone-insoluble samples in d_{10}-o-xylene at 120°C. Overall, the weight percent PCL in the resulting graft copolymer increased linearly with increasing ε-CL in the feed.

ADHESION STUDIES OF PP/AL AND PP/GLASS LAMINATES

The bondings between PP to glass or aluminum are both of scientific and commercial interest. The hydroxylated polypropylene was used as surface modifier to improve PP adhesion. The flexibility of hydroxyl groups located at the ends of side chains in PP-OH may enhance the interaction of PP-OH to substrates. On the other hand, the PP-OH may co-crystallize with PP. The co-crystallization of PP and PP-OH was revealed by DSC analysis as shown in Figure 3. A drawn PP film and PP-OH copolymer were melted together. The sample was reheated after cooling, co-crystallization of PP and PP-OH was shown by the presence of a single melting peak intermediate below the melting region of PP as shown in Figure 3(b). The other study directly used the laminated sample containing PP and PP-OH layers. As shown in Figure 3(c), the melting endotherm of the PP/PP-OH laminate sample consists of three peaks. The peak in the middle is at approximately the same position as the previously identified co-crystallization peak in Figure 3(b), whereas the other two peaks are indicative of pure PP and PP-OH. Accordingly, co-crystallization occurs at the PP/PP-OH interface in the laminate.

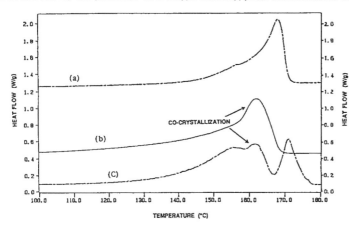

Figure 3. DSC curves of (a) PP, (b) PP/PP-OH blend and (c) PP/PP-OH laminate.

the laminate.

Table 3 summarizes the peel test results of PP/Al laminates. PP/Al laminates bonded by PP-OH were found to exhibit an extraordinary 7-10 fold increase in peel strength over acid etched samples. Contact angles of peeled Al and PP surfaces reveal typical hydrophobic surfaces. The same results were revealed in SEM studies. Both peeled Al and PP surfaces show similar morphologies. It is clear

Table 3. Peel strength of PP/Al laminates

Sample	Peel strength, N/m
Acid etched	
undrawn PP/Al	126±26
drawn PP/Al	130±34
PP-OH solution cast	
undrawn PP/Al	675±44
drawn PP/Al	1155±52

that cohesive failure occurs and gives rise to the high peel strengths observed for these PP/Al laminates. The same results were observed in the PP/Glass laminates.

PP/PC, PP/PCL, AND PP/PVC BLENDS

It would be extremely advantageous if the inexpensive, commodity PP could be effectively compatibilized in blends. The PP graft copolymers have been found to be very effective compatibilizers in polymer blends. One example of using PP-g-PCL for PP and Bisphenol-A polycarbonate, PC, blend was examined by polarized optical microscopy. As expected, a gross phase separation of the spherulitic PP and the amorphous PC phases is shown in a 70/30 blend of PP/PC. A blend containing 70/30/10 of PP/PC/ PP-g-PCL, where the graft contains 57 weight % PCL, shows the micrograph with only small distorted spherulites and a few very small distinct PC phases. The PP-g-PCL is clearly proven to be an effective compatibilizer for PP and PC blends. The similar results of PP/PVC/PP-g-PCL and PP/PMMA/PP-g-PMMA blends were also observed by optical microscopy.

CONCLUSION

The uses of borane containing monomers clearly present an effective and general approach in the functionalization of polyolefins. The success of this chemistry is based on the combination of advantages, (a) the stability of borane moiety to transition metal catalysts, (b) the solubility of borane compounds in hydrocarbon solvents (hexane and toluene) used in transition metal polymerizations, and (c) the versatility of borane groups, which can be transformed to a remarkable variety of functionalities and to free radicals for graft-form reactions. The functionalized polymers are very effective interfacial modifiers in improving the adhesion between polyolefin and substrates and the compatibility in polyolefin blends and composites.

ACKNOWLEDGMENT

Authors would like to thank for the financial support from the Polymer Program of the National Science Foundation.

REFERENCES

1.   M. D. Baijal in **Plastics Polymer Science and Technology**, *John Wily & Sons*, New York, 1982.
2.   E. C. Carraher, Jr. and J. A. Moore in **Modification of Polymers**, *Plenum*, Oxford, 1982.
3.   M. D. Purgett, **Ph.D. Thesis**, *University of Massachusetts*, 1994.
4.   K. I. Clark and W. G. City, **U.S. Patent 3 492 277** (1970).
5.   J. Boor, Jr. in **Ziegler-Natta Catalysts and Polymerizations**, *Academic Press*, New York, 1979.
6.   G. Ruggeri, M. Aglietto, A. Petragnani, and F. Ciardelli, *Eur. Polymer J.*, **19**, 863 (1983).

7.   T. Shiono, H. Kurosawa, O. Ishida, and K. Soga, *Macromolecules*, **26**, 2085 (1993).
8.   S. Marathe and S. Sivaram, *Macromolecules*, **27**, 1083 (1994).
9.   T. C. Chung, *Macromolecules*, **21**, 865 (1988).
10.   T. C. Chung and D. Rhubright, *Macromolecules*, **24**, 970 (1991).
11.   T. C. Chung and D. Rhubright, *Macromolecules*, **26**, 3019 (1993).
12.   T. C. Chung and G. J. Jiang, *Macromolecules*, **25**, 4816 (1992).
13.   T. C. Chung, G. J. Jiang, and D. Rhubright, *Macromolecules*, **26**, 3467 (1993).
14.   T. C. Chung and D. Rhubright, *Macromolecules*, **27,** 1313 (1994).
15.   T. C. Chung, D. Rhubright, and G. I. Jiang, **U.S. Patent 5 286 800** (1994).

SPS Crystalline Polymer: A New Metallocene-catalyzed Styrene Engineering Thermoplastic

Robert Brentin, David Bank, and Michael Hus
The Dow Chemical Company, USA

ABSTRACT

Using new developments in metallocene catalysis, a semicrystalline styrene polymer has been produced with polymer chains in a symmetrical configuration. This allows crystalline domains to form, yielding polymer properties in the engineering thermoplastics category.

This chapter focuses on polymer performance and processing characteristics of glass reinforced SPS crystalline polymer. The thermal, mechanical, and chemical resistance properties of SPS provide a unique combination and a good balance with respect to other glass-reinforced, semicrystalline resins.

INTRODUCTION

The semicrystalline SPS polymer provides good heat and chemical resistance, mechanical and electrical properties with low moisture sensitivity and easy processability. This set of properties makes glass-reinforced SPS resin well suited for industrial, automotive, electronics, electrical, and consumer applications (Table 1).

The significant interest in this new polymer arises from it being based on a well-established, widely available monomer, but with properties competitive with other engineering plastics and with some distinctive features.

POLYMER DESIGN AND MICROSTRUCTURE

Single site catalysts, such as metallocenes (containing transition metals), optimize polymer architecture to generate highly uniform molecules and even permit tailoring new categories of polymers, as is SPS. Single-site catalysis allows polymer units to be built in a highly selective manner. When the polymer's side chains are arranged in a symmetrical pattern on either side of the polymer backbone, it is described as "syndiotactic". SPS is a styrene polymer in which the styrene monomer units are arranged in a syndiotactic, or regularly alternating pattern. This arrangement allows crystallinity to develop at a rapid rate. The formation of crystalline domains in the melt results in a semicrystalline microstructure.

On a microscopic scale, SPS crystalline polymer appears as tightly packed spherulite structures. Compared to other common semicrystalline polymers, the microstructure is more uniform and

Table 1. SPS Crystalline Polymers - Typical Properties

Properties	ASTM Method	Units	30% glass	40% glass	30% glass, impact modified	30% glass, ignition resistant impact modified	40% glass, resistant impact modified
Tensile Strength	D 638	MPa	121	132	105	96	122
Tensile Modulus	D 638	MPa	10,000	11,170	7,580	9,650	10,340
Elongation	D 638	%	1.5	1.5	3.4	1.8	1.8
Notched Izod Impact (23°C)	D 256	J/m	96	112	117	70	91
Notched Izod Impact (-18°C)	D 256	J/m	96	112	96	70	91
DTUL@ 1.82 MPa	D 648	°C	249	249	232	221	240
DTUL@ 0.45 MPa	D 648	°C	263	263	263	263	263
Specific Gravity	D 792		1.25	1.32	1.21	1.44	1.47

relatively small. It is notable that the densities of the amorphous regions and the crystalline regions are both 1.05. These factors should favor more isotropic part shrinkage.

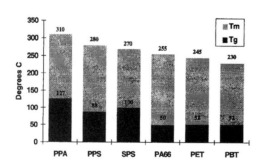

Figure 1. SPS heat resistance comparison. Glass transition and melting point temperature.

HEAT RESISTANCE

The heat resistance of SPS is attributed to its 100°C amorphous glass transition temperature and its 270°C crystalline melting temperature. In a molded part, glass fibers act to bridge the amorphous regions by tying together the crystalline spherulites to maintain modulus and dimensional integrity at higher temperatures. Above the glass transition temperature, polymers soften. The glass transition temperature of SPS at 100°C is higher than that of several other engineering thermoplastics in this category (Figure 1).

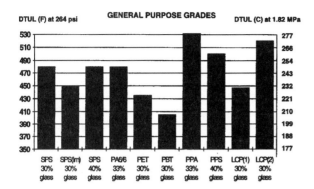

Figure 2. Heat deflection temperature of reinforced engineering thermoplastics.

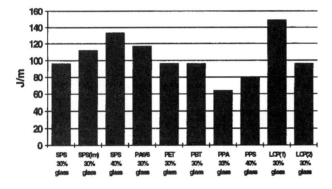

Figure 3. Notched Izod impact of glass reinforced engineering thermoplastics.

One measure of heat resistance is deflection temperature under load, DTUL. As illustrated in Figure 2, glass reinforced SPS heat resistance should be intermediate to that of glass reinforced polyesters and glass reinforced polyphthalamide/polyphenylene sulfide/liquid crystal polyester.

MECHANICAL PROPERTIES

The addition of 20 to 40% by weight of glass fiber reinforcement and/or mineral filler to SPS produces a range of products with high heat resistance and good dimensional stability. The most significant mechanical effect of adding glass fiber reinforcement to SPS is to raise tensile, flexural and impact strength. Impact modifier and flame retardant chemicals are added to other formulations to enhance certain properties.

Representative formulations include SPS with 30 and 40% glass fiber reinforcement and an impact modified SPS (designated SPS(im) in this paper) with 30% glass fiber. Tensile strength ranges from 100 to 130 MPa. Tensile modulus ranges from 7,500 to 11,000 MPa. Impact strength ranges from 70 to 120 J/m in the Izod test and 4 to 7 J in the Gardner impact test. In short, the mechanical properties of glass reinforced SPS are competitive with other glass reinforced engineering thermoplastics in this category.

Glass reinforcement of engineering thermoplastics tends to raise the impact resistance of some polymers and lower it for others. The result is a compressed range of impact strength, as measured by the notched Izod impact test (Figure 3). Dart drop testing also corroborates this leveling of impact strength. Mechanical property test results give direction in product design, however there is no substitute for testing molded parts under use conditions.

WATER RESISTANCE

SPS is a nonpolar polymer and does not attract appreciable moisture. In most cases, predrying before melt processing is not required. Moisture pickup by several glass reinforced engineering thermoplastics is shown in Figure 4. The feature of reduced or no drying allows for reduced inventory of in-process resin and lower energy consumption.

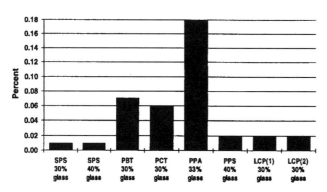

Figure 4. Water absorption of glass reinforced, ignition resistant thermoplastics. Water absorption after 24 h (50% RH, RT).

Since SPS is addition polymerized, as opposed to condensation polymerized, as are polyesters and nylons, the potential for molecular weight degradation due to reaction with water is removed. This allow greater operating flexibility for injection molders and serves to increase confidence in final part performance.

In applications involving water contact and water immersion, retention of physical properties over time becomes a factor. A pressurized hot water test was conducted to demonstrate the effect water has on SPS crystalline polymer. The results shown in Figure 5 indicate good retention of tensile strength after 3,000 hours immersion.

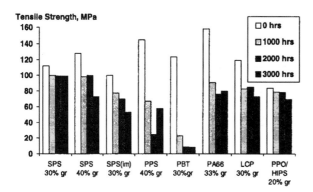

Figure 5. Hot water immersion, 200°F, 10 psi.

ELECTRICAL PROPERTIES

As a nonpolar polymer with low moisture absorption, SPS exhibits good electrical properties. Its dielectric constant of 3.1 as shown in Figure 6 remains relatively constant over a range of frequency and temperature. Other important electrical properties for 30% glass reinforced SPS having a UL94V0 rating at 0.8 mm include a dissipation factor of 0.0002 at 100 KHz, a dielectric strength of 23.6 kV/mm and a comparative tracking index of 570 volts. The electrical and without impact relative thermal index (RTI) for the 30% glass reinforced, impact modified, ignition resistant product is 130°C.

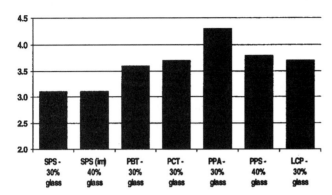

Figure 6. Dielectric constant of ignition resistant engineering thermoplastics @ 1 MHz

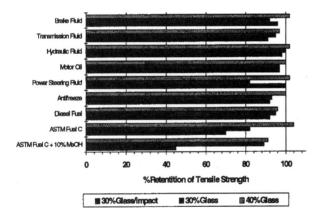

Figure 7. Chemical resistance - automotive fluids. 30 days immersion, RT.

CHEMICAL RESISTANCE

Anticipating reliability of parts exposed to a variety of chemical environments adds another layer of complexity to the task of plastic material selection. Semicrystalline engineering thermoplastics generally have good resistance to chemicals. Selection of which engineering thermoplastic to use depends on its performance in the intended chemical environment.

SPS has a wide range of chemical resistance, including acids, bases, and most organic solvents. Exceptions are organics with solubility parameters close to that of styrene polymer, such as benzene and toluene.

Figures 7-10 report retention of tensile strength for three SPS products after immersion in various fluids for 30 days. The standard tensile specimens were tested and compared to controls to determine the effect of the chemical reagent. Test results for automotive fuels, salt solutions, oil, battery acid and radiator fluid indicate broad suitability for SPS in automotive and industrial applications.

SPECIFIC GRAVITY

One feature of SPS that should have appeal across all applications is its relatively low specific gravity. Since the base resin has a 1.05 specific gravity, reinforced grades can be made at a density advantage over competitive materials. As shown in Figure 11, this specific gravity advantage results in a 10 to 20% greater part yield per pound of glass reinforced resin.

INJECTION MOLDING

Even a high performance polymer will realize only limited use if it is difficult to melt process. Low melt viscosity under process conditions is one measure of good processability. Capillary melt viscosity

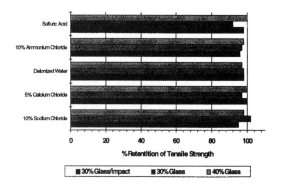

Figure 8. Chemical resistance - aqueous fluids. 30 days immersion, RT.

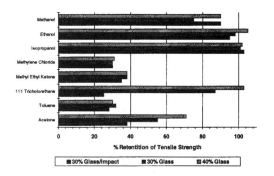

Figure 9. Chemical resistance - solvents. 30 days immersion, RT.

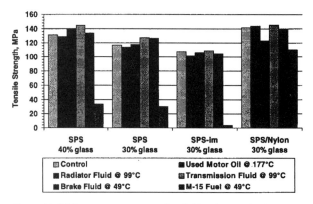

Figure 10. High temperature automotive fluid resistance. 30 days immersion.

measurements of several polymers at their processing temperatures (Figure 12) support the observations in field trials that SPS is able to fill thin sections and flow long lengths with modest injection pressure.

A 2D mold flow simulation of SPS and PBT indicates that parts designed with glass reinforced SPS should be able to incorporate greater flow lengths or reduced wall thickness, smaller runner diameters (decreased scrap and regrind), shorter molding cycle time, and lower injection pressure (more efficient use of molding presses).

CONCLUSIONS

The combination of a unique physical property profile, processing ease and exciting new catalysis technology makes SPS crystalline polymer a significant new development in polymer technology.

Glass reinforced SPS has the thermal, mechanical, chemical and electrical performance capabilities to compete in the engineering thermoplastics arena. SPS is now in the market development stage to validate its performance in a wide variety of applications and end use environments. Just how significant SPS becomes in the world of plastics remains to be seen. Based on a well-established monomer and possessing a unique combination of performance properties, SPS could become a workhorse engineering thermoplastic.

ACKNOWLEDGMENTS

The authors acknowledge all those at Dow Plastics and Idemitsu Petrochemical who have had a role in the research and development this new engineering thermoplastic.

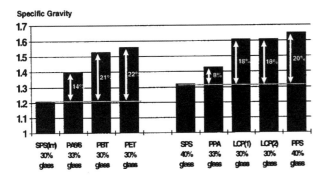

ARROW INDICATES % REDUCTION OF PART WEIGHT WITH SPS

Figure 11. Specific gravity of commonly used glass reinforced engineering thermoplastics.

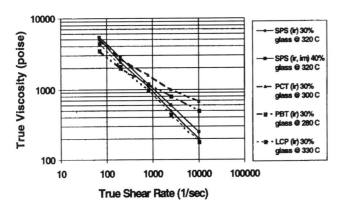

Figure 12. Capillary melt viscosity of ignition resistant engineering thermoplastics.

Ethylene/α-olefin Copolymers with Metallocene Catalysts in High Pressure Process

Akihiro Yano and Akira Akimoto
TOSOH Yokkaichi Research Laboratory, Japan

ABSTRACT

Metallocene catalysts usually produce low molecular weight ethylene/α-olefin copolymers at high polymerization temperature. We found some new metallocene catalysts which produced high molecular weight copolymers even at high polymerization temperature higher than 150°C. These catalysts showed good catalyst activity and good copolymerization reactivity. We adopted these catalysts at high pressure process and succeeded to develop production technology.Theses polymers produced in high pressure process showed narrow molecular weight distribution and narrow chemical composition distribution. Mechanical properties and physical properties of these polymers were superior to conventional LLDPE.

INTRODUCTION

In the last decade a lot of interest has been focused on the field of polymerization of a wide variety of monomers with metallocene catalysts such as ethylene, propylene, styrene, cyclic olefins.[1-4] Many companies are challenging for commercialization of these new metallocene polymers. Most progress has been achieved in the field of ethylene/α-olefin copolymers. EXXON has launched a range of ethylene/α-olefin copolymers since 1990 and DOW also produced these polymers. Ethylene/α-olefin copolymers are usually produced in gas phase, solution phase, slurry phase and high pressure process. In high pressure process, direct use of homogeneous metallocene catalysts is possible and this process is suitable for production of very low density polyethylenes and has the great advantage of carrying out grade change so easily. On the other hand, high polymerization temperature is needed to operate this process effectively. New metallocene catalysts are needed to produce high molecular weight polymers in high pressure process. In this paper, we describe the preparation of new metallocene catalysts and polymer properties yielded in high pressure process.

CATALYST DEVELOPMENT

We have to prepare the new catalyst which produces high molecular weight polymers at high pressure process.[5,6] It is clear that chain transfer reaction has to been controlled to yield high molecular weight polymers. Optimization of ligand structure of metallocene compound is important for production of high molecular weight polymers at high polymerization temperature. We optimized the structure of metallocene compound on the following two points.
1 Stereorigidity of metallocene compound
2 Electron-donating property of metallocene ligand.

We prepared metallocene compounds which had different bridge structure to investigate the effect of stereo rigidity of metallocene on catalyst performance.

Table 1 shows the results of ethylene polymerization using various metallocene compounds which have different bridge structures and two indenyl ligands. Ethylene-bridged metallocene compound produced the lowest molecular weight polymer and carbon-bridged and silicon-bridged metallocene compound produced higher molecular weight polymer than ethylene-bridged metallocene compound. These results show that increasing the stereorigidity of metallocene based on decreasing the number of bridged-atom is effective to obtain high molecular weigh polymers at high polymerization temperature.

Table 1. The results of ethylene polymerization with metallocene connecting indenyl ligand

Metallocene	Activity (kg/mmol.Zr)	M_w (10^{-4})
Et(Ind)$_2$ZrCl$_2$	40	2.0
Me$_2$C(Ind)$_2$ZrCl$_2$	6	3.2
Me$_2$Si(Ind)$_2$ZrCl$_2$	5	2.8
Ph$_2$Si(Ind)$_2$ZrCl$_2$	12	3.7

Polymerization temperature: 150°C, polymerization time: 20 min, ethylene pressure: 20 kg/cm^2

Table 2. The results of ethylene polymerization with metallocene connecting cyclopentadienyl and fluorenyl ligand

Metallocene	Activity (kg/mmol.Zr)	M_w (10^{-4})
Me$_2$C(Cp)(Flu)ZrCl$_2$	6	3.2
MePhC(Cp)(Flu)ZrCl$_2$	32	3.6
Ph$_2$C(Cp)(Flu)ZrCl$_2$	172	7.9

Polymerization temperature: 150°C, polymerization time: 20 min, ethylene pressure: 20 kg/cm^2

As said before, electron effect seemed to be important. Fluorene has a strong electron donating property among substituent cyclopentadienyl compounds. We prepared some metallocene compounds selecting carbon atom for bridge structure and fluorene for a ligand and investigated the effect of electron donation on catalyst performance.

Table 2 shows the results of ethylene polymerization using metallocenes having the fluorenyl ligand. Unfortunately dimethylmethylene cyclopentadienyl fluorenyl zirconium dichloride [Me$_2$C(Cp)(Flu)ZrCl$_2$] did not show high catalytic performance: activity was not so good and molecular weight was low. However by introduction of phenyl group instead of methyl group at methylene-bridged moiety, activity was enhanced by about 30 times and molecular weight of polyethylene has been also improved drastically. We think that electron donating effect due to the selection of fluorenyl ligand and stereo rigidity due to the selection of phenyl group at methylene-bridged substituents are important factors to produce high molecular weight polymers at high polymerization temperature.

COPOLYMERIZATION REACTIVITY

Table 3. The results of copolymerization with various metallocene catalysts

Metallocene	Activity (kg/mmol.Zr)	M_w (10^{-4})	T_m (°C)
Cp_2ZrCl_2	4	-	126
$Et(Ind)_2ZrCl_2$	40	2.0	124
$Ph_2C(Cp)(Flu)ZrCl_2$	170	7.6	116

Polymerization temperature: 150°C, polymerization time: 20 min, ethylene pressure: 20 kg/cm², hexene-1: 20 ml

Copolymerization reactivity at high polymerization temperature is one of important factors for the choice of metallocene catalyst. We checked the copolymerization reactivity of the catalyst using $Ph_2C(Cp)(Flu)ZrCl_2$. Table 3 shows the ethylene/1-hexene copolymerization results with various metallocene catalyst systems. These results show that the incorporation of higher olefins decreased in the following order: $Ph_2C(Cp)(Flu)ZrCl_2$ > $Et(Ind)_2ZrCl_2$ > Cp_2ZrCl_2. We think that the metallocene compounds which have wide reaction field are needed to prepare the good copolymerization reactivity catalyst system. From this point of view, $Ph_2C(Cp)(Flu)ZrCl_2$ has a small bite angle so this catalyst showed good copolymerization reactivity.

TEST RUNS IN HIGH PRESSURE PILOT PLANT AND ACTUAL PLANT

In general, high pressure processes are operated under high polymerization temperature and monomer pressure and polymerization time are very short. So at first we had to check if developed catalyst shows good catalyst performance under these special polymerization conditions.

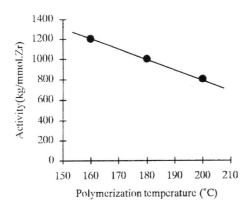

Figure 1. Effect of the polymerization temperature on the activity in actual plant.

We tried pilot tests with this catalyst system. This catalyst produced high molecular weight ethylene/α-olefin copolymers. The catalyst activity was very high and copolymerization reactivity was also very good in high pressure process. The catalyst activities were changed for polymerization conditions. The range of it was 200-800 kg/mmol.Zr. Copolymerization reactivity is very important feature to produce low density polymers in high pressure process. This catalyst showed good copolymerization reactivity, so it was possible to produce polymer which had density of 0.87 using this catalyst.

Consequently, we succeeded in ethylene copolymers production covering the wide range of density and the melt flow rates, MFR, regions in pilot plant.

Figure 1 shows the results of the actual plant test using the catalyst system containing $Ph_2C(Cp)(Flu)ZrCl_2$ compound. Catalyst activity was higher than the activity of pilot plant. Copolymerization reactivity was also as good as pilot test results. So far we succeeded to develop production technology of ethylene/α-olefin copolymers with these catalysts in high pressure process.

MOLECULAR CHARACTERIZATIONS OF COPOLYMERS PRODUCED IN HIGH PRESSURE PROCESS

Ethylene/α-olefin copolymers with metallocene catalysts usually show narrow molecular weight distribution and narrow chemical composition distribution. We characterized ethylene/α-olefin copolymers made by our metallocene catalysts in high pressure process. They showed narrow molecular weight distribution as expected for metallocene catalyst so the usual value of distribution width were in the range of $M_w/M_n \approx 2$. Chemical composition distributions was checked by temperature rising elution fractionation. These distribution curves showed narrow and only one peak. These results indicated that these polymers had a narrow chemical composition distribution as expected from metallocene catalyst.

Most important point is that these features hold for the sample of low density polyethylenes, so all products with our metallocene catalysts in high pressure process have very low extractables even for low density polyethylenes.

It is well known that polyethylene with long chain branch are produced by using some kinds of metallocene catalysts such as Constrained Geometry Catalyst, CGC, prepared by DOW. We can characterize the existence of long chain branch for investigating the relationship between the intrinsic viscosities [η] and the melt flow rates, MFR, and the relationship between the melt tensions and the melt flow rates, MFR. These samples showed high [η] and low melt tension against MFR. From these results it is clear that these copolymers do not have significant amounts of long chain branch. High ethylene concentrations are needed to operate high pressure process. Long chain branches are produced by copolymerization of polymer chain containing terminal double bond and ethylene. So it is very difficult to produce polyethylene with long chain branch in high pressure process.

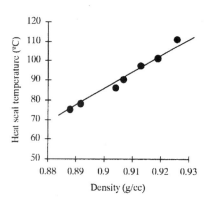

Figure 2. Relationship between heat seal temperature and density.

CAST FILM PRODUCT

During our actual plant trial, we successfully produced cast film grades. We evaluated these resins on conventional cast film equipment against conventional LLDPE. One important cast film property is heat sealing property. Figure 2 shows the relationship between heat seal temperature and polymer density. Heat seal temperature fell in proportion as polymer density decreased. We understood that heat seal temperature of polymer having density = 0.89 corresponded to that of EVA 20% from these results. As said before, metallocene LLDPE has a narrow chemical composition distribution, so lamellae thickness is thinner than that of conventional LLDPE. Thin lamellae breaks easily at lower temperature,

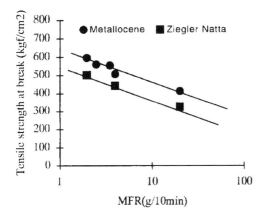

Figure 3. Tensile strength at break vs. density.

therefore melting point becomes also lower. This is a main reason that our new resins showed good heat sealing properties.

The clarity of cast films are usually improved because of quenching at cast film process. The clarity of our films higher than density = 0.91 were same as conventional LLDPE but in general blocking agents such as silica are used to inhibit blocking for conventional LLDPE. This causes decreasing the clarity of film. The clarity of our films lower than density = 0.91 was improved. This is due to the homogeneity of branch structures in our polymers.

Figure 3 shows tensile strength at break of our cast films. Tensile strength at break was superior to conventional LLDPE and ethylene/butene-1 copolymers indicated the lower value compared with ethylene/hexene-1 copolymers.

BLOWN FILM PRODUCT

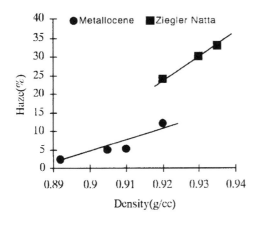

Figure 4. Haze vs. density.

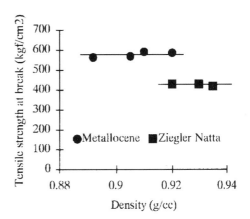

Figure 5. Tensile strength at break vs. density.

Blown film offered good strength properties and greatly improved optical properties compared with conventional LLDPE. Figure 4 shows the haze versus density of blown films. According as the density decreased the clarity of films was improved. These good optical properties are due to thin lamellae thickness and small size of spherulite, so in low density regions the clarity of films is especially improved.

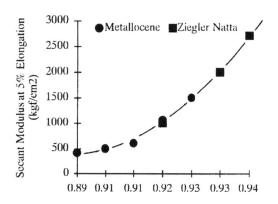

Figure 6. Modulus vs. density.

Figure 5 shows the tensile strength at break versus density. Tensile strength at break of our new resins was also superior to conventional LLDPE. Dart impact was also superior to conventional LLDPE. This feature was shown clearly at low density region especially less than 0.91 density. In general the ratio of tie molecules for metallocene LLDPE is higher than conventional LLDPE. We thought that it was a main reason to improve mechanical properties of our new materials.

Figure 6 shows modulus versus density. Modulus of our new resins indicated comparable performance to conventional LLDPE.

These results showed that our new resins had superior clarity and mechanical properties and similar modulus.

CONCLUSIONS

We prepared new catalysts which produced high molecular weight copolymers at high polymerization temperature. We adopted these catalysts to high pressure process and established metallocene LLDPE production technology. These resins showed good mechanical properties and optical properties. These features are due to homogeneity of polymer structures.

As mentioned before this catalyst system shows very high activity, so the residue of catalyst components in polymers is also very small. For example, the residue value of zirconium was about under 0.3 ppm. This feature is useful for opening up new avenues for use of theses materials such as medical and electric fields.

REFERENCES

1.   H. Sinn and W. Kaminsky, *Adv. Organomet. Chem.*, **18**, 99 (1980).
2.   W. Kaminsky, K. Kulper, H. H. Brinzinger, and F. R. W. P. Wild, *Angew. Chem. Int. Engl. Ed.*, **24**, 507 (1985).
3.   A. Ewen, R. I. Jones, and A. Razavi, *J. Am. Chem. Soc.*, **110**, 6255 (1988).
4.   N. Ishihara, T. Seimiya, M. Kuramoto, and M. Uoi, *Macromolecules*, **19**, 2464 (1986).
5.   A. Akimoto and A. Yano, Metallocene'94, 1994.
6.   A. Akimoto, Metallocene'95, 1995.

Semicrystalline Polyolefins - Narrow MWD and Long Chain Branching: Best of Both Worlds

João B.P. Soares

Department of Chemical Engineering, University of Waterloo, Waterloo, Ontario, Canada N2L 3G1

Archie E. Hamielec

Department of Chemical Engineering, McMaster University, Hamilton, Ontario, Canada L8S 4L7

ABSTRACT

Constrained geometry metallocene catalysts can be used to synthesize polyolefins with narrow molecular weight distributions and significant degrees of long chain branching. These polymers have excellent mechanical properties combined with good processability (good shear thinning, delayed melt fracture, and improved melt strength).

The bivariate distribution of molecular weight and frequency of long chain branching has been solved analytically for a continuous stirred tank reactor operating at steady-state. Numerical solutions of the population balance equations and Monte Carlo simulations are in perfect agreement with the analytical solution.

INTRODUCTION

Prior to the publications by Lai *et al.*[1,2] and Swogger and Kao,[3] there was no unambiguous experimental evidence in the literature that ionic catalyst systems such as Phillips chromium oxide, Ziegler-Natta, and metallocene catalyst systems, could be used to synthesize homopolyethylene or ethylene-olefin copolymers with long chain branches. Lai *et al.*[1] synthesized these branched polyolefins using a particular type of metallocene catalyst (constrained geometry catalyst). These polyolefins have a frequency of long chain branching in the range of 0.01 to 3 long chain branches per 1000 carbon atoms. The polymerization is done in a CSTR operating at steady-state and at high temperatures. An aliphatic solvent is used in the reactor and all of the polymer chains are in solution.

The most suitable catalyst types for long chain branch formation appear to be those with an "open" metal active center, such as the Dow Chemical constrained geometry catalysts. The active center of these catalysts is based on group IV transition metals that are covalently bonded to a monocyclopentadienyl ring and bridged with a heteroatom, forming a constrained cyclic structure with the transition metal center. Strong Lewis acids are used to activate the catalyst to a highly effective cationic form. This geometry allows the titanium center to be more "open" to the addition of ethylene and higher α-olefins, but also for the addition of vinyl-terminated polymer molecules.[4]

Lai *et al.*[1] presented some remarkable data on the effect of polydispersity on I_{10}/I_2 for linear polyolefins synthesized using classical heterogeneous titanium-based Ziegler-Natta catalysts and produced with a homogeneous constrained geometry catalysts. It is generally accepted that classical Ziegler-Natta catalysts have multiple active center types and consequently produce polyolefins with broad mo-

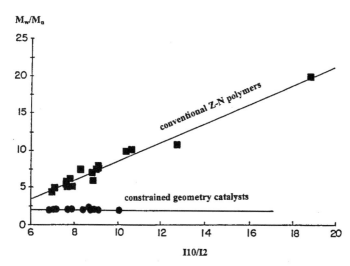

Figure 1. Relationships between polydispersity index and I_{10}/I_2 ratio for polyolefins produced with heterogeneous and homogeneous Ziegler-Natta catalysts and constrained geometry catalyst.

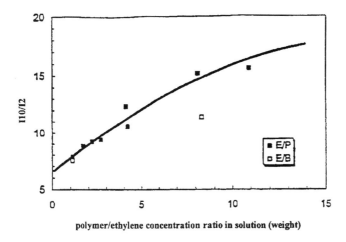

Figure 2. Calibration curve for I_{10}/I_2 ratio as a function of polymer/ethylene concentration ratio for a continuous stirred tank reactor.

lecular weight distribution.[5] Shear thinning increases as the molecular weight distribution broadens for polyolefins produced with these catalysts. On the other hand, polyolefins synthesized with constrained geometry catalysts have narrow molecular weight distribution, with polydispersity near the theoretical value of two for single-site type catalyst. However, the I_{10}/I_2 ratio can be increased at almost constant polydispersity, by increasing the long chain branching frequency. Figure 1 illustrates these remarkable results.

In fact, these authors have shown how to synthesize polyolefins with narrow molecular weight distribution and sufficient degree of long chain branching that combines the excellent mechanical properties of polyolefins with narrow molecular weight distribution (impact properties, tear resistance, environmental stress cracking resistance, and tensile properties) with the good shear thinning of linear polyolefins with broad molecular weight distribution. Polyolefins with narrow molecular weight distribution and containing no long chain branches generally have poor processability.

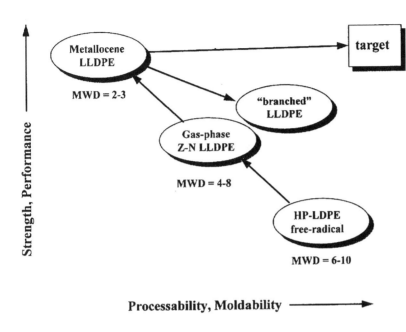

Figure 3. Balance of strength versus processability and moldability for polyolefins synthesized using different polymerization processes.

Sugawara[6] constructed the CSTR calibration curve shown in Figure 2 using data available in the patent application by Lai et $al.$[2] and then synthesized two ethylene-olefin copolymers (with densities of 0.906 and 0.912 g/cm^3). The melt index ratio I_{10}/I_2 correlates with long chain branching frequency (the higher the ratio, the higher the long chain branching frequency). Sugawara used this calibration curve to find CSTR operation conditions to produce a polymer with very low level of long chain branching (polymer A) and a polymer with a moderate level of long chain branching (polymer B). When compared with a HP-LDPE and a LLDPE, polymers A and B were superior in impact strength. Polymer B was superior to LLDPE but inferior to HP-LDPE in processability. Polymer B was significantly inferior to HP-LDPE in bubble stability. This superiority of HP-LDPE over polymer B was very likely due to the much higher levels of long chain branching in HP-LDPE polymers. Polymers A and B were inferior to both LLDPE and HP-LDPE in MD but superior to HP-LDPE in TD. Sugawara provided an interesting summary of property balances and the position of metallocene copolymers of ethylene-olefins in the balance of strength versus processability and moldability (Figure 3).

It seems that the target for polymers made with metallocenes is to produce polymer chains having narrow molecular weight distribution but to raise their level of long chain branching to further improve processability.

PROCESS FOR PRODUCING POLYOLEFINS WITH LONG CHAIN BRANCHES

Solution processes are preferred for long chain branch synthesis for several reasons. Macromonomers with terminal vinyl unsaturation are formed at higher rates at elevated temperatures. These dissolved polymer chains are mobile and have relatively large self-diffusion coefficients. Most importantly, at these higher temperatures the steric barrier for the addition of a macromonomer to the active center is relatively less important than at lower temperatures.

A potential negative effect of these elevated temperatures is the reduced lifetime of the catalyst active centers. Short residence times can be used with a CSTR to overcome this problem. Shorter reactor residence times would also have the advantage of making product grade transitions more efficient.

For continuous operation at steady-state, the residence-time distribution plays a very significant role in long chain branch generation. Consider the two extremes of plug flow tubular reactor, PFTR, with a narrow residence-time distribution, and of CSTR with a very broad residence-time distribution. With the PFTR, the polymer concentration increases while the monomer concentration decreases as one moves along the tube (similar to a batch reactor in time). High polymer concentrations are obtained near the exit of the PFTR, while much of the polymerizing mass in the reactor is at low polymer concentrations. With the CSTR, the entire reacting mass is at the same high polymer and low monomer concentration. Rates of long chain branch reactions are clearly higher at higher concentrations of polymer with terminal vinyl unsaturation, and rates of monomer consumption are clearly lower at the lower monomer concentrations, giving high levels of long chain branching per 1,000 carbon atoms. It is perfectly clear, therefore, that the optimal reactor type and operation conditions for maximum long chain branch formation is the steady-state operation of a continuous stirred tank reactor.

BIVARIATE DISTRIBUTION FOR MOLECULAR WEIGHT AND LONG CHAIN BRANCHING

The most likely long chain branch formation mechanism with metallocene catalyst systems is terminal branching, a mechanism which has been known in the free-radical polymerization literature for many years.[7] In free-radical polymerization, macromonomers are generated via termination by disproportionation and via chain transfer to monomer. With metallocene catalyst systems, the facile β –hydride elimination reaction appears to be responsible for in situ macromonomer formation. Other reaction types, such as β –methyl elimination and trans[8] may also generate dead polymer chains with terminal unsaturation. β –methyl elimination can actually be the most important transfer mechanism in propylene polymerization when $(Me_5Cp)_2Ti$, $(Me_5Cp)_2Zr$, and $(Me_5Cp)_2Hf$ complexes are used.[9] Therefore, these catalytic systems have the potential of producing polypropylene with long chain branches. These and other chain transfer mechanisms are summarized in Table 1.

The chain length distribution of polymer chains produced with metallocene catalysts that permit long chain branch formation via the terminal double-bond mechanism can be described analytically for each population containing a different number of long chain branches per polymer molecule.[10] The polymerization kinetic model involves the following steps:

$$P_{r,i} + M \rightarrow P_{r+1,i} \qquad k_p$$

$$P_{r,i} + D^=_{q,j} \rightarrow P_{r+q,i+j} \qquad k_{pLCB}$$

$$P_{r,i} + CTA \rightarrow D_{r,i} + P_{0,0} \qquad k_{CTA}$$

$$P_{r,i} \rightarrow D^=_{r,j} + P_{0,0} \qquad k_\beta$$

Table 1. Source of unsaturation and mechanisms of formation of dead polymer with terminal double-bond. (R - polymer chain, Cat - catalytic site, C_y - short chain branch containing y carbon atoms)

$$R\text{-}CH_2\text{-}CH_2\text{-}Cat \quad \rightarrow \quad R\text{-}CH=CH_2 + H\text{-}Cat$$

$$\text{(vinyl)}$$

$$R\text{-}Cat + \underset{\underset{C_y}{|}}{CH}=CH_2 \quad \rightarrow \quad R\text{-}\underset{\underset{C_y}{|}}{\overset{\overset{H}{|}}{C}}\text{-}\underset{\underset{H}{|}}{\overset{\overset{H}{|}}{C}}\text{-}Cat \quad \rightarrow \quad R\text{-}\underset{\underset{C_y}{|}}{C}=CH_2 + H\text{-}Cat$$

$$\text{(vinylidene)}$$

$$R\text{-}\underset{\underset{CH_3}{|}}{\overset{\overset{H}{|}}{C}}\text{-}\underset{\underset{H}{|}}{\overset{\overset{H}{|}}{C}}\text{-}Cat \quad \rightarrow \quad R\text{-}CH=CH_2 + CH_3\text{-}Cat$$

$$\text{(vinyl)}$$

$$R\text{-}Cat + CH_2=\underset{\underset{C_y}{|}}{CH} \quad \rightarrow \quad R\text{-}\underset{\underset{H}{|}}{\overset{\overset{H}{|}}{C}}\text{-}\underset{\underset{C_y}{|}}{\overset{\overset{H}{|}}{C}}\text{-}Cat \quad \rightarrow \quad R\text{-}\overset{\overset{H}{|}}{C}=\underset{\underset{C_y}{|}}{\overset{\overset{H}{|}}{C}} + H\text{-}Cat$$

$$\text{(trans)}$$

where: $P_{r,i}$ is a living polymer molecule of chain length r containing i long chain branches, $D^=_{q,j}$ is a dead polymer molecule of chain length q containing j long chain branches and having terminal vinyl unsaturation, $D_{q,j}$ is a dead polymer molecule of chain length q containing j long chain branches and a saturated chain-end, M is the monomer, CTA is a chain transfer agent, k_p is the propagation rate constant for monomer, k_{pLCB} is the propagation rate constant for dead polymer with terminal vinyl unsaturation, k_{CTA} is the rate constant for transfer to chain transfer agent, and k_β is the rate constant for β –hydride elimination.

Dead polymer chains having terminal vinyl unsaturation, $D^=_{q,j}$, can coordinate to the catalytic active site and insert in the growing chain, forming a trifunctional long chain branch. Dead polymer chains with saturated chain ends, $D_{q,j}$, cannot polymerize again. Observe that, by examining the mechanism of chain formation, one can conclude that the instantaneous molecular weight distributions of live polymer, dead polymer with vinyl chain-end unsaturation, and dead polymer with saturated chain-ends will be the same. However, the relative amount of dead polymer with and without terminal vinyl unsaturation is proportional to the rates of β –hydride elimination (producing dead polymer chains with terminal vinyl unsaturation), and transfer to chain transfer agent (commonly to hydrogen, producing dead polymer chains with saturated chain-ends).

The frequency distribution of chain length for polymer populations with n long chain branches per chain is given by:[10]

$$f(r,n) = \frac{1}{(2n)!} r^{2n} \tau^{2n+1} \exp(-\tau r)$$ [1]

where, r represents chain length, n represents the number of long chain branches per polymer molecule, and τ is given by:

$$\tau = \frac{R_\beta}{R_p} + \frac{R_{CTA}}{R_p} + \frac{R_{LCB}}{R_p}$$ [2]

where R is the rate of β-hydride elimination, R_p is the rate of monomer propagation, R_{CTA} is the rate of transfer to chain transfer agent, and R_{LCB} is the rate of macromonomer propagation or long chain branch formation.

The chain length averages of the polymer populations with different number of long chain branches per chain are related by the simple relationships:

$$\bar{r}_{n,i} = (1+2i)\bar{r}_{n,0}$$ [3]

$$\bar{r}_{w,i} = (1+i)\bar{r}_{w,0}$$ [4]

$$\bar{r}_{z,i} = (1+2i/3)\bar{r}_{z,0}$$ [5]

$$pdi_i = \left(\frac{1+i}{1+2i}\right)pdi_0$$ [6]

where, i indicates the number of long chain branches per chain, and $\bar{r}_{n,i}$, $\bar{r}_{w,i}$, $\bar{r}_{z,i}$, and pdi_i are the number, weight, and z-average chain lengths, and polydispersity, respectively.

Average values for long chain branching frequency are expressed as:

$$\bar{B}_N = \left(\frac{R_{LCB}}{R_\beta - R_{LCB}}\right)$$ [7]

where $\bar{B}_N$ is the average number of long chain branches per polymer chain. The number of long chain branches per 1000 carbon atoms, $\bar{\lambda}_N$, is given by:

$$\bar{\lambda}_N = 500 \frac{R_{LCB}}{R_p}$$ [8]

SIMULATION RESULTS

Figures 4a and 4b show the predicted weight chain length distributions for a polyolefin produced in a CSTR for a given value of τ. Table 2 shows the chain length averages and average degree of long chain branching predicted with the model.

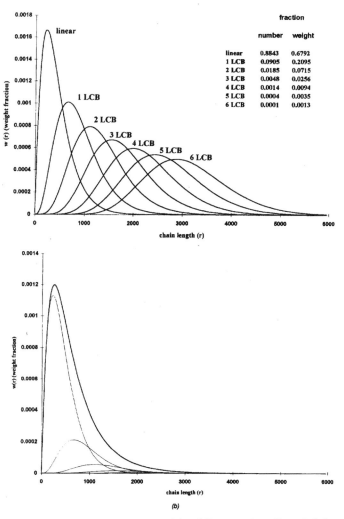

Figure 4. Weight chain length for polymer populations containing different number of long chain branches per polymer chain: (a) weight chain length distributions normalized with respect to the weight of individual populations; (b) weight chain length distribution normalized with respect to the total weight of the whole polymer. The global distribution (over all polymer populations) is indicated by the bold line. The table in the top right corner indicates number and weight fractions of each population.

Table 2. Comparison of chain length averages and degree of long chain branching calculated by numerical solution, Monte-Carlo simulation, and analytical solution. (r_n-number average chain length, r_w-weight average chain length, r_z-z-average chain length, pdi-polydispersity index, B_N long chain branches per polymer chain))

	Population balance	Monte-Carlo	Analytical solution
$\bar{r}_n$	289	289	288
$\bar{r}_w$	661	664	662
$\bar{r}_z$	1110	1117	1110
pdi	2.29	2.30	2.30
$\bar{B}_N$	0.15	0.15	0.15

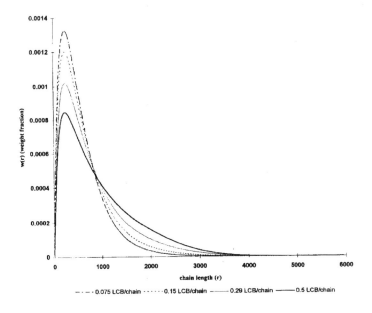

Figure 5. Effect of varying macromonomer addition rate on weight chain length distribution.

Table 3. Chain length averages and degree of long chain branching as a function of long chain branching incorporation. ($\bar{r}_n$-number average chain length, $\bar{r}_w$-weight average chain length, $\bar{r}_z$-average chain lenght, pdi-polydispersity index, $\bar{B}_N$-long chain branches per chain.)

$\bar{B}_N$	$\bar{r}_n$	$\bar{r}_w$	$\bar{r}_z$	pdi
0.075	268	580	928	2.16
0.150	288	662	1110	2.30
0.290	320	806	1378	2.51
0.500	363	952	1556	2.62

Figure 5 and Table 3 show the effect of varying the rate of long chain branch incorporation in the chain length distribution. The chain length distribution becomes broader as the long chain branching level increases, but the distribution is still very narrow as compared with the ones for polyolefins made with heterogeneous Ziegler-Natta catalysts.

It is clear that the analytical solution shown in Eq 1 can also predict the instantaneous chain length distribution of polymer produced during non-steady-state operation of a CSTR or in other reactor types.

The analytical solution was compared with the numerical solution of the polymer population balances and with chain length distributions generated with a Monte-Carlo simulation. The agreement with the analytical solution was excellent (Table 2).

The analytical solution for the chain length distribution of homopolymer with long chain branches, Eq 1, is also valid for copolymerization when appropriate pseudo-kinetic rate constants are used to calculate τ. Additionally, for the case of binary copolymerization, Stockmayer's[11] distribution can be used to obtain the chemical composition distribution of the copolymer. This is possible because the branched chains are formed by linear copolymer chains which follow Stockmayer's bivariate distribution.

CONCLUSIONS

Polyolefins with long chain branches produced with metallocene catalysts have remarkable physical properties, combining excellent mechanical properties with good processability.

The bivariate molecular weight and long chain branching distribution of these polyolefins can be described by a simple analytical mathematical expression. This distribution can be used to accurately predict the polymer microstructure synthesized with these catalysts.

REFERENCES

1.   Y. Lai, J. R. Wilson, G. W. Knight, J. C. Stevens, and P. W. S. Chum, **U.S. Patent 5 272 236**, 1993.
2.   S. Y. Lai, J. R. Wilson, G. W. Knight, and J. C. Stevens, **U.S. Patent Application WO 93/08221**, 1993.
3.   K. W. Swogger and C. I. Kao, *Polyolefins VIII, SPE RETEC*, p.14, Houston, February, 1993.
4.   T. K. Woo, L. Fan, and T. Ziegler, *Organometallics*, **13**, 2252 (1994).

5.   J. B. P. Soares and A. C.Hamielec, *Polymer*, **36 (11)** 2257 (1995).

6.   M. Sugawara, *SPO '94*, p. 37, 1994.

7.   A. E. Hamielec, J. F. MacGregor, and A. Penlidis, *Makromol. Chem., Macromol. Symp.*, **10/11**, 521, (1987).

8.   L. Resconi, F. Piemontesi, G. Franciscono, L. Abis, and T. Fiorani, *J. Am. Chem. Soc.*, **114**, 1025 (1992).

9.   J. Huang and G. L. Rempel, *Prog. Polym. Sci.*, **20**, 459 (1995).

10.   J. B. P. Soares and A. E. Hamielec, *Macromol. Theory Simul.*, 1995.

11.   W. H. Stockmayer, *J. Chem. Phys.*, **13**, 199, 1945.

Structure and Properties of Single Site Constrained Geometry Ethylene-Propylene-Diene, EPDM, Elastomers

D. R. Parikh, M. S. Edmondson, B. W. Smith, J. M. Winter
M. J. Castille, J. M. Magee, R. M. Patel, and T. P. Karajala
The Dow Chemical Company, Freeport , Texas 77541, USA

ABSTRACT

A new class of ethylene-propylene-diene, EPDM, elastomers has been derived utilizing single site constrained geometry catalyst technology. In this paper, microstructure and property relationships of various base polymers as well as formulated EPDM's are described. Influence of composition on melt rheology and solid state properties are also described.

INTRODUCTION

Since the introduction of ethylene-propylene elastomers, their importance has grown significantly in a wide range of applications.[1] This progress can be attributed to their unique end use performance in highly filled applications along with excellent weather resistance and thermal stability. EPDM catalysts and manufacturing process technologies have remained relatively stagnant since their introduction in the 1960's.[2]

The advent of single site single site constrained geometry technology has allowed for the production of polyolefins with novel molecular architecture. INSITE™ technology developed by The Dow Chemical Company has been used to produce a variety of polyolefins[3] including Engage® polyolefin elastomers and Nordel® IP hydrocarbon rubber.

Various applications require a balance of properties such as processability, efficient cure rate and ultimate physical properties. The balance of properties is achieved by the appropriate molecular composition, molecular weight and molecular weight distribution. Hence, if one can control and manipulate the molecular architecture and rheological behavior, the overall performance of EPDM for a given application and formulation can be optimized.

This paper describes the influence of molecular weight, crystallinity and diene level of EPDM elastomers derived from single site constrained geometry catalyst technology on their rheological and ultimate physical properties. This paper is an extension of the structure-property relationships of single site constrained geometry EPDM described earlier by the authors.[4]

EXPERIMENTAL

The selected EPDM elastomers are described below in Table 1.

Table 1. Description of targeted EPDM elastomers Mooney Viscosity (ML 1+4 @ 125°C)

Wt% C_2	Wt% C_3	Wt% ENB	Mooney viscosity
72	28	0	20
72	25	3	20, 40, 50, 60, 80
72	23	5	20, 40, 50, 60, 80, 90
72	19	6	20, 40, 50, 60, 80

Thermal characterization (crystallization and melting) of the EPDM samples was obtained using a differential scanning calorimeter (DSC), TA DSC-2920, equipped with the liquid nitrogen cooling accessory. The DSC was calibrated using Indium and a low temperature standard (deionized water and isopropanol). The samples were prepared in the form of melt-pressed films and placed in aluminum pans. The samples were heated at 10 to 180°C in the DSC and kept at 180°C for 4 minutes to ensure complete melting. The samples were then cooled at 10°C/min to -100°C using liquid nitrogen and heated to 140°C at 10°C/min. The glass transition temperature, T_g, was obtained from the heating curve using the inflection point method. The total heat of fusion was obtained from the area under the melting curve. Crystallinity was obtained by dividing the heat of fusion by 292 J/gm. Note that EPDM resins melt over a range of temperature and melting often immediately follows the glass transition.

RHEOLOGY

The dynamic melt rheological properties of the samples were studied using a Rheometrics, Inc. RMS-800 rheometer with 25 mm diameter parallel plates. The frequency was varied from 0.1 rad/s to 100 rad/s at an isothermal temperature of 190°C. According to the Cox-Merz rule, the complex viscosity versus frequency data measured using the dynamic shear rheology is considered equivalent to shear viscosity versus shear rate data. In general, low Mooney samples were run at 15% strain and higher Mooney (>50) samples were run at 8% strain. The Mooney viscosity (ML 1+4 @ 125°C) of the raw polymer was measured at 125°C whereas the final Mooney viscosity (ML 1+4 @ 100°C) of the formulated compound was measured at 100°C.

COMPOSITIONAL ANALYSIS

Determination of weighted amounts of ethylene and propylene was carried out via infrared analysis, FTIR, according to ASTM D3900. The ENB content was carried out in accordance with the method proposed by the ASTM.[5]

RESULTS AND DISCUSSION

MW FRACTIONATION

The unique combination of single site constrained geometry catalyst and solution process technology allows for a wide range of composition and molecular weights. Compositional consistency of EPDM elastomers produced using single site constrained geometry catalyst technology is investigated. There was an attempt made by Datta[6] to address the compositional consistency for EPDM blends. However, in order to investigate the compositional consistency of EPDM elastomers produced using single site constrained geometry catalyst, a representative sample was fractionated based on molecular weight. Table 2 shows the molecular weight, MW, molecular weight distribution, MWD, wt% ENB, wt% propylene and glass transition temperatures, T_g, for the base EPDM and the fractions.

Table 2. Composition of EPDM and its MW fractions

Sample	M_w(X1000)	MWD	Wt% ENB	Wt% C_3	Tg °C
Polymer	108.0	2.25	3.1	52.7	-49.9
MW1	34.4	1.7	3.0	52.2	-51.4
MW2	54.8	1.7	3.2	53.3	-48.5
MW3	108.0	1.8	2.9	53.7	-48.7
MW4	137.0	1.6	3.4	54.0	-47.9

The fractionation results shown in Table 2 indicate that molecular weight fractionation was achieved based on the MW data. Therefore, the MW fractions have lower polydispersity, MWD compared to the base EPDM. All fractions have similar amounts of ENB and propylene contents which indicate compositional consistency across the molecular weight range. The glass transition temperatures of the MW fractions are also within ±2°C compared to the base EPDM. These fractionation data indicate that there is a compositional consistency with the single site constrained geometry based EPDM which in turn can translate into consistent cure properties and ultimate physical properties. This has even more important implications for ultra fast cure EPDM based formulations.

THERMAL BEHAVIOR

It is known that as one changes the ethylene content of an ethylene interpolymer, crystallinity, melting temperature, crystallization temperature and hence ultimate mechanical properties are influenced.

Table 3. Relationship between mole % and Wt%. Comonomer content for various elastomers

Wt% C_2	Wt% C_3	Wt % ENB	Mole% C_2	Mole% C_3	Mole% ENB
72	25	3	80.6	18.7	0.8
72	23	5	81.4	17.3	1.3
72	19	9	83.0	14.6	2.4

Blends of ethylene interpolymers will exhibit different relationships. During the course of developing structure-property relationships, we analyzed several EPDM elastomers derived from a single site constrained geometry catalyst (described in Table 1) having about 72 wt% ethylene and different amounts

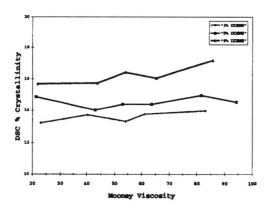

Figure 1. Influence of ENB on crystallinity at constant ethylene content (72 wt%).

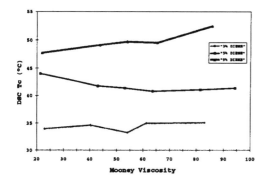

Figure 2. Influence of ENB on crystallization temperature at constant ethylene content (72 wt%).

of ENB and propylene comonomers. Figure 1 shows that at about 72 wt% ethylene, total crystallinity is independent of Mooney viscosity but the crystallinity increases with increasing ENB content. It could be hypothesized that a bulkier ENB molecule would be more effective in disrupting crystallinity and hence would result in lower crystallinity with an increase in ENB content at the same ethylene content. To explain this observed behavior, comonomer contents were compared based on mole rather than weight % (Table 3).

These data show that at constant wt% ethylene the actual mole% of ethylene increases as wt% ENB increases due to the higher molecular weight of ENB. Thus higher crystallinity as ENB increases can be explained based on the fact that there is higher mole % ethylene and hence a higher level of crystallinity. At the same time, it is expected that the crystallinity will remain constant over the Mooney viscosity range from 20 to 90, so long as the molecular composition remains unchanged. This is also shown in Figure 1.

It is known that a higher crystallinity in ethylene interpolymers also results in higher crystallization temperatures (Figure 2) as well as melting temperatures. Crystallization temperature is primarily determined by the ethylene sequence length and comonomer distribution along the polymer backbone. In a single component system, higher mole % ethylene would result in a longer ethylene sequence for homogeneous ethylene interpolymers which will show higher crystallization temperatures. However, this is not necessarily true for the blends in which overall crystallinity could be lower but a higher crystallization temperature. Crystallization temperature has an influence on solidification as well as green strength during fabrication processes such as calendering and injection molding. Therefore, it is important to understand the influence of microstructure and how to design the molecular structure to deliver optimum performance.

For semicrystalline thermoplastic polymers, crystallinity level strongly influences Young's modulus and tensile strength, semicrystalline EPDM's follow the same fundamental relationship. Higher crystallinity of the raw polymer has a strong influence even on the tensile properties of

Table 4. General purpose formulation

Ingredients	Parts per hundred, PHR
EPDM elastomer	100.0
Zinc oxide	5.0
Stearic acid	1.0
Carbon black, N550	80.0
Oil, Sunpar 2280	50.0
Butyl zimate	2.0
TMTD	0.5
MBT	1.0
Sulfur	1.5
Total	241.0

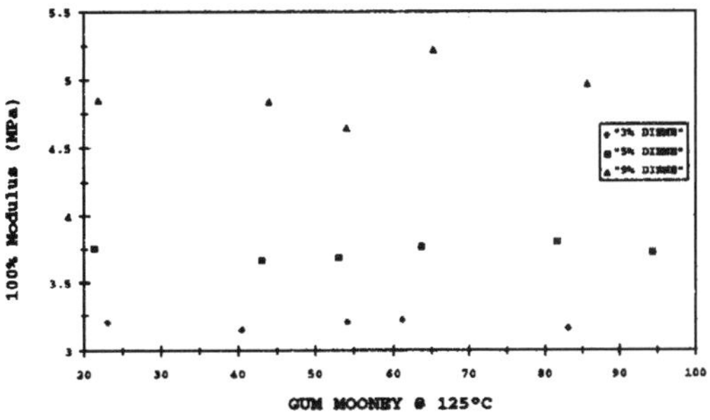

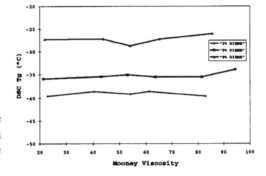

Figure 3. Relationship between 100% modulus and ENB content at constant ethylene content (72 wt%).

Figure 4. Influence of ENB content on glass transition temperature at constant ethylene content (72 wt%)

formulated compounds. To understand the relationship between the polymer composition and its compounded properties a general purpose formulation, as shown in Table 4, was evaluated.

Figure 3 shows that at constant (72 wt%) ethylene content, a higher 100% modulus was obtained as the wt% ENB was increased. As explained earlier, this was to be expected. It is noteworthy that 100% modulus was not dependent on Mooney viscosity.

The glass transition temperature, T_g, of single site constrained geometry derived semicrystalline EPDM's depends on the type and amount of diene, the ethylene amount,[1] the ethylene sequence length and the comonomer distribution. It is the ethylene mole content (see Table 3) which influences the crystallinity for homogeneous ethylene interpolymers which in turn affects the low temperature relaxation process. Figure 4 shows that the glass transition temperature is raised by about 12-15°C as the wt% ENB is increased from 3 to 9 (or crystallinity is increased from about 13 to 16 %). Influence of the EPDM's glass transition temperature was investigated in a general purpose formulation as well and similar trends were observed (Figure 5). The temperature retraction test determines the temperature at which a certain % retraction occurs for a pre-stretched specimen. In the present study, we measured the temperature for a 10% retraction of pre-stretched formulated EPDM samples. In this test, a lower retraction temperature would indicate improved low temperature properties. Thus a higher ethylene

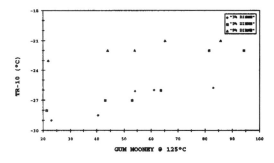

Figure 5. Relationship between low temperature performance and ENB content at constant ethylene content (72 wt%).

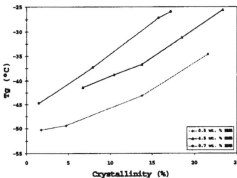

Figure 6. Relationship between glass transition temperature and crystallinity of various EPDM elastomers.

amount (by about 2 mole %) plays a key role in influencing low temperature properties of formulated EPDM.

A relationship between the amount of ENB and total crystallinity was also explored. Figure 6 shows a plot of T_g versus crystallinity for various EPDM elastomers having different compositions of ENB (and ethylene as well as propylene). The objective was to investigate the influence of ENB on the glass transition temperature at a constant crystallinity content. It is obvious from the plot that a higher amount of ENB raises the glass transition temperature of EPDM elastomers at the same level of crystallinity. It is interesting that the low temperature relaxation process is strongly influenced by the amount of ENB despite similar crystallinity and melting characteristics. This phenomenon could be explained based on the nature of the ENB molecule. The cyclic structure and reduced mobility of the side group (ENB molecule) would reduce the local segmental motion of the polymer chain. Therefore, the glass transition temperature of an EPDM elastomer would increase with the higher amounts of ENB.

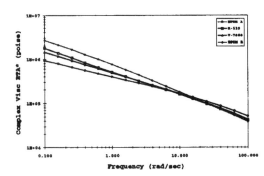

Figure 7. Shear rheology of metallocene EPDM with other 70 Money commercial EPDM.

RHEOLOGICAL BEHAVIOR

Generally, higher shear sensitivity is desired for most fabrication processes. Higher viscosity at a lower shear rates (e.g., 0.1 sec[-1]) provides melt strength whereas a lower viscosity at higher shear rates (e.g., 100 sec[-1]) enhances the melt processability. It is also known that the presence of the long chain branching and/or broad molecular weight distributions alters the shear thinning behavior.[7] Single site constrained geometry catalyst technology combined with solution process technologies alters the shear rheology of EPDM elastomers. Figure 7 shows the shear rheology of four

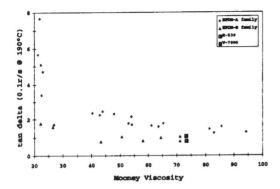

Figure 8. Enhanced melt strength at various Mooney viscosity due to altered molecular architecture.

EPDM elastomers with Mooney viscosity of approximately 70 and a diene content of 5 wt%. Two of these samples (EPDM-A and EPDM-B) were produced using single site constrained geometry catalyst technologies. These two samples were compared with two commercial EPDM elastomers. EPDM-B elastomer has more shear thinning behavior and it is expected to have improved melt processability. Shear rheology data indicate that the breadth of the single site constrained geometry catalyst and process technologies can allow one to produce EPDM's having a broad range of shear sensitivity while maintaining compositional consistency.

Melt strength is one of the key properties during various fabrication processes. It is expected that higher molecular weight (or Mooney viscosity) EPDM elastomers would have higher melt strength, however, manipulation of molecular architecture enables enhanced melt strength (determined using tanδ) even at a lower Mooney viscosity as shown in Figure 8. Tanδ is a ratio of loss modulus over storage modulus from dynamic mechanical spectroscopy and lower tanδ values are indicative of higher stored energy in the melt resulting in higher melt elasticity and melt strength. In the present example, it can be seen that over a wide range of Mooney viscosities, single site constrained geometry catalyst and solution process technologies allows for the production of EPDM elastomers having enhanced melt strength.

CONCLUSIONS

Compositional consistency of EPDM elastomers plays a key role in determining ultimate physical and cured properties. Metallocene catalyst and solution process technologies can produce EPDM elastomers having compositional consistency along with a wide range of melt processability combined with melt elasticity. Mole % ethylene content or crystallinity is a more appropriate measure of polymer performance than wt% ethylene content, and is a better explanation of the structure-property relationships based on the true composition.

ACKNOWLEDGMENTS

The authors wish to thank Debbie Mangold and Don Germano of DuPont Dow Elastomers for facilitating the fractionation and Bobbie Guilbeaux and Robin Yaws also of DuPont Dow Elastomers and Harolyn Perkins of The Dow Chemical Company for analytical characterization.

REFERENCES

1.   J. Riedel and R. Vander Laan, **The Vanderbilt Rubber Handbook**, 13th Edition, Ed. R. F. Ohm.
2.   Auchter, F. Stahel, and Y. Ishikawa, **Ethylene-Propylene Elastomers**, SRI Report, 1993.
3.   C. Pappas, Insite Technology: New Products and Rapid Global Progress, NPE, June, 1994.

4.   M. S. Edmondson, T. P. Karjala, R. M. Patel, and D. R. Parikh, 149th ACS Rubber Division Meeting, May, 1996.
5.   *Materiaux et Techniques*, December, 1991.
6.   S. Datta, N. P. Cheremisinoff, and E. N. Kresge, *Polymer Mater. Sci. & Eng.*, **68**, 1993.
7.   K. P. Beardsley and R. W. Tomilson , *Rubber Chem.Techn.*, **63**, 1990.

Enhancing Polyethylene Performance with INSITE® Technology and Molecular Design

Kaelyn C. Koch and Bill Van Volkenburgh
Dow Plastics, USA

ABSTRACT

Polyethylenes with new performance enhancements can be designed by using the tailoring ability of metallocene catalysts, an understanding of materials science and a molecular design approach. Molecular design begins with a clear understanding of performance requirements followed by the use of computer-aided design tools to "engineer" the desired polymer structure and manufacturing process conditions. Examples show that molecular design can be used to develop enhanced polyethylenes for diverse market applications including consumer and industrial films, high performance sealants, injection-molded containers and injection-molded lids.

INTRODUCTION

Many polyolefin design rules and structure-property relationships have been changed by INSITE® technology. For example, the processability of polyolefin plastomers and elastomers made with INSITE® technology can be decoupled from molecular weight and molecular weight distribution. The ability to incorporate a controlled level of long chain branching results in good processability even for narrow molecular weight distribution polymers.[1] INSITE® technology has now been leveraged to enhance the performance of polyethylene by using a molecular design approach to deliver performance combinations which were previously unavailable.

MOLECULAR DESIGN

Traditionally, many of the parameters used to design polyethylenes are constrained by the catalyst and manufacturing process. The conventional approach to polyethylene product design is to specify the melt index and density (or comonomer level) to satisfy the key performance requirements. These basic design parameters of melt index and density are typically related to a molecular microstructure which results from the particular catalyst, manufacturing process and comonomer type. Since the key parameters for controlling polymer performance (molecular weight, molecular weight distribution and comonomer type, level and distribution) are interdependent, the polymer designer is often required to make performance compromises.[2,3] For example, modulus and impact are inversely related for conventional Ziegler-Natta linear low density polyethylene, LLDPE, and thus high impact strength is not possible at higher modulus.

With the advancement of constrained geometry catalyst technology and polymers with controllable comonomer and molecular weight distributions, new capabilities are available for polymer modeling and materials science understanding. Molecular architecture control using INSITE® technology provides the capability to develop new products based on the performance required by the customer, rather than the traditional characteristics of melt index and density.[4] So, rather than beginning with old performance rules and the associated constraints and compromises, the molecular design process begins with an understanding of the performance requirements. Equipped with a clear understanding of the performance requirements, the resin designer uses computer-aided design tools and materials science understanding to define the desired polymer structure and to determine the best process conditions to manufacture the new product.

Many tools are available to the resin designer for building the desired molecular architecture. Broadly, the tools could be considered in five categories including raw materials, polymer microstructure, polymer macrostructure, manufacturing processes/catalysts and fabrication processes. The manner in which these tools are used will determine the final product attributes and performance.

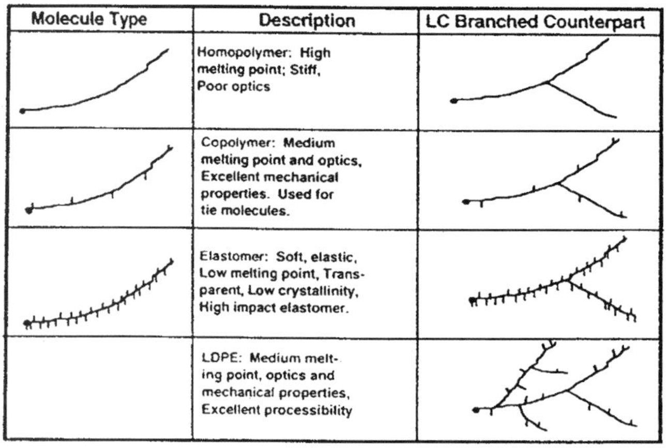

Figure 1 conceptually illustrates the basic polymer types which are the building blocks for a molecularly-designed polyethylene. Each polymer molecule offers specific, well-characterized behaviors. The objective is to select the combination of molecules which will provide the desired performance properties. The enhanced polyethylenes are based on the narrow molecular weight distributions, long chain branching and narrow composition distributions that result when polymers are produced with INSITE® technology; this technology offers the resin designer the opportunity to optimize the desired mechanical properties,

Molecule Type	Description	LC Branched Counterpart
	Homopolymer: High melting point; Stiff, Poor optics	
	Copolymer: Medium melting point and optics. Excellent mechanical properties. Used for tie molecules.	
	Elastomer: Soft, elastic, Low melting point, Transparent, Low crystallinity, High impact elastomer.	
	LDPE: Medium melting point, optics and mechanical properties, Excellent processibility	

Figure 1. Building blocks of polyethylene molecular design.

processability and sealing characteristics.

The remainder of this paper describes several enhanced-performance polyethylenes which were designed with the molecular engineering concept to deliver new performance combinations. The examples show that molecular design can be used to develop enhanced polyethylenes for diverse market applications including consumer and industrial films, high performance sealants and injection-molded containers and lids. In a separate paper, Kale is discussing the performance of an enhanced polyethylene designed for cable jacketing.[5]

EXPERIMENTAL PROCEDURES

BLOWN FILM

Fifty-one micron films were blown on a Gloucester film line equipped with a 64 mm extruder and a 152 mm die. Films were produced at 55 kg/hr with a 1.8 mm die gap using a 230°C melt temperature and a 2.25:1 BUR. Dart impact and other physical properties were determined according to standard ASTM procedures.

Sealing properties were determined on 89 μm coextrusions of nylon-6/6,6 copolymer/tie layer/sealant (28.5/28.5/43 layer ratio). These films were fabricated on a blown coextrusion film line equipped with two 24:1 L/D 63.5 mm extruders, one 24:1 L/D 51 mm extruder and a 203 mm three layer coextrusion die. The tie layer was an ethylene-acrylic acid copolymer (1.5 MI, 9.5% acrylic acid).

Heat seal measurements were made on a Topwave Hot Tack Tester using a dwell time of 0.5 seconds and a seal pressure of 0.275 MPa. Heat seals were pulled on an Instron tensiometer at 250 mm/min. Heat seal initiation temperature was defined as the temperature at which a seal strength of 1.7 N/cm was achieved.

INJECTION MOLDING

Injection-molded cups with a wall thickness of 0.68 mm were produced on a Huskey XL225P two-stage injection molder equipped with a four-cavity mold. Dynatup impact was measured on the injection molded parts according to standard ASTM procedures. Processing index is the apparent viscosity measured at high shear at a constant stress of 2.15×10^6 dynes/cm^2 on a gas extrusion rheometer.

Two percent flexural modulus was measured according to ASTM D790 on injection-molded plaques produced at 200°C on a 150-ton DeMag injection molding machine. The machine was equipped with a reciprocating screw and a seven-cavity ASTM plaque mold.

Heat sag was conducted by attaching injection-molded flex bars to a metal rack with five vertically aligned spring claps. The rack containing the suspended flex bars was placed into a forced-air convection oven at 100°C for 10 minutes. Each bar's configuration was compared to its original configuration before heating and the heat sag was measured as the change in configuration in centimeters.

RESULTS AND DISCUSSION

CONSUMER AND INDUSTRIAL FILMS

For many consumer and industrial film applications, excellent tensile, tear and impact properties are desirable for toughness and downgauging. For typical Ziegler-Natta LLDPE, one way of increasing dart impact is to lower product density or increase molecular weight; however, these options have the undesirable side-effects of reducing stiffness or processability, respectively. The molecular design approach was coupled with materials science understanding to develop a family of resins with exceptional toughness properties at higher modulus.

Tie molecules are the key for enhancing toughness in semicrystalline polymers such as polyethylene. A chain-folding model can be used to describe the basic crystalline morphology of conventional polyethylene with the polymer macrostructure being classified into three domains consisting of the crystal core (lamellae), the interfacial region and the amorphous region. The model also suggests that the crystalline lamellae are connected by tie molecules which are formed during crystallization. Tie molecules are formed when a chain segment is rejected by a crystalline lamella due to presence of side-chain branches (from the α-olefin comonomer) or due to crystallization kinetics when the polymer chain is long enough to be incorporated into more than one lamella.[6,7] Key polymer variables affecting

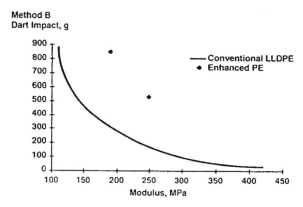

Figure 2. New performance combination: High dart impact and high modulus.

tie molecule formation and concentration are polymer chain length and co-monomer length, concentration and distribution. Tie molecules are generally believed to favorably affect mechanical strength.

In the molecular design for consumer and industrial film applications, tie molecule concentration was maximized to enhance strength properties while other polymer molecules were selected to control modulus. Figure 2 shows an enhanced polyethylene that delivers 530 g of dart impact at a modulus of 255 MPa (about 0.925 g/cc). This is about 200% higher dart impact than a conventional 0.920 g/cc LLDPE with a lower modulus of 189 MPa.

INJECTION-MOLDED HDPE CONTAINERS

Excellent processability is critical for injection molding applications, particularly when multi-cavity molds are used to achieve maximum output. Traditionally, high flow is achieved by decreasing the molecular weight and/or broadening the molecular weight distribution; however, these approaches compromise mechanical strength in the finished part.

The molecular design for this application also maximizes tie molecule concentration for toughness. The molecular weight and molecular weight distribution are selected to achieve the desired flow characteristics. Figure 3 shows that enhanced polyethylene can deliver an exceptional combination of strength and processability.

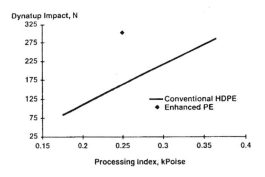

Figure 3. New performance combination: Flowability and impact strength for injection molding applications.

HIGH PERFORMANCE SEALANTS

The heat seal initiation temperature of a conventional LLDPE or ultra low density polyethylene, ULDPE, sealant is related to its density or modulus. Conventional polyethylenes, including LLDPE, ULDPE, ethylene vinyl acetate copolymers and low density polyethylene, are not able to offer stiffness or temperature resistance (e.g., for cook-in applications) while also delivering low heat seal and hot tack initiation temperatures. In the sealant molecular design, the objective was to decouple the relationship be-

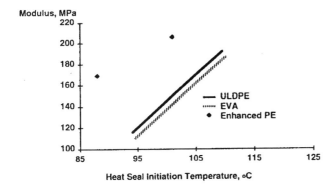

Figure 4. New performance combination: Low heat seal initiation and high modulus.

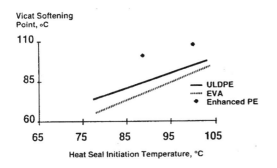

Figure 5. New performance combination: Low seal initiation and temperature resistance.

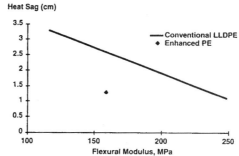

Figure 6. New performance combination: Flexibility and temperature resistance.

tween seal initiation temperature and modulus as well as the relationship between seal initiation temperature and upper service temperature.

Resins were designed to have higher modulus and higher Vicat softening points than would be expected for products having low heat seal and hot tack initiation temperatures. The resulting sealants offer easy handling characteristics, good packaging line speeds and excellent package integrity. Figures 4 and 5 show two products designed as high modulus alternatives to ULDPE. Additionally, the enhanced polyethylenes have low blocking and excellent mechanical strength.[3]

INJECTION-MOLDED LIDS

Conventional LLDPE and ULDPE used for injection-molded lids do not simultaneously offer high flexibility for easy opening along with high temperature resistance for dish washing and microwaving. Molecular design was used to deliver a higher melting/softening temperature than would be ordinarily expected for a lower modulus resin. Figure 6 shows that the enhanced polyethylene has the temperature resistance of a conventional 0.925 g/cc density LLDPE at a modulus that corresponds to a lower density polyethylene (i.e., 0.915 g/cc density).

CONCLUSIONS

Enhanced polyethylenes can be designed by using the tailoring ability of constrained geometry catalyst technology, an understanding of materials science and a molecular design approach. Molecular design begins with a clear understanding of the performance requirements followed by the use of computer-aided design tools to "engineer" the right polymer structure and to select the best process conditions. New property combinations which have been described include: (a) high impact and high modulus, (b) high impact and exceptional processability, (c) low temperature sealability and higher upper service temperature, (d) low temperature sealability and high modulus, and (e) flexibility and higher upper service temperature.

ACKNOWLEDGMENTS

The authors wish to recognize Steve Chum for his leadership in applying molecular design and materials science principles to enhance polyethylene performance.

REFERENCES

1.   G. W. Knight and S. Lai, *RETEC*, February, 1993.
2.   K. W. Swogger and G. M. Lancaster, *SPO*, September, 1993.
3.   N. F. Whiteman, J. A. deGroot, L. K. Mergenhagen, and K. B. Stewart, *RETEC*, February, 1995.
4.   G. Lancaster, J. Daen, C. Orozco, and J. Moody, *METCON*, May, 1994.
5.   L. T. Kale, *ANTEC*, May, 1996.
6.   A. Keller, *Phil. Mag.*, **2**, 1171 (1957).
7.   P. H. Geil in **Polymer Single Crystals**, *John Wiley*, New York, 1993.

Morphological Investigation on Very Low Density Ethylene-Octene Metallocene Copolymers Using SAXS

Paul J. Phillips and Kenneth Monar
Department of Materials Science and Engineering
The University of Tennessee, Knoxville, TN, 37996-2200

ABSTRACT

The most advanced analytical methods currently available for SAXS studies of particle-like systems have been applied to a non-spherulitic very low density ethylene-octene copolymer, useful information being obtained from one dimensional correlation function and direct invariant analyses. The morphological detail emerging from the broadest interpretation of the data presented here is not that of a fringed-micellar system, but is better described as system of plate-like structure of unknown lateral extents. First estimates from a new method for deriving the invariant of the corresponding sharp interface structure render crystal densities in the plates which suggest an "imperfect Swiss cheese-like structure" to the plates. The crystals are approximately 5 nm thick and have interfacial layers which are each about 2 - 3 nm thick.

INTRODUCTION

The development of new synthetic routes for ethylene-octene copolymers via DOW Chemical's constrained geometry catalyst technology (INSITE® CGCT) has produced a range of low density polyethylenes with unique molecular properties such as high degree of randomness in the comonomer distribution, narrow molecular weight distribution, and controlled quantity and length of long chain branching. They have resulted in copolymers with densities ranging from 0.9600 to as low as 0.855 g/cc. It has been observed that a morphological transition occurs at ca 0.8750 g/cc (ca 37-42 Me/1000C) at both the light and electron microscopy levels from an open lamellar system characterized by spherulites akin to an LDPE[1] at higher densities, to one consisting of an ill-defined morphology which has been described as a granule-like structure of low aspect at lower densities, based on ruthenium tetraoxide stained films, which may show a few isolated lamellae depending on thermal treatment.[2] It has recently been suggested[3] that this morphology at very low densities consists of fringed micellar crystals.[4] The lack of any discernible UV light scattering pattern evidences the low haze exhibited by films from these low density copolymers.[2] Concomitant with these morphological changes is a distinct change in deformation behavior in uniaxial tension from a mode typical for semicrystalline polymers consisting of a yield point followed by some degree of cold drawing dependent upon crystallinity to one with no yield point, low modulus, and the high permanent set indicative of elastomers.[5,2]

The use of small angle X-ray scattering, SAXS, is an important tool in polymeric investigations at the nanometer scale. It does not suffer from the problems of submitting organic materials to quantita-

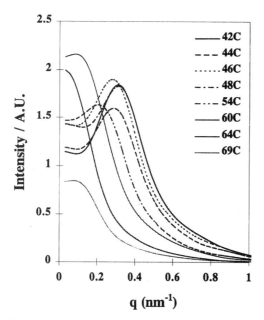

Figure 1. Corrected absolute intensity vs. angle.

tive analyses under electron beams. Potentially it is possible to obtain very detailed information on morphological structures at the lamellar and sub-lamellar levels provided that appropriate models for the scattering can be developed. While the analysis known as the one-dimensional correlation function (6, 1DCF) has been a stalwart of modern semicrystalline structure analyses, especially with regard to the 2 phase approximation, it has been well known to suffer from its various assumptions (e.g. 7). Nonetheless, it has been thought that the 1DCF will give the correct information so long as the distances considered are small in comparison with the curvature of the phase boundary[8] and/or the amorphous heterogeneities have domain sizes either fully within or outside the experimental range in more dilute lamellar systems.[9]

EXPERIMENTAL

A single copolymer system is examined having a number average molecular weight of ca. 20,000 with a polydispersity of ca. 2.0. It had 39 hexyl side chains per 1000 carbon atoms based on infrared analyses. The specimens were either crystallized to an equilibrium degree of crystallinity isothermally or were quenched to temperatures up to 8K below the crystallization temperature. The samples were allowed to anneal at room temperatures and were subsequently submitted to SAXS analysis at or very close to the crystallization temperature. SAXS studies were carried out using the 5 m and 1 m cameras at the Center for Small Angle Scattering Research at Oak Ridge National Laboratory.

RESULTS

The azimuthally averaged scattering patterns for a series of samples are shown in Figure 1 where absolute intensity is plotted against the momentum transfer given as:

$$s = 2\sin\theta/\lambda \quad or \quad q = 2\pi s \qquad [1]$$

Figure 2 shows raw, unsmoothed data plotted in what is know as a Porod Plot which is based on the premise that the scattered intensity will approach a maximum value as the scattered light originates from probing intraparticle and not interparticle properties according to:

$$\lim_{q \to limit} I(q)q^4 = 16\pi^4 Kp \qquad [2]$$

As is typical for semicrystalline polymers, this region is not seen in its entirety because of other intervening effects. The curves quickly rise with an approximate q^2 dependence as a result of thermal fluctuations and inhomogeneities within each of the component phases. The nature of the pseudo maximum in these plots near the beginning of this region and the strong downturn in intensity resulting from the presence of an interface are well known problems that complicate the determination of the background and also of the correction of the tail end of the curve. In this regard, the relatively recent methodology of Sobry et al.[10] was employed, permitting recovery of the Kp term and the determination of the interfacial thicknesses between the two phases.

It is now known that features can occur in the Porod region of the scattering pattern which are well expected from the scattering theory which produce subsidiary maxima between the first minimum and first maximum in the

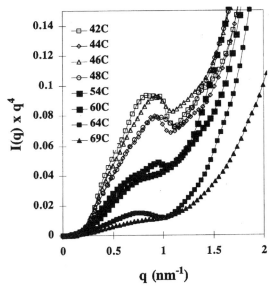

Figure 2. Porod Plot on the raw data.

1DCF.[10] These features undermine the determination of both the Specific Surface and the Invariant, Q; two quantities related to the quantity of interface per volume of scattering entity and to the total scattering power. This has been the topic of a recent paper[11] based on earlier work showing the effect of edges on the scattering pattern[12] the details of which had been provided earlier.[13] However, it is now known that even this view is incomplete and that it is the limited edges and their vertices which are responsible.[12]

We will examine the results from the 1DCF and the corresponding sharp boundary structure to better understand the morphological changes occurring during the changeover from well-defined lamellar stacks to the new morphology discussed above. This will be examined in a more comprehensive manner as a function of branch content, undercooling, and kinetics at a later date.[14]

The 1DCF is shown in Figure 2 and was produced from the 3-dimensional correlation function, 3DCF:

$$\gamma(r) = \frac{\int\limits_{0}^{\infty} s^2 I(s)\dfrac{\sin(2\pi sr)}{2\pi sr}ds}{\int\limits_{0}^{\infty} s^2 I(s)ds} \qquad [3]$$

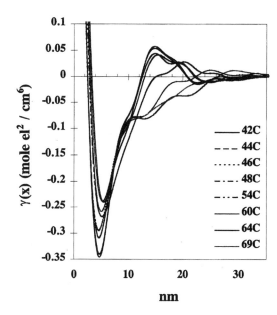

Figure 3. One-dimensional correlation function.

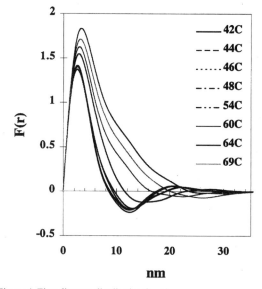

Figure 4. First distance distribution function.

where the term in the denominator is the invariant. Two related functions are the 1st distance distribution function F(r) and the 2nd distance distributions P(r) which are related to the 3DCF as the first and second moments rγ(r) and r²γ(r). The 1DCF is related to 3DCF as the first derivative of the F(r) function. As compared to the raw data in Figure 1, it is related through the inverse Fourier transform from reciprocal space to real space coordinates where the dimensions can have physical meaning.

Figures 3 and 4 show several interesting and expected features for these very low crystallinity systems (less than approximately 6% based on DSC at the crystallization temperature). First as expected, the quenched samples (isothermal crystallization temperatures of 42-48°C) all show the maximum at approximately the same value of the distance plotted on the x-axis (ca. 5 nm). Samples crystallized at higher temperatures (i.e., lower undercoolings) show a continuous increase to higher values, the maximum observed being ca 6 nm. It is believed that the dimension quoted refers to crystal thickness, since the crystallinity is less than 0.5.

The undulations observed in the 1DCF (Figure 3) are not artifactual and are present in the raw data. At the present it is not possible to assign the physical cause of the undulations, however, they should be the result of either multiple crystal size populations or higher order terms in the extended Porod Law.[10] DSC curves, not shown here, show clearly bimodal melting curves for the quenched samples and so for these specimens the undulations may be caused by crystal thickness populations. For samples crystallized at higher temperatures such multiple melting behavior is not observed, so in these cases the source of the undulations has not been identified.

An interesting feature of the 1CDF when applied to non-lamellar stack systems are the high values observed for the crystallinity (ϕ), obtained from the first minimum (see Table 1). It is noted that these values correspond to approximately the squares of the DSC crystallinities. Also presented in the table are the interfacial thicknesses derived from the background correction, using the method of Sobry. Note that these values are in the range of 2.2 - 3 nm, which is about half of the crystal thickness. The combined interfacial content of a crystal is therefore as large as that of the crystal itself!

Table 1. Data from SAXS analyses of specimens as a function of crystallization temperature

T_c °C	E_{VI}nm	$\frac{<\eta^2>}{10^{-5}}$	$\Delta\eta$	ρ_a	ρ_c	ϕ_1%	$(\phi_1)^{1/2}$%	ϕ_c, DSC %
42	2.82	2.45	0.0136	0.841	0.861	25.5	5.0	5.9
44	3.88	2.85	0.0126	0.840	0.862	23.4	4.8	8.0 (?)
46	2.28	2.09	0.0109	0.838	0.858	22.7	4.76	5.2
48	3.08	2.63	0.0180	0.837	0.858	25.3	5.0	4.9
54	2.38	1.09	0.00809	0.834	0.847	21.1	4.6	2.7
60	2.23	0.912	0.0074	0.830	0.843	20.5	4.5	1.34
64	1.73	0.349	0.0046	0.827	0.827	19.3	4.3	0.915
69	2.57	0.257	0.0041	0.824	0.832	19.0	4.3	0.911

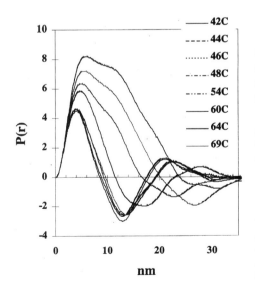

Figure 5. Second distance distribution function.

Interestingly, the F(r) plots (Figure 4) show a distinct lamellar character, as evidenced by the modeling studies of Glatter.[16] Likewise, the P(r) function (Figure 5) also shows a lamellar-like character, with the exception of the Porod region intensity anomaly.

Additional information can be gleaned from the application of the integrated scattering intensity (i.e. the invariant) corrected with the Porod parameter obtained during background analysis:

$$\langle \eta^2 \rangle = C * Q = \phi_1 \phi_2 \Delta\eta^2 \qquad [4]$$

where C is a constant and the subscripts refer to the crystalline and amorphous phases respectively. The term $\Delta\eta$ is the electron contrast between the two phases. Referring to Table 1, the electron contrast has been calculated using the crystallinities from the first minimum in the 1DCF curve of each specimen and the data of Richardson et al.[17] for the amorphous density. These values are very low and suggest that the scattering crystals are of very low average density. This in itself may be interpreted to mean that the crystals are discontinuous in nature, containing many trapped amorphous regions. A simple analogy might be to regard the scattering as arising from stacks of slices of Swiss cheese in which the holes of each slice are randomly arranged relative to the holes of adjacent slices.

CONCLUSIONS

Correlation function analyses show that both particle and dense phase structural information is contained in the SAXS. To the extent of our current analyses on the familiar 2 phase approximation, the following information can be summarized. The measurable "particle" dimension ranges from 5 nm in quenched samples to 6 nm in samples crystallized at low supercoolings. It is suggested that this dimension is the thickness of individual lamellae of localized and very limited continuous lateral dimensions. The gross picture emerging of the morphology, as based on the "crystal" bulk densities (i.e., inclusive up to the midplane of the interface) is a one of randomly oriented discontinuous stacks of planar or gently undulating crystals of small nonspecific lateral dimensions. The interfaces of the crystals are individually about a half of the crystal thickness. The crystals also appear to contain many amorphous islands within their bounds and need not be aligned parallel to one another. The comparison of the baseline to the DSC crystallinities seems to hold better for the more concentrated systems (up to 54°C) than those crystallized at the highest temperatures. This however may be more related to finding the baseline in the thermogram (i.e., what the heat capacity is doing) since there are shorter chains in these systems which will crystallize at temperatures close to the T_g (ca. 55°C). This concept of how much crystal to assign the heat of fusion will be quantified using WAXD at the crystallization temperature and is the subject of current research.

These conclusions are subject to further scrutiny of the observed behavior in the Porod region. There is additional structural information contained therein which cannot currently be extracted from the data. This aspect of the analysis is the subject of current research by several groups.

ACKNOWLEDGMENT

The authors wish to thank The Dow Chemical Company for their support in this research and Dr. Michael D. Heaney at Dow for support in the SAXS analyses.

REFERENCES

1.   G. R. Strobl et al., J. Polym. Sci., Polym. Phys., **18**, 1361 (1980).

2.   Y.-C. Hwang, et al., Proc. ANTEC '94, 3414 (1994).

3.   P. S. Chum et al., Plastics Eng., June, 21(1995).

4.   P. J. Flory, J. Am. Chem. Soc., **84**, 2857 (1962).

5.   S. Bensasson et al., Proc. ANTEC '95, 2256 (1995).

6.   C. G. Vonk and G. Kortleve, Koll. Zeit. Zeit. Polym., **19**, 220 (1967); ibid., 124.

7.   C. G. Vonk, J. Appl. Cryst., **11**, 541 (1978).

8.   C. G. Vonk and F. J. Balta-Calleja in **X-ray Scattering of Synthetic Polymers**, Elsevier, NY, 1989.

9.   G. R. Strobl and M. Schneider, J. Polym. Sci. Polym. Phys., **18**, 1343 (1980).

10.   R. Sobry et al., J. Appl. Cryst., **24**, 516 (1991).

11. T. Albrecht and G. Strobl, *Macromolecules*, **28**, 5267 (1995).

12. J. Mering and D. Tchoubar, *J. Appl. Cryst.*, **1**, 153 (1968).

13. S. Ciccariello et al., *Phys. Rev.*, **B26**, 6384 (1982).

14. R. Sobry et al., *J. Appl. Cryst.*, **27**, 482 (1994).

15. K. Monar and P.J. Phillips (1996) *to be published*.

16. O. Glatter, (1979), *J. Appl. Cryst.*, **12**, 166 (1979).

17. M. J. Richardson et al., (1963), *Polymer*, **4**, 221 (1963).

Further Studies on Metallocene ULDPE/PP Blends Impact-Morphology Relationships

S. P. Westphal, M. T. K. Ling, S. Y. Ding, and L. Woo
Baxter Healthcare, Round Lake, IL 60073, USA

ABSTRACT

In a previous study, metallocene polymerized ultralow density polyethylenes, ULDPE, was found to be effective impact modifier for rigid matrices such as polypropylene, PP. This is mainly due to the very high level of comonomer incorporation with homogeneous short chain branching distributions, SCBD, brought about by the single site metallocene catalysts. However, a very heterogeneous Ziegler-Natta copolymer of 0.88 density was found to be surprisingly effective in impact modification efficiency.

In this study, detailed control of the phase morphology of the impact modifier/matrix was sought. Using a variety of characterization techniques: differential scanning calorimetry, DSC, dynamic mechanical analysis, DMA, and low voltage scanning electron microscopy, LVSEM. It was found that the modulus difference between the two phases, degree of coupling as evidenced by morphology at the phase boundaries, and crystallizable fraction of the rubber phase all played an important role in impact modification.

INTRODUCTION

One of the most immediate consequences of metallocene catalysts ability to homogeneously incorporate high levels of comonomers is the significantly depressed β-relaxation, which at high contents, resembles a glass transition, T_g. These very low T_gs led to superior subambient impact performance compared to conventional low density polyethylenes, LDPE, and linear low density PEs from Ziegler-Natta catalysts. Additionally, these metallocene copolymers, despite their exceptionally low crystallinity, remain free flowing due to their narrow composition distributions. These two properties combined allow them to be used as very efficient impact modifiers for rigid matrices such as polypropylene.[1,3,11] However, in one of our earlier studies,[1] a very heterogeneous Ziegler-Natta ULDPE, although very difficult to process by itself, was found to have surprisingly good low temperature impact efficiency in polypropylene. It was proposed that the extremely broad compositional distributions led to enhanced coupling between the rigid matrix and the elastomeric domains. It is the goal of this study to determine if this interface can be tuned through deliberate broadening of the short chain branching distribution, SCBD. Perhaps, through this understanding, we can further optimize the impact modification efficiency of metallocene ULDPEs.

Available literature on impact modification indicate that the morphology plays an important role on effectiveness, hence in this study we characterized these blends by thermal analysis and electron microscopy to determine the likely origin for the performance difference and attempt to construct a plausible model that fits most observations. It was hoped, through studies like this, additional insights can lead to further performance improvements.

MATERIALS

Matrix polypropylene (PP w/0.91 density, MFR=2)
Dispersed phase ZN ultralow density polyethylene-A (ULDPE w/0.88 density and MI= 0.8)
Dispersed phase homogeneous ULDPE-B (ULDPE w/ 0.87 density and MI=5)
Modifier A: High density polyethylene (HDPE w/0.96 density, 20 MI)
Modifier B: Homogeneous medium density polyethylene (MDPE w/ 0.935 density, 2.5 MI).
Modifier C: Homogeneous very low density polyethylene (VLDPE w/ 0.902 density, 1 MI)
Polyethylene modifier blends were first compounded and pelletized with the 0.87 density base modifier B in a 1:4 ratio of PE to ULDPE. The resultant modified ULDPE pellets as well as base ULDPE resins were blended with the PP material and compounded on a 1.5 inch Davis Standard extruder with a two-stage screw and a Maddox mixing section at a melt temperature of 230°C and extrusion casted into film. The target composition was 75% PP and 25% modified ULDPE or straight ULDPE. The base blend for interface modification was made from ULDPE B. The modified composition was targeted to contain 5% interface modifiers.

CHARACTERIZATION

Thermal analysis was carried out on a TA Instruments 2100/2910 Differential Scanning Calorimeter, DSC, at a cooling rates of 10°C/min. to -30°C after equilibration at 200°C to erase the previous thermal history. The 2nd melt was done at a heating rate of 10°C/min. Dynamic Mechanical Analysis, DMA, was done on a Seiko DMS-110 Dynamic Mechanical Analyzer over a temperature range of -150 to 150°C over a frequency range of 0.5 to 100 Hz. Electron microscopy was carried out on the film samples microtomed at -100°C in a Reichert FC4E Cryo-ultramicrotome and etched in n-heptane at ambient temperature in a sonic bath for 30 min. The film samples were first relaxed at 180-200°C for 5 min before microtoming. Imaging was carried out on a JEOL 6300 LVSEM or a JEOL 35CF SEM. Rheological measurements were done on a Rheometrics Fluids Rheometer at 200°C with a parallel plate geometry over a frequency range of 0.1-200 rad/sec.

Impact studies were performed on a Dynatup (Model 8200) Instrumented Impact Tester with a 2 cm semi-spherical tup interfaced with personal computer with LabView automation software. A 14 cm circular film holder was used for sample mounting. Samples were conditioned and tested in a computer controlled, liquid N_2 cooled, environmental chamber.[2]

RESULTS AND DISCUSSIONS

Instrumented impact data are shown in Figures 1-4. The reference blend with the metallocene modifier has a very low brittle/ductile transition of about -35°C. After rising steeply with temperature after the B/D transition, the impact energy moderated somewhat before rising again at about -15°C to a very strong plateau at ambient temperature. In comparison, the Ziegler-Natta heterogeneous modifier gives a surprisingly low B/D transition of -35°C. After steady increases at a slightly lower rate than the reference, it begins to decrease at about -15°C and ending at room temperature at a significantly lower value than the reference. As discussed in the previous paper, this was probably caused by the larger particle size of the dispersed phase.

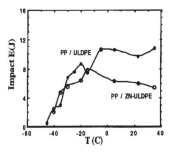

Figure 1. Impact comparison.

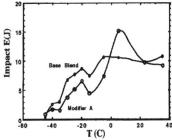

Figure 2. Modifier A (HDPE) impact.

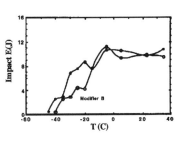

Figure 3. Modifier B (0.935 density).

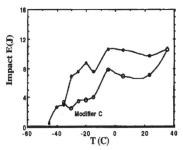

Figure 4. Modifier C (0.902 density).

The modifier A with 5% of a highly crystallizable high density polyethylene, achieved a B/D transition at -35°C, equaling that of the reference blend. However, between -40°C and -20°C, the impact performance was slightly but detectably lower than that of the reference, presumably from the slightly lower ULDPE content (20% versus 25 %). However, at about -20°C, after the β-relaxation of the ULDPE, the impact energy rises sharply, significantly surpassing that of the reference blend before settling down to near equality at about room temperatures. This strong rise and high ambient impact clearly indicated the effectiveness of the interface coupling the modifier was designed for, validating the hypothesis set forth in the previous paper.

For modifier B, a 0.935 medium density homogeneous polyethylene, the B/D transition was slightly elevated to about -30°C, and the rise at about -20°C was still evident. However, the impact performance above -20°C pretty much matched that of the base reference. This indicated a moderate level of the modification of the interface.

For modifier C, it appears an anti-synergism was at work: the modifier was too high of a crystallinity to improve deep subambient impact, and yet too low of a crystallinity to modify the domain interface. As a result, the impact performance stayed significantly below the reference material throughout the temperature range.

DYNAMIC MECHANICAL ANALYSIS

Dynamic mechanical analysis, DMA, uses low amplitude sinusoidal stresses to probe various molecular relaxation processes in the polymer at various temperatures. The gamma relaxation at about -120°C for polyethylenes was assigned the so-called "crankshaft" motion[4] of more than four methylene sequences. In addition, there is a strong inverse relationship between the crystallinity and the intensity of the relaxation. Evidently, at higher crystallinities, the crystalline matrix constrains and restricts the crankshaft motion at these temperatures. In addition, earlier DMA studies[5] correlated the

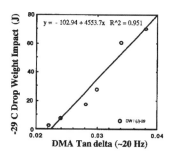

Figure 5. Correlation of gamma intensity with impact data from ref. 5.

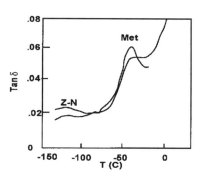

Figure 6. DMA comparison of ZN and metallocene-modified PP.

intensity of the gamma transition with subambient impact performance of PP/EPR blends of varying rubber content and processing conditions (Figure 5). In this case, the rigid matrix is the crystalline PP phase, while the rubber domains the dispersed phase. An analogy may also exist where the freedom of motion for the rubbery phase is modulated by the surrounding rigid matrix. Here, the coupling at the domain interface must be the source of the modulation.

For pure ULDPE's the γ-relaxation is quite active and as a result the brittle-ductile transition is located at temperatures far below the β-relaxation, the main relaxation for the rubber phase. For example, the base metallocene modifier with a β maximum of -44°C exhibits a B/D transition at -75°C.

For impact modified polymers, in the region between the γ maximum and the β-relaxation, the mechanical coupling between the rubber phase and the rigid matrix is important in transferring stresses from the brittle matrix to the rubber domains. This coupling is reflected in the intensity of the γ –transition for similar rubber content systems.

In comparison, as shown in Figure 6, the Ziegler-Natta modified polypropylene exhibited a significantly more active γ-relaxation compared with the metallocene ULDPE with a slightly lower crystallinity.

MORPHOLOGY

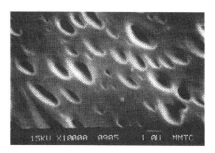

Figure 7. n-heptane etched morphology metallocene ULDPE/PP blend.

It is well known that particle size and dispersion are important factors in achieving optimal product performance and that particle size is a function of the chemical structure of the components and the differences in rheological behavior.[9-11] Rheology data indicates that the polypropylene matrix exhibits a greater shear thinning response than the dispersed phase ULDPEs. Among the ULDPEs, the Ziegler-Natta sample has the greatest shear sensitivity due to its much broader molecular weight distribution. The viscosity match at processing temperature and shear rates determines the particle size and distribution of the dispersed phases.

Sonication with n-heptane has virtually no effect on crystalline polypropylene but dissolves primarily

Figure 8. n-heptane etched morphology
Z-N ULDPE/PP blend.

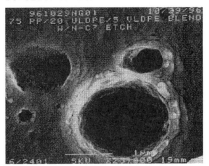

Figure 9. n-heptane etched morphology of
VLDPE modified ULDPE/PP blend.

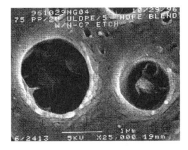

Figure 10. n-heptane etched morphology of
HDPE modified ULDPE/PP blend.

amorphous and low crystallinity fractions in olefin polymers and leaves a cavity where the extracted material was located. This process creates the topographical contrast for characterization of multiphase materials with electron microscopy. For example, Figures 7-8 compare the etched morphology of the metallocene modified PP with that of the Ziegler-Natta modified PP. The homogeneously polymerized ULDPE domains were nearly completely extracted from the matrix. In strong contrast, the very heterogeneous Z-N modifier exhibited far less extractability at very similar crystallinities. In addition, the extracted interface also showed numerous fibrils connecting the ULDPE rubbery domain to that of the polypropylene matrix. In our opinion, this is increased coupling across the domain interface that led to the surprising performance.

The morphology of the polyethylene modified ULDPE is shown in Figures 9-10. When linear polyethylene was present within the ULDPE there was residue in the cavity after the heptane extraction, residue which was connected to the matrix with microfibrils. This did not occur at all with the VLDPE modified material. When VLDPE was present within the ULDPE there was virtually nothing remaining in the cavity after the n-heptane extraction. The greater the crystallizability of the polyethylene the greater the coupling between the matrix and the dispersed phase.

SUMMARY

Previously, it was established that the ductile-brittle transition temperature of a polypropylene modified with a homogeneous ULDPE decreases linearly with the density of the ULDPE. This work has shown that the presence of a more crystallizable polyethylene within the ULDPE domains strengthens the interface between the particle and the matrix and reduces the ductile-brittle transition compared with homogeneous modifiers of equal density. The toughening occurs through the development of crystalline microfibrils of polyethylene connecting the ULDPE with the PP matrix. The extent of the coupling increases as the crystallizability of the polyethylene increases. This work points to the direction with which impact modification can be further optimized.

REFERENCES

1. L. Woo, T. K. Ling, and S. P. Westphal, *ANTEC'95,* 2284 (1995).
2. L. Woo and M. T. K. Ling, *ANTEC'90*, 1116 (1990).
3. T. C. Yu and G. J. Wagner, *Polyolefins VIII*, 539, Houston, TX, 1993.
4. N. G. McCrum, B. E. Read, and G. Williams in **Anelastic and Dielectric Effects in Polymer Solis**, p. 366, *John Wiley*, New York, 1967.
5. P. S. Gill, R. L. Blaine, and R. P. Carter, *ANTEC'79*, 816, (1979).
6. L. Woo, S. Westphal, and M.T. K. Ling, *Polym. Eng. Sci.*, **34(5)** 420 (1994).
7. B. Ohlsson and B. Tornell, *J. Appl. Polym. Sci.*, **41**, 1189 (1990).
8. J. S. Trent, J. I. Scheinbeim, and P. R. Couchman, *Macromolecules*, **16**, 589 (1983).
9. B. D. Favis and J. M. Willis, *J. Polym. Sci., Polym. Phys.*, **B28**, 2259 (1990).
10. J. White and K. Min, *Adv. Polym. Techn.*, **5(4)** 225 (1985).
11. T. C. Yu, *ANTEC'95*, 2374 (1995).

Morphology of Low to Very Low Density Ethylene-Octene Metallocene Copolymers

P. J. Phillips and K. Monar

Department of Materials Science & Engineering
University of Tennessee, Knoxville, TN 37996-2200, USA

ABSTRACT

The melting phenomena of random, hexyl-branched ethylene copolymers have been examined for branch contents ranging from 4-53 SCB/1000C at near constant molecular weight with low polydispersity (ca. 2.0) corresponding to densities of ca. 0.94-0.866 g/cm^3. We have found that, irrespective of morphology, melting after isothermal crystallization and bulk isothermal crystallization phenomena can be grouped into two types of behavior: Group 1 (G1), nominally with up to 2% defects (as branches) and Group 2 (G2) with greater than 2% and up to 8.5% defects. The first group is characterized by single peak melting behavior above T_c and classical annealing via thickening as evidenced by T_m−log time relations, at least for short times. G1 follows the Avrami equation up to ca 90-98% of the developed crystallinity with growth as spherulites. Group 2 is generally characterized by two melting peaks above T_c (one narrow, one broad). These samples show distinctive two-stage Avrami crystallization curves with lower than expected initial Avrami index where spherulitic forms are growing (up to 0.88 g/cm^3) and even lower indices where no spherulitic forms are observed (below 0.872 g/cm^3).
Three unexpected phenomena are observed for the melting of isothermally crystallized G2 samples: (a) the narrower peak melts at a lower temperature than an underlying, broad peak; (b) the early peak is absent at short times (less than 1 m) and develops with increasing time and shifts to higher melting temperatures with time and; (c) the early melting peak appears representative isothermally crystallized samples, at lower undercoolings, insofar as these melting points show parallel T_m-T_c behavior in a Hoffman-Weeks Plot whereas the broad melting peak (observed at short times) exhibits more conventional behavior. The use of WAXD, DSC, and to a lesser extent, density are used on slow cooled and quenched bulk crystallized samples to gain some insight into this process.

INTRODUCTION

Random octene-ethylene copolymers with densities ranging from 0.9600 to as low as 0.855 g/cc are commercially available via DOW Insite® metallocene technology which possess high degree of randomness in the comonomer distribution, narrow molecular weight distribution, and controlled quantity and length of long chain branching, LCB.

It has been observed that a morphological transition occurs at ca 0.875 g/cm^3 (ca 37-42 Me/1000C) which has been characterized as fringe-micellar[1] based on stained, ultra-thin sections and electron microscopy. However, some lamellae have been shown to exist in these systems, depending on thermal treatment.[2] Surprisingly, the most discernible lamellae were observed at 15°C/m cooling while none are observed under isothermal conditions and quenching may indicate a few broken up lamellae. The case for the non-lamellar structure shows the reverse trend, appearing more prominent for the isothermal condition and becoming increasingly coarse. A morphological/mechanical model was developed[3] characterized by a distinct change in deformation behavior in uniaxial tension from one consisting of a yield point followed by some degree of cold drawing dependent upon crystallinity, to one with no yield point, a low modulus, and high permanent set indicative of elastomers.[3,2] These have been correlated with deformation modes for crystalline homopolymer, spherulites with a open lamellar system akin to HP-LDPE,[4] to a mixed system of open lamellae and fringe micelles to a fully non-lamellar system. Moreover, the mechanical α-relaxation associated with viscous transport from lamellar crystals[5] was present in a slowly cooled 0.882 g/cm^3 sample, but not a 0.870 g/cm^3 one.[6] These authors conferred a bundle model for the lowest densities based on the melting temperature, T_m, and the Gibbs-Thomson equation for the melting of thin crystals:

$$T_m = T_m^0 \left(1 - \frac{2\sigma_e}{l(\Delta H_m^0 \rho_c)}\right)$$

(where l is the crystal thickness ΔH_m^0 is the specific equilibrium heat of fusion, ρ_c the crystal density and σ_e the surface energy) without adjustment for the equilibrium melting point, T_m^0, depression[7] or ΔH_m^0 for melting far from T_m^0 where it is defined. Now the bundle model[8] assumes, as a minimum, the value of σ_e quoted for relatively short segments (2,000) increasingly dramatically for the number of nuclei a (much) longer chain would participate in. The critical temperature for massive primary nucleation in linear polyethylene or LDPE is ca 87°C ($\Delta T=58$°C)[9,10] which would have resulted in much higher melting temperatures than observed under the suggested undercooling.

Earlier we had reported that particle analysis using small angle x-ray scattering from a 0.87 density system had substantial platelet character with interfacial regions having a density approaching that of the crystal at the lowest undercooling.[11] Crystal thicknesses of ca. 45Å were noted with total attached interfacial content at least as thick. Also, strong two-stage bulk crystallization was observed for non-spherulitic samples at shorter times with Avrami indices of 1-2 as varying from low to high undercooling.[12] Such bulk crystallization curves had been known for some time from LDPE with indices of 1-2.5.[13] SAXS analyses[14] indicated that thickening was inoperative and that [$(\rho_c - \rho_a)$=constant] which implied that the branching frequency changes with T_c; implying lamellar space filling within spherulites can diminish with time, which has been documented[15] and that the constant (low) indices are indicative of isolated lamellae. This result is not consistent with the TEM evidence for VLD-MI-LLDPE's even though qualitatively similar. Neither is the suggested model[1] (possibly with massive nucleation) consistent with the change in Avrami index during crystallization from higher to lower values.

Thermodynamically, some balance must be struck between enthalpy and crystal thickness with the effective ethylene run length and the high σ_e of a bundle crystal without defects, or the work of chain folding and fold surface energy plus defect incorporation. This is not an insignificant point as the difference in tear strength in MI-LLDPEs of octene are greater by a factor of seven over similarly prepared butene systems at their maximum strengths, nominally occurring at the same comonomer contents.[16] It is important to note here that the hypothetical volume of the dominant defect in PE crystals (2g1) is estimated to be 60 Å^3 (17; we calculate butene, 63 Å^3 and octene 172 Å^3) and that the defect incorporation expansion coefficient (a measure of lattice disruption, 18) is greater by a factor of three for methyl and ethyl branches relative to butyl branches.

Table 1. Sample characterization

Sample	SCB[1]	PD	$X_b\%$[2]	$C(X_b)^3$,A
L04	4.0	2.2	0.4	319
L24	24.0	2.2	2.9	53
L42	42.0	2.1	5.9	30
L39	39.6	2.0	5.5	32

[1]per 1000C; [2]branching frequency as CH_2; [3]estimated core crystal thickness with tilt

EXPERIMENTAL

The samples used are listed in Table 1 together with the branching frequency, maximum sequence length, and minimum melting temperature. The short time crystallization/melting studies, STC, were performed with ca. 7 mg using a Perkin-Elmer DSC which was also used to melt bulk specimens (1.6 mm thick) crystallized in oil baths (T_c ±0.4°C). WAXD data were collected on a Scintag PAD diffractometer using Cu-K$_\alpha$ radiation and a Kevex electronic detector with an electronic window of ca. 600 eV centered on 8.04 keV. Data were collected for slow cooled (SC, 5°C/m) and quenched samples (FQ, sCO$_2$ cooled ethanol) for comparative purposes.

RESULTS

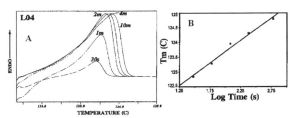

Figure 1. (A) Melting of L04 from T_c = 113°C at 10°C/min after annealing for indicated times. (B) Melting and thickening show linear increase with log(time).

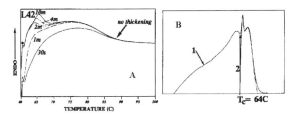

Figure 2. (A) Melting of L04 from T_c = 60°C at 10°C/min after annealing for indicated times. '*' indicates initial C_p transient. (B) Melting of L42 after annealing at Tc = 64°C for 5 min; (1) Quench to T_g and (2) Heat from T_c; both at 10°C/min.

Figures 1 and 2 illustrate the different annealing phenomena observed in isothermally crystallized samples for G1 and G2. In G1 the crystal thickness (highest melting point, '1A') increases with the logarithm of time (1B) for sample L04 annealed at 113°C, whereas for L42 the highest melting point does not change (2A,arrow) with annealing time at the crystallization temperature. Both were crystallized at the highest accessible undercooling. Additionally, the G2 samples (e.g., L42) show the appearance of a lower melting peak, T_{ml}, during crystallization which shows a shift toward higher melting temperatures with time. It is readily distinguishable from the C_p transient in shifting from isothermal crystallization to heating (cf. 2A,'*') and is absent at the shortest times. Melting from T_c renders any recrystallization from quenched material mute, but as can be seen in (2B) this mechanism appears inoperative in this sample.

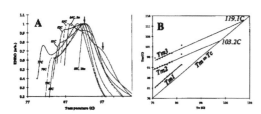

Figure 3. (A) STC melting behavior (from T_c) for L24. (B) Hoffman-Weeks plot of melting data in 'A' as described in the text.

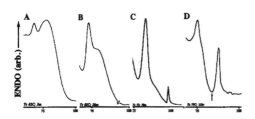

Figure 4. Melting behavior for L42 bulk crystallized samples at indicated T_c.

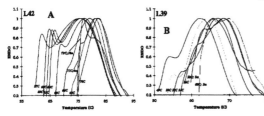

Figure 5. STC melting behavior for L42 (A) and L39 (B).

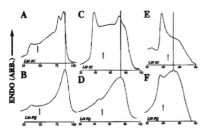

Figure 6. DSC melting curves of quenched (FQ) and slow cooled (SC) bulk crystallized samples. L24 (A,B), L42 (C,D), L39 (E,F).

The STC behavior for sample L24 is shown in Figure 3A as a function of crystallization temperature for times of 2 m, unless indicated otherwise. The data are normalized for comparative purposes and the amount of data is limited by the low heats of fusion, for short times. Several points are noted: (a) the more narrow T_{m1} peak is fairly well resolved from the broader melting T_{m2} peak; (b) the two peaks tend to become less differentiated at lower undercoolings and; (c) the T_{m1} peak eventually becomes the dominant peak. The latter point is supported on two phenomenological observations: first, the rather narrow character of the T_{m1} peak and; second, from the behavior of the T_{m1} peak on a Hoffman-Weeks Plot, HWP. The latter is illustrated in Figure 3B where the T_{m1} peak shows parallel T_m-T_c behavior while T_{m2} or T_{m3} both extrapolate to the T_m-T_c line and begin to deviate at longer times. The melting peaks at the lowest undercoolings likewise show the parallel trend. Figure 4 illustrates the G2 melting phenomenon in bulk crystallized samples for L42 showing an increase in T_{m1} relative to T_{m2} with time and temperature (A,B,C). Additionally, another melting peak begins to be observed (B,C,'*') and at the highest crystallization temperatures (D), only a single peak is observed above T_c (arrow). For some STC studies, T_{m1} appears much diminished relative to T_{m2} for selective temperature ranges (5A) but does develop in time (e.g. 72C; 10 m) and gives some indication that this is at the expense of the highest melting point crystals. For non-spherulitic L39 STC studies (5B) yield a more uniform behavior for the two peaks but also reveal the tendency shown in (4A-D) for sufficient time.

The annealing effects observed isothermally prepared samples and their weakness in a slow cooling/heating experiment, led us to compare some structural information on the basis of slow cooled and fast quenched samples. Figure 6 shows a comparison of the melting curves for the slow cooled samples (SC, top) and the quenched samples (FQ, bottom). L24 (A, B) shows that the T_{m1} peak development is at the ex-

Table 2. Melting, fusion, and density for SC and FQ

Sample	SC Fusion (J/g)	FQ Fusion (J/g)	SC Density (g/cm³)	FQ Density (g/cm³)
L24	94	69	0.899	0.894
L42	48	31	0.883	0.881
L39	31	32	0.971	0.872

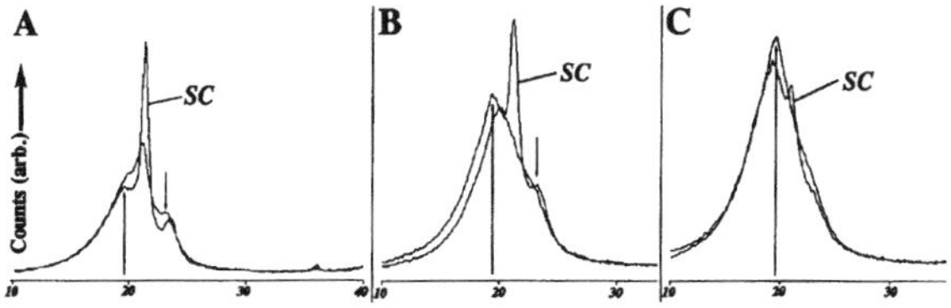

pense of the higher melting temperatures and that the peak temperature is higher for the FQ sample and occurs at ca. 33% lower crystallinity (determined above the arrow to the melt on a constant ΔH_f^0 basis). The same trend is observed for L42 (C, D), but to a lesser extent. The case of L39 (E, F) shows the least change for T_{m2}. Both L42 and L39 FQ samples have enthalpies which are at least as great as SC samples (Table 2). Room temperature WAXD for the same three comparative series are shown in Figure 7 for constant conditions, normalized between sample types but not between SC and FQ subtypes. In all of the cases, the SC treated samples have substantial more orthorhombic character than the corresponding FQ treated sample, however, some additional factors underlie this expected situation. First, for all SC samples, there is

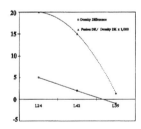

Figure 7. WAXD intensity vs. degrees 2θ (unnormalized) for SC and FQ L24 (A), L42 (B), L39 (C).

Figure 8. Density difference vs. fusion difference for SC & FQ samples.

more liquid-like polyethylene as revealed by the sub-20°(2ϑ) amorphous scattering while FQ samples possess more 20°(2ϑ) character (characteristic of amorphous fraction in homopolymer PE). Secondly, all SC samples possess a more ordered lattice relative to the comparable FQ sample; especially L24 and L42. The unit cell dilation (evidenced in the (200) reflection) is greater in the L24 SC sample than for the FQ sample. L42 shows many of the same trends with substantially greater disruption of the orthorhombic phase and, like L24, the SC sample would appear to possess a greater quantity of liquid-like PE than the FQ sample. In the non-spherulitic L39 sample there is liquid scattering for the FQ sample, even though substantial fusion energy is evolved (6F). A comparison of the differences in the heat evolved on melting in SC and FQ treatments and density is shown in Table 2 and Figure 8. Note in (8) that the macroscopic density actually increases slightly for L39-FQ and fusion is nearly identical. Also, when normalized by the density difference, the fusion difference is not a constant, but decreases rapidly to ca 1.0 in this region.

CONCLUSIONS

The classical thickness/log time relation observed for L04 is not surprising but we would expect to find a limiting value relative to inter-crystalline structural factors and, more importantly, the side branches. It has recently been shown that a low branch content metallocene LLDPE inhibited the growth of the hexagonal phase at high pressure with thickness' limited ultimately by the ethylene run length[19] defining the necessity of chain folded nuclei to reach this condition.

Now the ca. $20°(2\vartheta)$ amorphous maximum in PE must, necessarily, be considered indicative of mostly interfacial material as the liquid scattering of PE, PE-copolymers, and n-alkanes has been defined.[20] This is not surprising given the tendency for adjacently folded-chain crystals. The disruption of the lattice in the FQ series indicates higher presence of incorporation of comonomer. We will report later on the small maximum appearing at the apex of the liquid peak in (7B, C arrow) which is observed to disappear with increasing temperature and is correlated with the disappearance of T_{ml} in (6C, E). The liquid-like diffraction pattern for quenched VLD-PE is known for molten PE as well where it disappears at the "so-called" extrusion window for PE[21] and its origin has been described as a thermodynamic state change.[22]

The delineation of the liquid-like and constrained amorphous material[23] combined with the WAXD data, and the isothermal crystallization and melting processes above, and knowledge of the effects of branch incorporation[19] lead us to propose a model of initial incorporation of comonomer branches followed by temperature and time dependent rejection. As such, T_{ml} may represent a surface/interfacial/intercrystalline enthalpy, a concept which has precedence[23] and has recently been proven with reversible surface melting in LPE.[24]

This concept would clarify several matters: the two stage crystallization;[12] the residual plateletlike structure for the crystallites and thick interfaces indicated by SAXS[11] suggested as "swiss-cheese" like (or, better, surface craters); the disordered surfaces observed in all single crystals of PE copolymers; the greater quantity of liquid-like material in SC than in FQ material as indicated by WAXD while possessing more perfect crystals, and; the presence/lack of lamellae or coarseness of the TEM ultrastructure dependent upon thermal history in VLD samples.[2]

The difference in tear strength with comonomer type (16; strongly related to amorphous content) may be related to incorporation/rejection issues on the one hand and competing surface energy increases and lattice strain with concomitant dilation on the other. Obviously, this competition is complex as illustrated in (3A, 5A, 5B) but would predict that higher molecular weight material, at similar branch content, would always have more of the T_{ml} material initially (more intercrystalline links, tethered by chain ends).

Note that the mixed morphological model characterizing the lamellar to fringed micelle transition for L- and VLD-MI-LLDPEs[3,6] is not consistent with several pieces of definitive evidence. It would not be expected to show substantial differences in SC and FQ treatments in SAXS and WAXD nor annealing behavior producing even lower melting temperatures. Instead, a transformative process between metastable states (noting that the case for the reverse process could be made) is argued on the basis of several lines of experimental evidence. We will consider the form shown for the Gibbs-Thomson equation combined with a suitable consideration for the melting point depression and lattice constants to come to a reasonable accounting for the melting point.[25]

REFERENCES

1.   P. S. Chum, C. I. Kao, and G. W. Knight, *Plastics Eng.*, June, 21 (1995).

2.   Y.-C. Hwang, S. Chum, R. Guerra, and K. Sehanobish, *Proc. ANTEC '94*, 3414 (1994).

3.   S. Bensason, J. Milnick, A. Moet, A. Hiltner, and E. Baer, *Proc. ANTEC '95*, 2256 (1995).

4.   G. Strobl, M. J. Schneider, and I.G. Voigt-Martin, *J. Polym. Sci., Polym. Phys.*, **B18**, 1361 (1980).

5.   R. H. Boyd, *Polymer*, **26**, 1123 (1985).

6.   S. Bensason, J. Milnick, A. Moet, A. Hiltner, and E. Baer, *J. Polym. Sci. Polym. Phys.*, **B34**, 1301 (1996).

7.   P. J. Flory, *J. Chem. Phys.*, **17**, 223 (1949).

8.   H. G. Zachman, *Pure Appl. Chem.*, **38**, 79 (1974).

9.   R. L. Cormia, F. P. Price, and D. Turnbull, *J. Chem. Phys.*, **37**, 1333 (1962).

10.   J. A. Koutsky, G.A. Walton, and E. Baer, *J. Polym. Sci., Polym. Phys.*, **B5**, 185 (1967).

11.   P. J. Phillips and K. Monar, *Proc. ANTEC '96*, 1996.

12.   P. J. Phillips and K. Monar, *Proc. ANTEC '95*, 1995.

13.   A. J. Kovacs, *Ricerce Sci.*, **251**, 669 (1955).

14.   G. R. Strobl, T. Engelke, E. Maderek, and G. Urban, *Polymer*, **24**, 1585 (1983).

15.   S. Hosoda, K.Kojima, and M. Furuta, *Makromol. Chem.*, **187**, 1501 (1986).

16.   L. T. Kale, T. A. Plumley, R. M. Patel, and P. Jain, *Proc. ANTEC '95*, 2249 (1995).

17.   W. Pechold and S. Blasenbrey, *Koll. Z. Z. Polym.*, **235**, 216 (1967).

18.   F. J. Balta-Calleja, and C. G. Vonk in **X--Ray Scattering of Synthetic Polymers**, *Elsevier*, Amsterdam, 1989.

19.   J. A. Parker, D. C. Bassett, R.H. Olley, and P. Jaaskelainen, *Polymer,* **35**, 4140 (1994).

20.   D. C. McFadden, K. E. Russell, G. Wu, and R. D. Heyding, *J. Polym. Sci., Polym. Phys.*, **B31**, 175 (1993)

21.   A. J. Waddon and A. Keller, *J. Polym. Sci., Polym. Phys.*, **B30**, 923 (1992).

22.   H. M. M. van Bilsen, H. Fischer, J. W. H. Kolnaar, and A. Keller, *Macromol.*, **28**, 8523 (1995).

23.   M. G. Peterlin, *J. Polym. Sci., Polym. Phys.*, **B5**, 613 (1967).

24.   R. Mutter, W. Stille, and G. Strobl, *J. Polym. Sci., Polym. Phys.*, **B31**, 99 (1993).

25.   K. Monar, **Ph.D. Dissertation**, *The University of Tennessee*, 1997.

Morphological Studies of Metallocene Plastomer Modified Polypropylenes

N. R. Dharmarajan and Th. C. Yu

Exxon Chemical Company, Baytown, TX 77522, USA

ABSTRACT

The dispersion of metallocene plastomers in homopolymer polypropylene varying from low to high melt flow rates were examined by low voltage scanning electron microscopy, LVSEM, using ruthenium tetroxide stained cryomicrotomed samples. Micrographs obtained were used to count the cross-sectional area of several hundred particles. The particles areas were next converted in equivalent diameters. At up to 20 melt flow rate of the base polypropylene, the average particle size remained to be 0.5 μm. Only at high melt flow rate, i.e., 35 MFR did the base polypropylene alter the plastomer dispersion. In this instance, the average particle size was increased to 0.8 μm, but still less than the 1 μm threshold necessary for cold temperature impact modification of polypropylene. Solubility data for ethylene-butene plastomers at varying comonomer content were measured using small angle neutron scattering, SANS. The interfacial tension between plastomer and polypropylene could be then calculated from measured solubility parameters. Low interfacial tensions between ethylene-butene plastomer and polypropylene promoted submicron dispersion of plastomer in both low and high melt flow polypropylenes. The application of Wu's model for predicting plastomer dispersion was not successful.

INTRODUCTION

The recently introduced metallocene plastomers have been found to possess plastic as well as elastomeric attributes. The free flowing plastomers allow them to be handled on existing plastics equipment. Uniform comonomer incorporation along the ethylene backbone was achieved by the single site catalysts, so that at relatively low comonomer content, the ethylene-butene plastomer exhibits many elastomeric attributes such as low melting point, low crystallinity, and filler acceptance. The ethylene-butene plastomer is being used in the next generation of high flow thermoplastic olefins, because the final compound shows a much higher flow rate than the traditional ethylene-propylene based products. In a previous mixing study,[1] both a twin screw extruder and a single screw extruder fitted with a mixing screw were found to be the effective compounders for plastomer-polypropylene blends. In this study, an ethylene-butene plastomer was melt blended with five low, medium, and high flow homopolymer polypropylenes, PPs, on a twin screw extruder. Scanning electron micrographs for these blends were obtained from ruthenium tetroxide stained specimens. These micrographs were next used to conduct detailed particle size estimates and particle size distribution. Rheological characterization of both plastomer and polypropylene together with estimates of their interfacial tensions allowed us to model plastomer dispersion at different compounding shear rates.

EXPERIMENTAL

A 30 mm ZSK twin screw extruder was used to compound 30 wt% EXACT 4033 into five homopolymer polypropylenes with melt flow rates ranging from 1.7 to 35. They are shown in Table 1. Sample pellets were next injection molded into 3-1/2" and 1/8" thickness test plaques using a 75 ton Van Dorn injection molding machine.

Table 1. Effect of melt rheology on plastomer dispersion (iso-PP 70 wt% EXACT™ 4033 30 wt%)

	MFR
Escorene PP 1042	1.7
Escorene PP 1012	5.3
Escorene PP 1154	13.0
Escorene PP 1074 E-1	20.0
Escorene PP 1015 E-1	35.0

IMAGE ANALYSIS

Image analysis of low voltage electron scanning micrographs, LVSEM, was used to determine the average particle size and particle size distribution of EXACT 4033 in the five polypropylene matrices. The LVSEM micrographs were scanned at 200 to 300 pixels per inch on an Apple laser scanner to create a digital dot matrix file. The scanned image was next analyzed using Image (Version 1.22), a software developed by Wayne Rasband (NIH), operating on an Apple Macintosh IIcx personal computer. Through phase contrast, this program was able to distinguish particles of the dispersed EXACT phase from the surrounding polypropylene matrix. The area of the dispersed phase, the major and minor axes and orientation of each modifier particle were calculated. Typically about 80 particles were analyzed from each micrograph, and three to four images of every EXACT 4033/PP combination were examined for statistically meaningful results. Data analysis was performed on the same PC using KaleidaGraph 2.1 software.

The following parameters were calculated:

$$D_{avg1} = (D_{major} + D_{minor})/2 \qquad\qquad\qquad [1]$$

$$D_{avg2} = (4\ Area/\pi)^{1/2} \qquad\qquad\qquad [2]$$

$$D_n = \sum n_i d_i / \sum n_i d_i^2 \qquad\qquad\qquad [3]$$

$$D_v = \sum n_i d_i^3 / \sum n_i d_i^2 \qquad\qquad\qquad [4]$$

$$AR = D_{major} / D_{minor} \qquad\qquad\qquad [5]$$

The average particle diameter was calculated in two ways:
1. Average length of the particle major and minor axes lengths (D_{avg1} in Eq 1).
2. A circle diameter (D_{avg2} in Eq 2) with an area equal to the measured area of the dispersed particle.

In most cases D_{avg1} was found to be nearly equal to D_{avg2}. The exceptions occurred when the particles were significantly oriented. In this case D_{avg1} was somewhat greater than D_{avg2}. D_{avg1} was used in all analyses. The number average diameter, D_n, and the volume average diameter, D_v, were calculated using Eqs 3 and 4, respectively. The aspect ratio, AR, was defined as the ratio of major to minor particle axis (Eq 5) and it is equal to 1 if the particles are perfectly spherical.

The diameters reported in this study were based on projections onto a two dimensional surface from the cut surface of the individual particle. This method excluded three dimensional effects and consequently did not represent the true particle diameter. There were several methods proposed in the literature to estimate the actual diameter from two dimensional projections. In a recent study, Scott and Macosko[2] concluded that applying these corrections only provided an 8% increase in D_n values, and produced a small shift in the overall distribution to lower diameters. These corrections were considered insignificant and were not applied in our analysis.

RESULTS AND DISCUSSION

The left side of Figure 1 shows the dispersion of EXACT 4033 in a 5 MFR PP and the corresponding particle size distribution is shown on the right side. The average particle size is 0.5 μm, and the particles range in diameter from 0.1 to 2.3 μm. The particles appeared oriented in the 5 MFR PP matrix. The average aspect ratio of the dispersed particle is calculated to be 2.8. There is also some indication of particle coalescence. Similarly, the left side of Figure 2 shows the dispersion in a 35 MFR PP, and the right side of Figure 2 shows the corresponding particle size distribution. The average particle size increased to 0.78 μm, and the particle diameter ranged from 0.1 to 2.3 μm. The particles appeared less oriented. The average aspect ratio was also reduced to 1.8.

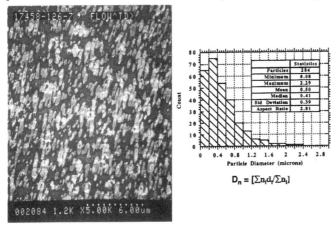

Figure 1. EXACT™ 4033 in 5 MFR isotactic polypropylene.

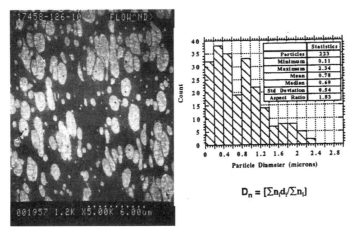

Figure 2. EXACT™ 4033 in 35 MFR isotactic polypropylene.

Figure 3 contains the cumulative particle size distribution of EXACT 4033 in PP resins ranging in melt flow rate, MFR, from 1.7 to 35. The cumulative size distribution is very nearly the same in the PP matrices varying in MFR from 1.7 to 20. However, in the 35 MFR PP, the distribution tends towards increased particle size. As an example, from Figure 3 we observe that in a 20 MFR PP over 85 % of

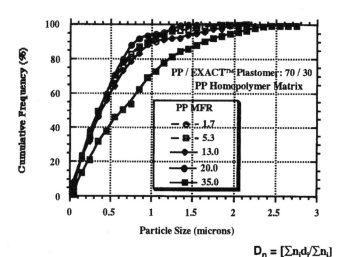

Figure 3. Cumulative frequency plot.

$$D_n = [\Sigma n_i d_i / \Sigma n_i]$$

the EXACT particles have a diameter less than 1 μm, while in the 35 MFR PP matrix only 70 % of the particles are smaller than 1 μm.

Figure 4 shows a log-normal plot of the EXACT particle size distribution in the various polypropylenes. This type of distribution has been commonly applied by other researchers in particle size analysis of multi-component polymer blends.[3,4] A linear fit can be applied to the data points in Figure 4 confirming that the distribution is log-normal. As expected, the particle distribution converges to a single curve for PP MFRs ranging from 1.7 to 20 while in the 35 MFR PP matrix the distribution is towards larger particles.

MODELING OF PLASTOMER DISPERSION

The viscosity ratios between the EXACT 4033 and the five polypropylene base resins were experimentally determined at several shear rates. These measurements were performed at 225°C, which is a typical temperature used in melt compounding EXACT 4033 with polypropylenes. The experimental data of viscosity, η, variation with shear rate, γ, was modeled according to the Cross[5] equation given by:

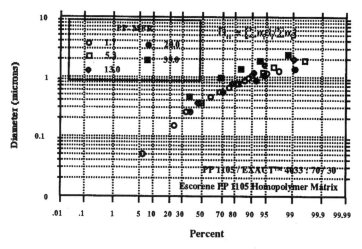

Figure 4. Log normal distribution.

$$\eta = \eta_o / (1 + (\lambda\gamma)^{\alpha}) \qquad\qquad [6]$$

where η_o is the zero shear, λ is the relaxation time and α a constant. The rheological properties of EXACT 4033 and polypropylenes fitted the Cross equation well.

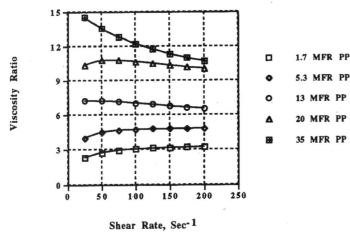

Figure 5. Viscosities ratios for ethylene-butene plastomer and isotactic polypropylene. Test temperature: 225°C.

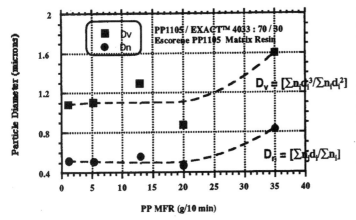

Figure 6 Effect of melt flow rate of polypropylenes on plastomer particle size.

Figure 5 shows the viscosity ratio dependence on shear rates for the five plastomer polypropylene blends. It is desirable to minimize this viscosity ratio to achieve optimal dispersion of the EXACT modifier in PP matrix. This can be implemented by suitably adjusting the shear rate during compounding. As seen in Figure 5, the viscosity ratio increases with decreasing PP molecular weight (or increasing PP melt flow). For polypropylene resins ranging from 1.7 to 20 MFR the viscosity ratio was found to be almost independent of shear rate, while for the 35 MFR PP the viscosity ratio decreased continuously with shear rate. This suggests that an intensive shear environment is not required for good plastomer dispersion in low melt flow (up to 20 MFR) polypropylene resins. In a 35 MFR PP matrix, a high intensity mixer (e.g., twin screw extruder) that can reduce the viscosity ratio would be required.

Figure 6 shows the change in particle diameter with polypropylene melt flow rates. The particle diameter is represented using both the number average value, D_n, and the volume average value, D_v. The dispersity index, defined as D_v/D_n, is nearly 2.0 in all the systems. Both D_n and D_v remain constant at 0.5 μm and 1.1 μm, respectively, up to 20 MFR. In the 35 MFR PP, D_n and D_v values increase to 0.8 and 1.6 μm, respectively.

MODEL SELECTION

We have compared our experimental data showing the effect of viscosity ratio on particle breakup with models proposed in the literature. Taylor[6] first studied this phenomenon for a single droplet breaking up under the influence of a simple shear field. By balancing surface tension forces that resist droplet deformation with the matrix shear forces that facilitate this breakup, Taylor was able to determine the maximum droplet size that would be stable. Taylor's equation is expressed in the form:

$$Ca = 8(\eta_\tau + 1) / (19\eta_\tau + 6) \qquad [7]$$

where Ca represents the Capillary number and η_τ the viscosity ratio. The Capillary Number is defined as

$$Ca = \gamma \eta_m D_n / (2\Gamma) \qquad [8]$$

where Γ represents the interfacial tension, η_m the matrix viscosity and γ the shear rate. We can then calculate the droplet size D_n using Eqs 7 and 8. Droplet break up occurs when $Ca > Ca_{crit}$. Taylor's equation is valid only for small deformations in Newtonian fluids. Another correlation that has been proposed and applied in modified nylon systems is by Wu.[7] His model is of the form:

$$D_n = 2Ca(\eta_r) \pm p \qquad p = +0.84 \quad for \quad \eta_r > 1 \qquad [9]$$

$$p = -0.84 \quad for \quad \eta_r < 1$$

SOLUBILITY PARAMETER AND INTERFACIAL TENSION

For applying either Taylor or Wu's model, the interfacial tension, Γ, between EXACT 4033 and isotactic polypropylene should be either experimentally determined or estimated otherwise. Lohse[8] has gathered extensive data on solubility parameters, δ, of ethylene butene at varying butene comonomer content using small angle neutron scattering measurements, SANS. Figure 7 shows the temperature dependence of solubility parameter for several ethylene-butene plastomers. The interfacial tension can be calculated from solubility parameters using the equation

$$\Gamma = [B_{12}\rho RT\{< Rg^2 > / M\}]^{1/2} \qquad [10]$$

where B_{12} is the interaction energy density, ρ the density, R is the gas constant, T the absolute temperature, Rg^2 the mean square radius of gyration and M represents the molecular weight. For a polymer pair B_{12} can be written as

$$B_{12} = (\delta_1 - \delta_2)^2 \qquad [11]$$

Using Eqs 10 and 11, the interfacial tension is calculated. For EXACT 4033, a value of 1.68 mN/m which corresponds to Γ at 225°C is used in the model calculations (see Table 2).

Figure 8 compares the model predictions with the experimental data. Both Taylor and Wu's models are used in the calculation. Taylor's model severely underestimates the particle size. The limiting value identified as Taylor limit in Figure 8 is about 0.05 μm. This is anticipated since the Taylor model

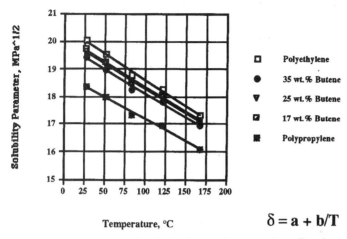

Figure 7. Solubility parameters for ethylene-butene plastomers via small-angle neutron scattering, SANS.

$$\delta = a + b/T$$

is developed for deformation of viscous droplets, while polymers are viscoelastic. Further, the Newtonian fluid criteria specified by Taylor is also not applicable as both EXACT plastomer and PP are non-Newtonian at high shear deformations. Wu's correlation, which is phenomenological, appears to fit the data at low viscosity ratios, but in the higher melt flow polypropylenes this model overestimates the observed particle size. Both models ignore coalescence effects

Table 2. Interfacial characteristics of EXACT 4033/PP blends (Γ of PE/PP @ 225°C is 2.05 mN/m)

Temperature, °C	EXACT 4033 solubility parameter, δ, MPa 0.5	EXACT 4033/PP blend interfacial tension, Γ, mN/m
200	16.84	1.70
225	16.59	1.68
250	16.36	1.77

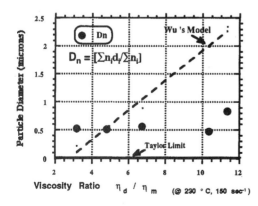

Figure 8. Model correlations.

which occur at high modifier concentrations. These effects would provide further deviations in the comparison between the experimental data and model calculations.

CONCLUSIONS

An ethylene-butene plastomer was melt blended with five low, medium, and high flow isotactic polypropylenes on a twin screw extruder. Low voltage scanning electron micrographs for these blends were obtained from ruthenium tetraoxide stained specimens. Image analysis shows that 0.5 μm dispersion of plastomer particles was in both low and medium flow rate polypropylenes. Only in high flow polypropylene did the plastomer particle increase to 0.8 μm, but still less than the 1 μm threshold necessary for cold temperature impact modification of polypropylene. Log-normal distribution of plastomer particles was determined. Rheological characterization of both plastomer and polypropylene together with estimates of their interfacial tensions allows us to generate model plastomer dispersion at different shear rates during compounding. The interfacial tension between the ethylene-butene type of plastomers and polypropylene was found to be very low so that submicron dispersion of plastomers has been achieved even in high flow polypropylene. The application of Wu's model for predicting plastomer dispersion was not successful.

ACKNOWLEDGMENT

The authors would like to thank Kelli Brightwell, Rose Singleton, Tony Flores and the Polymer Application Laboratory in Baytown for their support of this study. We are indebted to D. J. Lohse, Exxon Research, for providing solubility parameter measurements based on small angle neutron scattering. Finally , D. S. Davis and L. G. Kaufman reviewed the manuscript and made several valuable suggestions

REFERENCES

1.   T. C. Yu, SPE ANTEC Proceedings, 1996
2.   C. E. Scott and C. W. Macosko, *Polymer*, **35**, 422 (1994).
3.   B. D. Favis, *J Appl. Polym. Sci.*, **39**, 285 (1990).
4.   S. Wu, *Polymer*, **26**, 1855 (1985).
5.   M. M. Cross, **Advances In Polymer Science And Technology**, *S. C. I. Monograph*, 22, 381, (1967).
6.   G. I. Taylor, *Proc. R. Soc. Lond.*, **A138**, 41, (1932).
7.   S. Wu, *Polym. Eng. Sci.*, **27**, 335, (1987).
8.   D. J. Lohse, personal communication.

Comparison of the Crystallization Behavior of Ziegler-Natta and Metallocene Catalyzed Isotactic Polypropylene

E. B. Bond and J.E. Spruiell

Center for Materials Processing, University of Tennessee, Knoxville, TN 37996-2200, USA

ABSTRACT

The crystallization behavior of three metallocene catalyzed isotactic polypropylenes were compared to that of a Ziegler-Natta catalyzed resin. The differences in behavior between the two type of resins are suggested to be a result of the differences in molecular microstructures of the resins. Characterization of the molecular microstructures suggest that the miPP resins have a more uniform defect distribution from chain to chain, compared to the zniPP resin.

INTRODUCTION

The study of stereoregularity in polypropylene has long been an area of active scientific research. The research has shown that the isotacticity level is the parameter that most determines the crystallization temperature, T_c, and melting temperature, T_m, as well as the crystallinity.[1-6] As the percent isotacticity decreases, these thermal characteristics decrease due to the rejection of imperfect stereoirregular chains and segments from the crystals.

Polypropylene resins have been produced from Ziegler-Natta catalysts for over 30 years now. These Ziegler-Natta catalysts exist in several variations: supported, unsupported, heterogeneous, etc. Results from research on these resins have shown catalyst type has an influence on the stereostructure.[7-9] Recently, the polymerization process has been studied to determine the types of polymer chains formed and stereoirregular distribution within each type of chain.[10] The results have suggested that the Ziegler-Natta catalysts have more than one type of active site, perhaps as many as three or more.[9] Typically in these mechanisms, one site will produce highly isotactic polypropylene and the others will produce atactic polypropylene (this atactic material can also contain chains and segments that are highly syndiotactic).[10]

There have been several techniques used to characterize the molecular architecture of isotactic polypropylene, iPP. These techniques are infrared analysis, IR, carbon-13 nuclear magnetic resonance, cNMR, and various fractionation methods.[1-10] IR and cNMR can be used to estimate the average stereoregularity of a resin, with cNMR considered the definitive method of tacticity determination.[10] The fractionation methods are used as a technique of analytically separating the various tactic chains by crystallization. The highly isotactic chains will crystallize as the solvent is cooled, allowing for a separation based on the isotactic content, with the atactic and low molecular weight isotactic material remaining in solution.[1-3] In the present paper, a study of the crystallization behavior of three metallocene catalyzed resins are compared to that of a comparable Ziegler-Natta catalyzed resin. The observed differences are discussed in terms of the molecular microstructures resulting from the two types of catalysts.

EXPERIMENTAL

MATERIALS

The basic characteristics of the four resins studied in this paper are listed in Table 1. The large, M, stands for metallocene catalyzed iPP and large, ZN, for Ziegler-Natta catalyzed iPP. The large, N, indicates that 1 wt% nucleating agent was added to the M35 resin. All the resins were unfractionated homopolymers that had a nominal 35 melt flow rate, mfr, except for one metallocene resin that has an mfr of 22. The miPP resins have very narrow molecular weight distributions, nMWD, a result of the single site metallocene catalyst system that produces these distributions, without viscosity breaking from the addition of peroxides. This fact alone, the generation of nMWD resins, is a major breakthrough in catalyst technology that allows custom generation of molecular chains that are similar in length, and as will be shown, tacticity. The zniPP resin was prepared with the desired narrow MWD and 35 mfr by visbreaking a reactor resin with higher molecular weight and broader MWD.

Table 1. Material characteristics

Sample Code	Catalyst	MFR	Polidispersity	M_n	M_w
M35	Metallocene	35.0	1.93	71,146	137,277
ZN35	Ziegler-Natta	35.0	2.56	59,577	152,332
M22	Metallocene	22.0	2.01	76,687	154,210

CRYSTALLIZATION RATES

The crystallization rates were investigated using differential scanning calorimetry, DSC. The isothermal crystallization studies were carried out using a Perkin Elmer DSC-7.

All samples tested were first prepared by melt pressing resin pellets into films at 210°C to minimize any sample to sample variations.

ISOTACTICITY DETERMINATIONS

FTIR was used to determine the percent isotactic content. Sample preparation is very important for tacticity determinations when using FTIR. Reference 12 was followed for sample preparation and evaluation. The cNMR tacticities were estimated from the FTIR results by a calibration curve based on previous measurements where both cNMR and FTIR measurements were carried out on the same zniPP resins.[13] The two measurements obviously do not measure precisely the same quantity, though there seems to be a reasonable proportionality between them.

A crystallization fractionation was also performed on these resins to determine the percent soluables. Xylene was the solvent, and ASTM standard D5492-94 was followed for the experimental procedure and results calculations. The percent soluables in cold xylene is generally accepted to be a measure of the atactic and low molecular weight fraction.

RESULTS AND DISCUSSION

The melting and crystallization temperatures determined by heating and cooling in the DSC are reported in Table 2. The results show that the ZN35 resin crystallizes and melts at significantly higher temperatures than the nonnucleated metallocene isotactic polypropylene, miPP, resins. Table 3 shows the enthalpy of melting and crystallization and the estimated crystallinities measured on heating. As the data from Table 2 and Table 3 show, there is very little difference in the thermal behavior of M22 and M35. Since the zniPP resin had the highest melting temperature and crystallinity of all the resins studied, this would suggest that it has a higher isotactic content, based on previous studies of the influence of isotacticity on crystallization and melting behavior.[1-5,12,13]

The equilibrium melting temperatures, T_m^o, were then determined using the Hoffman-Weeks method,[14] as shown in Figure 1. The ZN35 resin had the highest T_m^o value, followed by M22 and M35. These results confirmed what was found in the DSC cooling and heating curves. The results in Figure 2, the isothermal crystallization half-time plot, show that the metallocene catalyzed resins crystallize more slowly than the ZN35 resin at a given temperature. The T_m^o data indicates that the crystalline lamellae produced by the metallocene resin are not as stable as the lamellae produced by the Ziegler-Natta resin, thus, they melt at lower temperatures. It is known that the T_m^o value of polypropylene will decrease as the isotactic content decreases, due to the increasing amount of defects in the developing

Table 2. Thermal Characteristics of Resins*

Sample Code	T_m (Peak) °C	T_m(Onset) °C	T_c(Peak) °C	T_c(Onset) °C
M35	151.1	144.6	100.7	106.8
ZN35	159.7	152.8	104.8	110.6
M22	149.6	143.9	100.7	106.7

*heating and cooling 20°C/min using Perkin Elmer DSC7
T_m-melting temperature; T_c-crystalization temperature

Table 3. Crystallinity and Enthalpy Measurements*

Sample Code	Melting Enthalpy J/g	Crystallization Enthalpy J/g	Crystallinity %**
M35	84.787	-86.246	51.4
ZN35	93.574	-95.328	56.7
M22	84.759	-82.979	51.4

*heating and cooling rates 20°C/min; **determinated upon the melting endotherm at 20°C

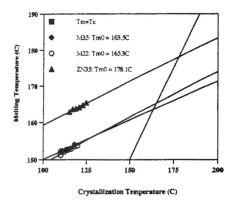

Figure 1. Hoffman-Weeks plot.

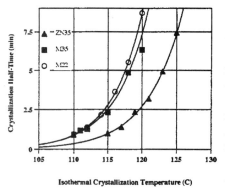

Figure 2. Isothermal half-times.

crystal. This may explain why the M22 resin has a slightly higher T_m^o than does the M35 resin, since its isotactic content is slightly higher (see below). It also suggests that the isotacticity of the crystalline fraction is higher in the zniPP resin than in the miPP resins.

In the introduction, a description of Ziegler-Natta catalyst polymerization was briefly covered. The results from investigations on these catalysts showed that several active sites may be possible. In a typical two site type catalyst, one site will produce highly isotactic chains while the other produces atactic chains.[1,15] However, from previous cNMR work, it is also known that within the highly isotactic chains, there can be stereo inversions that lower the isotactic content within the highly isotactic chains. These inversions are typically syndiotactic[16] sequences. Knowing these facts, when a fractionation study is performed on Ziegler-Natta catalyzed iPP resin, there is a clear separation of highly isotactic chains crystallizing and precipitating, while the atactic chains remain in solution until a further precipitation step is carried out.[17] The results of measurements of the xylene soluables on these resins is shown in Table 4. These results indicate that the ZN35 resin contains about 5.7% atactic and low molecular weight fraction, while the metallocene resins have very little, if any.

The FTIR results given in Table 4 suggest that the average isotactic contents of the resins differ only slightly, with the tacticity of M35 being slightly lower than that of either M22 or ZN35. The FTIR technique is very sensitive to crystallinity; this apparently arises from the fact that the vibrations used in the measurements arise from the close packing found in polymer crystals. A recent paper[12] suggests that the observed A998/A973 ratio can vary by as much as 37%, depending on the crystallization and annealing conditions of the sample films. The samples used in this study were prepared by slowly crystallizing films in a hot stage at a cooling rate of 0.5°C/min. A nitrogen gas purge was used to minimize thermal degradation. The crystallinity values for these films are given in the second column of Table 4, as measured in a density gradient column. Notice that all the resins increased in crystallinity when compared to the crystallinity values shown in Table 3. This is due to the slower crystallization rate, which would allow the crystal defects to be removed and minimized during crystallization. The cNMR isotactic contents were then calculated from previous work performed on similar polymers, where the isotactic content had been measured by cNMR. Since the isotacticity values measured by FTIR are relatively close to one another, these results alone do not fully explain the differences found in the

Table 4. Stereoregularity Determination

Sample Code	FTIR % Tacticity[a]	% Crystalinity[b]	% Xylene Solubles[c]	cNMR % Tacticity[d]
M35	87.3	51.9	0.19	91.24
ZN35	89.5	57.5	5.66	93.47
M22	90.0	57.1	0.21	93.98

[a]IR method by Luongo and Sundell; [b]slow crystallization in hot stage (0.5°C/min); [c]determination from ASTM standard D5492-94; [d]calculated from FTIR and previous cNMR work

thermal behavior of these resins, especially the differences between the zniPP resin and the miPP resins. We note however, that the results can be explained by assuming that the distribution of chain defects is different between the two types of resins. For the ZN35 resin, the solubility data show that 94.3% of the resin crystallized and precipitated during the fractionation. This value is roughly equivalent to the isotactic content of the resin as determined by FTIR and extrapolated to its cNMR value, a value of 93.5%. These values are in very good agreement with each other, considering the possible sources of error. The results from the fractionation study, along with the isotactic content by FTIR, suggest that the precipitated chains in the zniPP sample have a relatively high degree of isotacticity. These observations can explain the differences observed in the crystallization studies if we account for the difference in the distribution of the defects.

The suggested explanation for the crystallization behavior of the resins studied in this paper goes as follows: During isothermal and slow cooling rate crystallization, the zniPP resin can reject the atactic chains from the growth front, creating lamellae that are relatively free of defects due to the high isotactic content of chains that are crystallizing at the interface. The miPP resins have the stereoirregular defects in each chain. Since the defects are in each chain, they are likely incorporated into the growing lamellae in whole, or in part, depending on whether the inversions are grouped together or isolated. This will lead either to thinner lamellae or to lamellae that have defects contained within them. Such lamellae would be less stable than lamellae formed from highly isotactic chains, and they would require less thermal energy to melt.[18]

CONCLUSIONS

The miPP resins have a different molecular architecture than the zniPP resin, due to the defects being more uniformly distributed from chain to chain, whereas the zniPP has chains that are highly isotactic in nature with other chains highly atactic. This results in less perfect lamellae formation for the miPP resins, that melt at lower temperatures and consequently crystallize at lower temperatures.

The metallocene catalyst technology has produced molecules that are more uniform, in molecular weight and molecular architecture, that have not been encountered before. The differences between the zniPP and miPP resins are only now beginning to be understood.

ACKNOLEDGMENTS

The authors wish to thank Dr. Khaled Mezghani and Dr. Bernhard Wunderlich for their many fruitful discussions concerning the crystallization behavior of these miPP resins. They also thank Exxon Chemical Company for supplying the resins studied and for support of the research.

REFERENCES

1.   R. Paukkeri and A, Lehtinen, *Polymer*, **34(16)** 4083 (1993).
2.   E. Martuscelli, M. Pracella, and L. Crispino, *Polymer*, **24**, 693 (1983).
3.   E. Martuscelli, M. Avella, A. Segre, E. Rossi, G. Drusco, P. Galli, and T. Simonazzi, *Polymer*, **26**, 259 (1985).
4.   D. Burfield and P. Loi, *J. Appl. Polym. Sci.*, **41**, 1095 (1990).
5.   J. Janimak, S. Cheng, A. Zhang, and E. Hsieh, *Polymer,* **33(4)** 729 (1992).
6.   J. Luongo, *J. Appl. Polym. Sci.*, **3**, 302 (1960).
7.   Y. Inoue, Y. Itabashi, R. Chujo, and Y. Doi, *Polymer*, **25**, 1640 (1984).
8.   V. Busico, P. Corradini, L. De Martino, F. Graziano, and A. Iadicicco, *Macromol. Chem.*, **192**, 49 (1991).
9.   M. Kakugo, T. Miyatake, Y. Naito, and K. Mizunuma, *Macromolecules*, **21**, 314 (1988).
10.   Y. Hayashi, Y. Inoue, R. Chujo, and Y. Doi, *Polymer*, **30**, 1714 (1989).
11.   T. Sundell, H. Fagerholm, and H. Crozier, *Polymer*, **37(15)** 3227 (1996).
12.   R. Ramamurthy, **Masterís Thesis**, *University of Tennessee--Knoxville*, 1996.
13.   S. Williams, Z. Ding, E. Bond, and J. Spruiell, *Proc. SPE ANTEC,* 1996.
14.   J. Hoffman and J. Weeks, *J. Res. Natl. Bur. Stand.*, **A66**, 13 (1962).
15.   C. Wolfsgruber, G. Zannoni, E. Rigamonti, and A. Zambelli, *Makromol. Chem.*, **176**, 2765 (1975).
16.   A. Zambelli, D. Dorman, R. Brewster, and F. Bovey, *Macromolecules*, **6(6)** 925 (1973).
17.   A. Lehtinen and R. Paukkeri, *Macromol. Chem. Phys.*, **195**, 1539 (1994).
18.   B. Wunderlich, *Personal Communication*, October, 1996.

Explaining the Transient Behaviors of Syndiotactic Polypropylene

W. R. Wheat

Fina Oil and Chemical Company, P.O. Box 1200, Deer Park , TX 77536, USA

ABSTRACT

In a previous paper, some of the more unusual behaviors of metallocene-based syndiotactic polypropylene were explained, relating its inherently higher entanglement density to such things as crystallization rates and rheological responses. In this work, this basic understanding will be extended to describe time-dependent characteristics. The room temperature aging of this new polymer is compared to that of standard isotactic polypropylenes.

INTRODUCTION

Highly syndiotactic polypropylene, sPP, has recently been made possible by metallocene catalyst advances and is being made on commercial-scale equipment by Fina Oil and Chemical. Merely by alternating the steric addition of propylene monomers in their production, these polymers gain some very unusual behaviors. These behaviors, such as slow crystallization rates, different molecular weight-viscosity relationships and rheological responses, are in some cases significantly removed from those of sPP's chemical cousin, standard isotactic PP.

An explanation for many of these differences was described previously.[1] The more flexible backbone chain and higher molecular entanglement density of sPP in the melt (by a factor of 2 or more) begins to give the understanding needed to improve the prediction of properties and the effects of processing parameters. With more chain segments anchored by physical "crosslinks" throughout the material, there is significantly more resistance to large-scale molecular movements, such as found in crystallization processes and material flow, in general.

Solid state behaviors reflect this increased entanglement density as well, with crystallization expected to trap more tie molecules and entanglements in the amorphous regions. For instance, comparing the two polypropylene types, sPP can be a much more elastic solid and retain a much higher shrinkage in an oriented part. Extending this idea, the amorphous regions would be expected to have less order on the molecular scale. That fact, together with the lower molecular weight of sPP, makes it a polymer with a different free volume character where permeation can occur to a greater extent and entanglement across an interface in heat sealing is easier to effect.

Still, it has been difficult to predict some behaviors. With processing differences and testing differences to contend with, getting a complete and accurate picture is sometimes harder than it sounds. One of the more significant variables to control has been the age of the article to be tested, as sPP has appeared to change quite significantly with time in some tests. The goal of this investigation was to study changes with time, and describe them in comparison to standard polypropylenes for an improved general understanding.

AGING THEORY

Being a standard semicrystalline polymer, isotactic polypropylene, iPP, has long been known to change with time at room temperature, although not to the extent more recently observed with the syndiotactic form. With iPP, aging has been thought of in the industry to be mainly a result of secondary crystallization—to include further crystallization after the quench and crystal perfection mechanisms.[2]

However, certain authors[3,4,5] have also described the aging of iPP at room temperature to be more of a "physical" aging—ultimately, a result of the slow collapse of free volume with time, from a non-equilibrated quenched state toward a more structurally-relaxed one. The free volume collapse in the amorphous region comes from a long relaxation time process based on segmental motions in the solid polymer. Usually, this would not happen above the glass transition temperature of an amorphous polymer. But in this case, the constraints on the amorphous phase caused by the crystalline portions of the polymer act to hold the amorphous region in this non-equilibrium state. This pseudo-higher glass transition has been measured in iPP fibers using a thermally stimulated current technique.[6] Collapse of free volume itself during aging has been measured using positron annihilation lifetime spectroscopy on a variety of amorphous and semicrystalline polymers.[7]

It is not obvious which process(es) are at work controlling the aging of sPP.

EXPERIMENTAL

Samples of commercially-produced Fina polypropylenes listed in Table 1 were collected for study. Included were standard Ziegler-Natta catalyzed isotactic PP's, both homopolymers and random copolymers with ethylene, as well as metallocene-based iPP and sPP. All tests were done on compression-molded samples, prepared by quenching quickly from above 200°C to room temperature in an ice water bath.

Table 1. Range of Polypropylenes Used in this Study

PP Type	MFR g/10 min 230°C	Xylene solubles wt %	MWD	Ethylene wt %
Normal iPP	2	4	-	-
BMWD iPP	3	4	Broad	-
High cryst. iPP	12	<1	-	-
RDP iPP	2	4	-	3
met-iPP	30	2	Narrow	-
sPP	4	2	Narrow	-

Densities were measured using one end of a standard Izod bar shape (1/8" x 1/2") by weighing, first in air and then in methanol. Methanol density was monitored closely as required for accuracy. A 30 mm length from the other end of the same bar was used to test the torsional modulus (G*) using a Rheometrics Dynamic Mechanical Analyzer, so there would not be any difference in quench between the samples in these two tests. The DMA tests were performed at room temperature, using a nondestructive 0.1 % strain with a bar-clamping fixture. Data were taken over the frequency range of 50 to 500 radians per second, so the testing at each time only took a matter of seconds rather than minutes.

Separate samples were prepared for X-ray diffraction, although an effort was made to duplicate the above procedure.

With all these tests, measurements with aging time were repeatedly made on the same, exact sample. Care was taken not to damage the sample in any way. Samples were always re-loaded with exactly the same orientation into the DMA or XRD instruments. After testing, the samples were returned to a drawer in the lab, which was controlled at all times between 20 and 24°C.

RESULTS

The change of densities with time for these materials is compared in Figure 1. Clearly, the lowest density materials are the ones expected to have the lowest crystallinity. For all polypropylene types, densities rose consistently throughout the 10 week (100,000 minute) measurement period. However, the density rise is fairly small—between 0.003 and 0.006 units. The largest rise in density came from the lowest crystallinity sample, sPP.

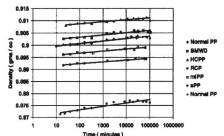

Figure 1. Changes in density with time.

A typical family of plots from the DMA is included as Figure 2 for the high-crystallinity Ziegler-Natta iPP, HCPP. This information was consolidated by taking the average modulus within the tested range as a single modulus and re-plotting this value versus the logarithm of time. All materials are shown compared in Figure 3. As would be expected, all samples get stiffer during room temperature aging. Perhaps unexpectedly, all materials appear to stiffen with approximately the same slope against log (time).

These are interesting results, as it would appear iPP and sPP age in exactly the same way! However, as you will note, the effect is much stronger on the initial properties of sPP—it is 2½ times its original stiffness in 10 weeks, whereas the HCPP only becomes 25 % stiffer in the same amount of time. Figure 4 shows this more clearly.

Percent crystallinity from the X-ray data is shown for all samples in Figure 5 versus log (time). There was some scatter of measurements here, but the general trend was to just slightly higher crystallinity with time—on the

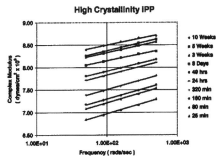

Figure 2. Raw DMA torsional modulus data on one bar aging with time.

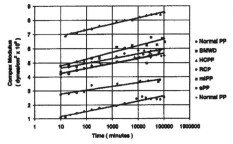

Figure 3. Torsional modulus of all the polypropylenes changing with time.

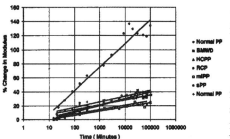

Figure 4. Percent change in modulus with age.

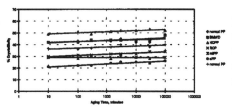

Figure 5. Crystallinity by X-ray diffraction with time.

order of 3 to 6 %. Although this accounts for a good portion of the density rises seen, it is believed the significant rise in mechanical properties must be coming from a more significant mechanism—densification in the amorphous regions, fitting the description of free volume collapse reported for iPP. (In fact, some of the references cited do not show any crystallinity rise at all for the iPP's they tested.)

None of the polypropylenes showed any real changes in their X-ray patterns, indicating any significant changes in crystal form. Figure 6 shows the before and after scans for sPP. Peak widths and placement were essentially unaffected with room temperature aging.

FURTHER DISCUSSION

It makes sense that the same mechanisms reported in the literature for the aging of iPP should be valid for sPP. Segmental motions are only slightly more difficult in sPP due to entanglements, as seen by the slightly higher flow activation energy at zero-shear viscosity.[1] The molecular architecture is really very much the same.

With the significantly larger amount of amorphous phase present in sPP, there should be a larger effect of any change there. Mechanical properties are noticeably changed to the hand within the first few hours. Other properties that rely on the amorphous phase should also be affected—like barrier properties and heat seal temperatures, as well as slip bloom in films. In fact, they are. What might have started out as a high permeability film made from sPP ages to the point that, after one year, has the barrier of a 2 % ethylene random copolymer of iPP. Likewise, fresh sPP-based film is easy to seal, giving low seal temperatures relative to other PP's. But after aging, this same film tightens up in the amorphous regions, requiring more temperature to provide mobility for entanglement across the interface. Particularly low coefficients of friction have been seen in sPP-based films.[8] The mo-

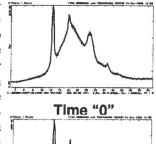

Time "0"

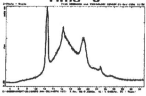

10 Weeks

Figure 6. XRD scans of sPP at 10 minutes and 100,000 minutes.

lecular motions giving free volume collapse are, at the same time, squeezing out the slip additives and making this process very efficient.[9]

CONCLUSIONS

Syndiotactic polypropylene, although a very different polymer in its properties and processing characteristics, behaves very much like isotactic polypropylene with respect to the physical aging processes at room temperature. It is expected that both secondary crystallization and crystal perfection processes are at work. But the larger effect on properties is expected to come from the slow collapse of free volume in the amorphous regions. Although above the material's glass transition temperature, crystallinity restricts molecules from achieving an equilibrium conformation during a real quenching process.

It has sometimes appeared that sPP is metamorphic—going through large, previously unexplained changes with time. However, the physical aging explanation provides a new understanding, offering potential production and application solutions.

ACKNOWLEDGMENTS

The author would like to thank Fina Oil and Chemical and Mitsui Toatsu Chemicals, Inc. for allowing this paper to be published and the supporting data included.

REFERENCES

1.   W. R. Wheat, *Proc. SPE ANTEC'95*, 2275 (1995).
2.   B. Wunderlich, *Macromol. Phys.*, **2**, 169 (1976).
3.   C. P. Buckley and M. Habibullah, *J. App. Polym. Sci.*, **26**, 2613 (1981).
4.   L. C. E. Struik, *Polymer*, **28**, 1521 (1987); *ibid*, **30**, 799 (1989).
5.   Y. C. Bhuvanesh. and V.B. Gupta, *Polymer*, **36(19)** 3669 (1995).
6.   M. Galop and R. Callahan, *Thermal Trends*, **3(2)** 18 (1996).
7.   D. M. Bigg, *Polym. Eng. Sci.,* **36(6)** 737 (1996).
8.   L. Sun, J. Schardl, S. Kimura, and R. Sugimoto, *Proc. Polyolefins IX* , p. 563, February, 1995.
9.   G. W. Schael, *J. App. Polym. Sci.*, **10**, 653 (1966).

Equilibrium Melting Temperatures of Ethylene Copolymers: An Appraisal of the Applicability of the Flory and Sanchez-Eby Approaches

Man-Ho Kim and P. J. Phillips

Department of Material Science & Engineering
University of Tennessee, Knoxville, TN, 37996-2200, USA

ABSTRACT

The equilibrium melting temperatures of linear polyethylenes and ethylene/1-octene random copolymers were determined as a function of molecular weight and branch content. This systematic study makes it possible to evaluate two equilibrium melting temperature depression equations for the olefin type random copolymers, the Flory equation and the Sanchez-Eby equation, as a function of defect content and molecular weight. An empirical equation suggested in this study for predicting the equilibrium melting point of linear polyethylene as a function of molecular weight is more sensitive to molecular weight than the Flory-Vrij equation. The range over which the two equations can be applied depends on the defect content, after molecular weight correction.

INTRODUCTION

The equilibrium melting temperature, T_m^o is an important physical parameter in polymers and may be determined from extrapolation techniques such as the Hoffman-Weeks plot and the Thompson-Gibbs plot. The more reliable method is to use the Thompson-Gibbs relationship of melting temperature, T_m, as a function of lamellar thickness, l.

The T_m^o of linear polyethylene is one of the controversies in polymer physics, many studies having been reported. However, there is inconsistency among literature values (Table 1), although only the values obtained from the plots of T_m^o versus $1/l$ are listed. In the case of random copolymers the equilibrium melting point is

Table 1. Equilibrium melting temperatures determined from the plot of T_m versus $1/l$ for linear polyethylenes.[12]

T_m^o °C	Samples [b]	Molecular weight	References
145.8	-	-	13,14
143.5±0.5	-	-	15
142.0±0.3[c]	various	various	16
141±0.5	various	various	17
139.0[d]	Marlex 6009	-	18
138.15	Marlex 6001[e]	-	19

[b]bulk crystallized; [c]our analysis showed 140.84°C instead of 142.0°C when the data in reference 16 was reanalyzed; [d]obtained from the plot of T_m versus reciprocal long period (not lamellar thickness). Thus, this value is not recommended for comparison, since the analysis is not correct based on Thompson equation; [e]suspension crystallization

Table 2. Characteristics of the as-received linear polyethylenes and ethylene/1-octene random copolymers and the equilibrium melting temperatures, $(T_m^o n,p)$

Code[a]	$\overline{M}_w$ g/mol	$\overline{M}_n$ g/mol	$\overline{M}_w / \overline{M}_n$	Density g/cm³	CH₃/ 1000C	T_m °C	$T_m^o(n,P)$ °C[f]
H-LPE	101,300	53,900	1.879	0.9540	0.00[b]	132.25	142.65
M-LPE	77,600	38,200	2.031	0.9510	0.00[c]	131.43	141.33
H7-O	-	43,600	-	0.9180	6.84	113.19[e]	140.44
H7'-O	98,400	44,800	2.196	0.9180	7.32	113.25	139.30
H10-O	-	48,800	-	0.9126	9.50	107.54[e]	138.64
H16-O	-	45,200	-	0.9003	16.34	94.54[e]	133.86
H17-O	102,700	48,700	2.108	0.9003	16.92	96.19	134.10
L4-O	59,900	27,300	2.194	0.9365	3.98	118.81	139.33
L10-O	-	23,700	-	0.9195	9.83	109.21[e]	134.55
L13-O	51,800	25,000	2.070	0.9110	12.93[d]	105.00	-
L18-O	-	24,900	-	0.9027	18.17	94.46[e]	131.92
L24-O	46,700	21,800	2.151	0.8975	24.04	90.24	127.51
L39-O	48,000	22,800	2.105	0.8700	38.62	48.23[e]	-

[a]H. M. L.- high, medium, and low molecular weight; LPE - linear PE; numbers - branch content; O - octene comonomer
[b]trans/1000C=0.003; vinyls/1000C=0.135; vinylidenes/1000C=0.013
[c]trans/1000C=0.027; vinyls/1000C=0.128; vinylidenes/1000C=0.023
[d]trans/1000C=0.298; vinyls/1000C=0.045
[e]measured at a DSC heating rate 10°C/min
[f]determined from the plot melting temperature versus lamellar thickness

depressed. This may be due to the effect of the chemical structure on the chemical potential of the amorphous phase in the case of complete exclusion of defects (Flory Theory).[1,2] It can also be depressed in the case of partial or complete inclusion of comonomers in the crystal due to the effect of defects on the crystal (Sanchez-Eby Theory).[3-5] Neither of these theories has been tested experimentally for olefin type random copolymer due to the difficulty of obtaining T_m^o.

Our purpose of this study is to report a systematic study of the T_m^o of linear polyethylenes and the ethylene/1-octene random copolymers as a function of molecular weight, MW, and branch content, and to evaluate the effectiveness of these two theories in predicting the experimental behavior.

EXPERIMENTAL

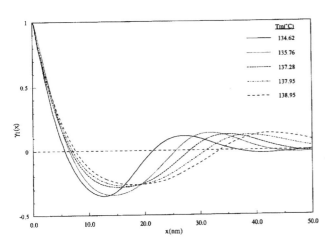

Figure 1. Normalized one-dimensional correlation function behavior of the H-LPE with different melting temperatures.

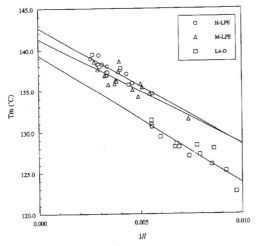

Figure 2. Plot of melting temperature against reciprocal lamellar thickness.

The polymers prepared using metallo-cene catalysts were supplied by the Dow Chemical Co. The details of as-received samples are listed in Table 2. Here, the code H, L, number and O stand for high molecular weight, low molecular weight, methyl groups per 1000 carbon (i.e., SCB or branch content) and octene comonomer, respectively.

The isothermal crystallizations were carried out in a silicone oil bath for at least 24 h. Sealed samples were melted at 150°C for 10 minutes to remove the thermal history, and then transferred to the oil bath, preset at the crystallization temperature. The isothermal crystallization studies were carried out as a function of crystallization temperature and time.

The 10 m small angle X-ray scattering, SAXS, facility of Oak Ridge National Laboratory was used (sample thickness: 1.0 mm). Two geometries, 5 m for lamellar thickness and 1 m for background, were used. Intensity was measured at room temperature for linear polyethylenes and at crystallization temperatures for EO copolymers. Lamellar thickness was estimated from the analysis of the one-dimensional correlation function, $\gamma_i(x)$, of the absolute intensity.

Melting temperatures were determined at a heating rate of 10°C/min under a nitrogen atmosphere using a Perkin Elmer DSC-7 (sample thickness: 0.1 mm). The deviation from the melting of indium before and after experiment was within ±0.01°C and ±0.15°C, respectively.

RESULTS

An example of the $\gamma_i(x)$ function of linear polyethylene is shown in Figure 1. As long period, L, increases, the T_m increases. Lamellar thickness, l, was obtained from the $l = Lv_c$ for linear polyethylene,

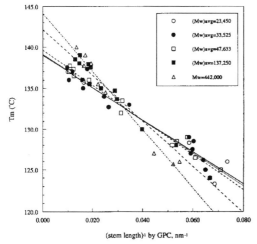

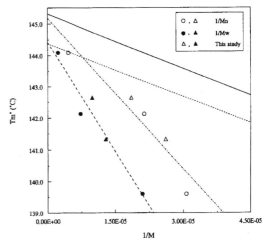

Figure 3. Plot of melting temperature against reciprocal stem length as indicated molecular weights, reanalyzing the data by Basset *et al.*[7]

Figure 4. Equilibrium melting temperatures against reciprocal molecular weights. Solid and dotted lines are the values calculated from Flory-Vrij equation and Eq 2, respectively.

where v_c is the volume crystallinity and from the analysis of the self-triangle zone for copolymers. The Thompson-Gibbs plot of T_m and lamellar thickness for linear polyethylene are shown in Figure 2. The T_m^o of linear polyethylene with $M_w = 77,600$ is 141.3°C and that with $M_w = 101,300$ is 142.7°C, demonstrating the known MW dependence. The T_m^o's of the linear polyethylenes and EO copolymers determined from the intercepts at infinite lamellar thickness are shown in Table 2.

DISCUSSION

An extensive study of the melting temperatures and average stem lengths of linear polyethylene fractions was reported by Bassett *et al.*[6,7] They obtained $T_m^o = 142\pm0.3$°C from the plot of T_m versus reciprocal stem length using all melting temperatures, disregarding the effect of MW on T_m^o. Their data were reanalyzed to obtain T_m^o as a function of MW. To obtain enough data points, similar molecular weights were grouped and the average MW of each group was taken. The corresponding plots of T_m against reciprocal stem length are shown in Figure 3. The intercept reflects T_m^o. T_m^o as a function of MW is shown in Figure 4, including the values of this work. The empirical equation to estimate T_m^o as a function of $\overline{M}_w$ is as follows:

$$T_m^o(^oC) = 144.416 - 2.302 \times 10^5 \frac{1}{M_w} \qquad [1]$$

The T_m^o obtained from the intercept is 144.42°C within the error range of Flory & Vrij's value,[8] obtained from the analysis of n-alkanes. Despite the similarities of the T_m^o values, the MW dependence of T_m^o is quite different for the Flory-Vrij equation and Eq 1. The values calculated from both the Flory-

Vrij equation (with 145.35±1°C) and the modified Flory-Vrij Eq 2 (with 144.42°C) to give better fitting at high molecular weights

$$T_m^o(n) = 417.57 \frac{n-3.0352}{n+0.6064\{1.987\ln(n)+2.45-0.65(1-T_m^o(n)/417.57)\}}$$ [2]

are about 1~5°C higher than those of this study dependent on MW. In the MW range used in this study, the T_m^o estimated from Flory & Vrij's semi-empirical equation levels off, while Eq 1 is more sensitive to molecular weight.

Shown in Figure 5 are the T_m^os of the EO copolymers as a function of the MW and branch content, together with those of the linear polyethylenes. The T_m^o of EO copolymers decreases slowly with branch content up to 1% and sharply beyond 1%. It also shows a strong MW dependence as well as a branch content dependence at low branch content. As branch content increases further, the T_m^o for high and low MW falls on the same extrapolated fitting line (broken line). This suggests that the MW effect is weak and the defect effect dominates the T_m^o of the copolymers.

Figure 5 makes it possible to evaluate Flory's melting point depression equation. Flory's equation can be expressed to correct the MW effect as follows:

$$\frac{1}{T_m^o(n,p)} - \frac{1}{T_m^o(n)} = -\frac{R}{\Delta H_u}\ln p$$ [3]

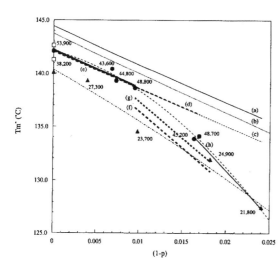

Figure 5. Equilibrium melting temperatures of linear polyethylenes and ethylene/1-octene random copolymers as a function of branch point mole fraction at the indicated number average molecular weights.

where p is a sequence probability of crystallizable structure units A in the copolymer system. When the equilibrium melting temperature $T_m^o(n\rightarrow\infty) = 144.42°C$ of infinite MW was used, the theoretical values give the highest deviation, a, from the experimental values, e. When the Flory & Vrij equation to correct MW was used, the calculated values b are still higher than the experimental results, e. However, when the values from Eq 1 are used instead of those from the Flory-Vrij equation, with $T_m^o(n)=142.12°C$ for the averaged $(\overline{M}_w)_{avg} = 100,550$ of a series of high molecular EO copolymers, or the $T_m^o(n)$'s obtained as a function of each MW, the calculated $T_m^o(n,p)$'s, d and c, coincide with the experimental values, e, up to 0.01 branch sequence probability $(1 - p)$ (i.e., less than 1.0% branch point mole%). As branch content increases further, the values calculated from Eq 3 deviate from the experimental values. Flory's equation was derived under the assumption of the com-

was derived under the assumption of the complete exclusion of defects from the crystalline phase. Since the hexyl branch is flexible and it is possible to rotate around single bonds, branch chains may obtain crystallographic register.[9,10] Unpublished results show that a small number of defects can be dissolved in the crystalline phase and the amount of inclusion increases as the branch content increases, similar to results of Hosoda et al.[10] On the basis of the assumption that the extrapolation of the high supercooling region can predict the T_m^o for complete inclusion at the intersection of the T_m versus T_c plot,[4,11] the possible maximum numbers of defect with Eq 4 were estimated for $\beta \approx 1$ as follows: H10-O has 1.8 defects per 1000 carbons and H16-O has as much as 4.1 defects per 1000 carbons in the crystalline phase. Inclusion lower than that of H10-O is small enough to maintain a well-ordered crystalline phase and to satisfy the exclusion model. Therefore, for less than 1.0% branch content, the Flory equation can be used. As the branch content of the copolymer increases, more defects enter the crystalline phase violating the approximate exclusion condition. The T_m^o behavior at the higher branch content permits the Sanchez-Eby Eqs 3 and 4 to be tested. The reexpressed Sanchez-Eby equation incorporating molecular weight is as follows:

$$\frac{1}{T_m^o(n)} - \frac{1}{T_m^o(n,p)} = -\frac{R(1-p)}{\Delta H_u}\left\{\frac{\varepsilon\beta}{RT_m^o(n,p)} + (1-\beta) + \beta\ln\beta\right\} \qquad [4]$$

where ε and β stand for the excess free energy due to the defect inclusion and the degree of the defect incorporation. The values of ε were calculated from the Hoffman-Week's plot and the method used by Morra et al.[11] Here $(1-p)$ is the B-sequence probability that represents the branch point mole fraction. The calculated values using the $T_m^o(n\rightarrow\infty) = 144.2°C$ from Eq 1 and the $T_m^o(n\rightarrow\infty) = 145.35°C$ of Flory & Vrij are shown as f and g respectively. If we use the equilibrium melting temperature of a finite chain length n, the calculated values will be lowered from f or g by as much as the difference between $T_m^o(n)$ and $T_m^o(n\rightarrow\infty)$, causing deviation from experimental values. Here 10-O was calculated for both models. The slope has roughly a similar tendency to that of the experimental values. However, there is still a difference between experimental and calculated values.

CONCLUSIONS

The MW dependence on T_m^o of linear polyethylene has been determined and it does not correspond to the Flory-Vrij equation. An empirical Eq 1 is more sensitive to MW than the Flory-Vrij equation. The T_m^o for linear polyethylene at infinite molecular weight, $T_m^o(n\rightarrow\infty)$, was found to be 144.4°C. The $T_m^o(n,p)$, of the ethylene/1-octene random copolymers are a function of the MW and the defect at low defect content $((p-1)\% \leq 1.0\%)$. $T_m^o(n,p)$ is a weak function of MW and a strong function of the defect content at high defect content $((1-p)\% \geq 1.0\%)$.

The Flory equation can predict the $T_m^o(n,p)$ for the ethylene/1-octene random copolymer at $(1-p)\% \leq 1.0\%$ when the MW effect is corrected using Eq 1. In the range of high defect content, the experimental values are always higher than the calculated values obtained using the uniform inclusion model, which does give a prediction that is closer than that of the Flory equation.

ACKNOWLEDGMENT

This research has been supported by the National Science Foundation grants DMR 91-07675 and 94-08678.

REFERENCES

1. P. J. Flory, *Trans. Faraday Soc.*, **51**, 848 (1955).
2. P. J. Flory, *J. Chem. Phys.*, **17(3)** 223 (1949).
3. I. C. Sanchez and R. K. Eby, *Macromolecules*, **8**, 639 (1973).
4. E. Helfand and J. I. Lauritzen, Jr., *Macromolecules*, **6(4)** 631 (1973).
5. J. Martinez-Salazar, M. S. Cuesta, and F. J. Balta' Calleja, *Colloid. Polym. Sci.,* **265(3)** 239 (1987).
6. D. C. Bassett, A. M. Hodge, and R. H. Olley, *Proc. Roy. Soc. London*, **A377**, 25 (1981).
7. D. C. Bassett, A. M. Hodge, and R. H. Olley, *Proc. Roy. Soc. London*, **A377**, 39 (1981).
8. P. J. Flory and A. Vrij, *J. Am. Chem. Soc.*, **85**, 3548 (1963).
9. J. N. Hay and Xiao-Qi Zhou, *Polymer*, **34(5)** 1002 (1993).
10. S. Hosoda, H. Nomura, Y. Gotoh, and H. Kihara, *Polymer*, **31**, 1999 (1990).
11. B. S. Morra and R. S. Stein, *J. Polym. Sci. Polym. Phys. Ed.*, **B20**, 2243 (1982).
12. L. Mandelkern, G.M. Stack, and P. J. M. Mathieu, *Anal. Cal.*, **5**, 223.
13. H. E. Bair, T. W. Huseby, and R. Salovey, *Amer. Chem. Soc. Polym. Preprints*, **9**, 795 (1968).
14. J. D. Hoffman, G. T. Davis, and J. I. Lauritzen, **Treatise on Solid State Chemistry**, Ed. N. B. Hannay, Chap. 7, *Plenum*, NY, 1976.
15. R. G. Brown and R. K. Eby, *J. Appl. Phys.*, **35**, 1156 (1964).
16. D. C. Bassett, A. M. Hodge, and R. H. Olley, *Proc. Roy. Soc. London*, **A377**, 39 (1981).
17. K. H. Illers and H. Hendus, *Makromol. Chem.*, **113**, 1 (1968).
18. R. Martuscelli and M. Pracella, *Polymer*, **5,** 306 (1974).
19. J. Runt, I. R. Harrison, and S. Dobson, *J. Macromol Sci. Phys. Ed.*, **B17**, 99 (1980).

Melt Rheology and Processability of Ethylene/Styrene Interpolymers

Teresa Plumley Karjala, Y. W. Cheung, and M. J. Guest

The Dow Chemical Company, Freeport, Texas 77541, USA

ABSTRACT

The melt rheology and processing characteristics of ethylene/styrene interpolymers, ESI, are correlated with the styrene content and molecular weight of the interpolymers. The complex viscosity, mastercurves, activation energy, high shear rate processability, and melt strength are discussed for four well-defined ESI. Comparisons are also made to polyethylene and polystyrene where appropriate.

INTRODUCTION

The development of single site constrained geometry catalysts (The Dow Chemical Company's INSITE[TM] Technology)[1,2] have provided a route to produce ethylene/styrene interpolymers, ESI. As a result of ^{13}C NMR analyses showing successive head-to-tail styrene chain insertions to be absent even at high levels of styrene incorporation, these interpolymers have been described as "pseudo-random" and can contain up to about 50 mol % (~ 80 wt%) styrene content. Structure, thermal transitions, and mechanical properties of ESI have been described elsewhere.[3]

In this paper, the melt rheology and processability of ESI will be examined. These melt properties are important, because they give information on 1) the thermal stability of these materials under typical processing conditions, 2) the processability of these materials, 3) the effect of molecular parameters such as styrene content and molecular weight on processing, and 4) the behavior of these materials in elongational type flows.

EXPERIMENTAL

The materials studied are described in Table 1 in terms of the total wt% styrene (wt% S) contained in the polymer, the weight average molecular weight, M_w, melt index, I_2, melt density, ρ_m, and glass transition temperature, T_g. The molecular weight distributions of all samples were narrow (~2.3). These polymers contain a small amount of atactic polystyrene, typically less than 4 wt%.

Low frequency melt rheological measurements were performed on a Rheometrics RMS-800 with 25 mm parallel plates in the dynamic mode under nitrogen. Frequency/temperature sweeps were performed over the range 0.1-100 rad/s at temperatures of 190, 170, and 150°C. Mastercurves were created at a reference temperature of 190°C using a horizontal shift mode with the Rheometrics Rhios V4.4 software. Activation energies for flow were determined from an Arrhenius dependence of the horizontal shift factors. Thermal stability data were determined in air at 1 rad/s at 230°C over 30 min.

Table 1. Structural description of materials*

Materials	wt % S	M_w	Melt index	$\delta_{m\ g/cc}$	T_g (°C)
ESI-1	74.2	340,000	0.18	0.952	30
ESI-2	71.5	250,000	1.14	0.916	24
ESI-3	76.5	140,000	12.40	0.929	35
ESI-4	29.8	240,000	0.03	0.805	-8

*Comonomer composition, expressed as weight % styrene, was determined using 1H NMR. Molecular weights, M_w, were obtained by gel permeation chromatography, GPC. The melt index, I_2, and melt density, δ_m, were determined with indexer, according to ASTM D1238 Procedure A, Condition E. The glass transition temperature, T_g, was determined from tan δmaximum by solid state dynamic mechanical spectrometry.[3]

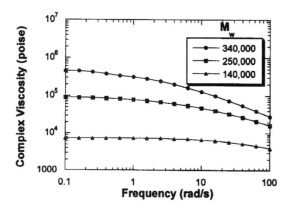

Figure 1. The effect of molecular weight on complex viscosity at 190°C of ethylene/styrene interpolymers containing approximately 75 wt% styrene.

Higher shear rate (100 - 4642 s⁻¹) viscosities were measured on a Goettfert Rheograph 2003 with a 1 mm diameter/20 mm length die with a 180° entry angle and a 1000 bar pressure transducer.

Melt strength measurements were taken on a Goettfert Rheotens at 190°C. The full force/velocity curve was examined as well as the melt strength, or the plateau of the force/velocity curve. A die diameter of 2.1 mm with an aspect ratio of 20 was used at an apparent shear rate of 33 s⁻¹. The air gap between the capillary rheometer outlet and the Rheotens was 100 mm, while the initial wheel velocity was 10 mm/s and the wheel acceleration was 2.4 mm/s².

RESULTS AND DISCUSSION

THERMAL STABILITY

The thermal stability testing indicated that ESI show little degradation of viscosity with time at elevated temperatures. For example, the change in viscosity for sample ESI-1 was 0.5% at 230°C over 30 minutes, showing that the ESI are not notably susceptible to crosslinking or degradation at processing temperatures.

COMPLEX VISCOSITY/RELATIONSHIP TO M_w

Figure 1 shows the effect of weight average molecular weight (M_w = 140,000 - 340,000) on complex viscosities at 190°C for interpolymers containing approximately 75 wt% styrene. With an increase in

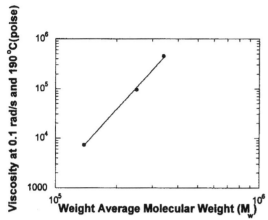

Figure 2. Relationship between zero shear viscosity and weight average molecular weight for the samples in Figure 1.

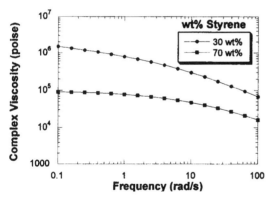

Figure 3. The effect of wt% styrene on complex viscosities at 190°C of ethylene/styrene interpolymers of approximately 250,000 MW.

molecular weight, the viscosity increases as expected. The zero shear viscosity, η_o, was estimated as the viscosity at 0.1 rad/s, since these interpolymers showed essentially Newtonian behavior at low frequencies. Figure 2 presents the relationship between the zero shear viscosity and the weight average molecular weight for these samples. Zero shear viscosities were also estimated by a Cross model with almost identical results. As shown in Figure 2, this relationship can be expressed as:

$$\eta_{0.1}^* \, (poise) = 4.07 \times 10^{-20} \, M_m^{4.52} \quad [1]$$

For linear polymers above their entanglement molecular weight, M_e, this relationship is generally expressed as $\eta_o = K(M_w)^a$, in which $a = 3.4 - 3.55$ and K is a constant. Elevated values of the molecular weight exponent a are typically found for samples which have long chain branching.[6,7] Recent work has also shown an elevated exponent (a=4.2) for metallocene-catalyzed polyethylenes.[8] The increased dependence of the zero shear viscosity on molecular weight for ESI may be due to the effect of the bulky phenyl side branch in the ethylene chain, which provides increased resistance to flow as compared to a linear polyethylene.

Figure 3 shows the effect of the wt% styrene on complex viscosity. At approximately equivalent $M_w = 250,000$, the viscosity of the 30 wt% styrene interpolymer is over an order of magnitude higher than that of the 70 wt% styrene interpolymer. As the wt% styrene incorporation is increased, the molecular weight between entanglements, M_e, for the interpolymer will increase (typical values are $M_e = 1,100$ for polyethylene and $M_e = 18,700$ for polystyrene[6]). Relationships have also been proposed for copolymers relating zero shear viscosity to the monomeric friction coefficient, ξ_o, M_w, and M_e:[9]

$$\eta_o = K\xi_o M_w (M_w / M_e)^{2.5} \qquad [2]$$

At equivalent molecular weight, as the styrene level increases, M_e increases, which is associated with a decrease in viscosity. A further consideration is that higher T_g's are found for higher styrene-containing interpolymers,[3] as is evident in Table 1. For isothermal testing of similar materials, a higher T_g polymer should have decreased molecular motions, and thus an increased ξ_0 coefficient, and correspondingly higher viscosity as shown in Eq 2. However, the M_e effect is considered to dominate for ESI as shown in Eq 2 and Figure 3.

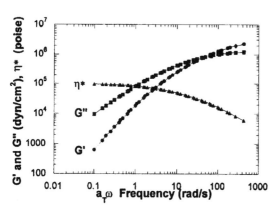

Figure 4. Mastercurve of ESI-2 at a reference temperature of 190°C.

MASTERCURVES/ϖ_C/G_N/E_A

Rheology data for the interpolymers exhibited excellent time-temperature superposition; an example is shown in Figure 4 for ESI-2. At low frequencies or long times, the viscous behavior, ($G''(\omega)$ or the loss modulus) dominates, while at higher frequencies or shorter times the elastic behavior ($G'(\omega)$ or the storage modulus) dominates. The crossover frequency, ω_c, at which the storage and loss moduli are equivalent, is shown in Table 2. This inverse of the ω_c is indicative of a characteristic relaxation time.[10] As M_w increases, $1/\omega_c$ increases, and the relaxation times increase. Longer relaxation times are also found at lower styrene contents, as discussed previously, since lower styrene contents influence viscous behavior in the same manner as do higher molecular weights. The plateau modulus, G_N, was estimated with the Rheometrics software by fitting storage moduli data, G', to:

Table 2. Description of rheology parameters determined by dynamic mechanical spectroscopy at 190°C*

Materials	$\eta^*(0.1)$ poise	ω rad/s	G_N dyn/cm²	M_e calc.
ESI-1	4.51E+05	5.9	2.68E+06	13,675
ESI-2	9.61E+04	44.2	2.28+06	15,467
ESI-3	7.36+03	456.9	-	-
ESI-4	1.50E+06	7.3	1.03E+07	3,009

*Values are the complex viscosity at 0.1 rad/s, $\eta^*(0.1)$, the frequency, ω_c, at which the storage and loss moduli crossover, the estimated plateau modulus, G_N, and the calculated entanglement molecular weight, M_e.

$$G' = G_N + C_1 / \omega^{C_2} \tag{3}$$

in which C_1 and C_2 are constants. The G_N determined by this method are in Table 2. The entanglement molecular weight, M_e, was evaluated from rubber elasticity theory as:[5,6]

$$M_e = \rho_m RT / G_N \tag{4}$$

in which R is the gas constant, T is temperature, ρ_m is the melt density, and G_N is the plateau modulus. The melt densities used are listed in Table 1. The G_N were reasonable for all samples except ESI-3, the lowest molecular weight sample, which was not fit well by Eq 3. The calculated M_e for the interpolymers given in Table 2 are intermediate between values for polyethylene, $M_e = 1,100$, and that of polystyrene, $M_e = 18,700.$[6] Other methods could be used to estimate the G_N which would refine the M_e predictions.[5]

The activation energy, E_a, or temperature dependence of the viscosity is also expected to be a strong function of the styrene level; base polymers of high density polyethylene are typically quoted with $E_a = 6.5$ kcal/mol and polystyrene of $E_a = 23$ kcal/mol.[11] An Arrhenius dependence of the horizontal shift factor, a_T, was used to estimate E_a for interpolymers, which was justified since $T > T_g + 100$:[5]

$$a_T = K e^{E_a / R(1/T - 1/T_o)} \tag{5}$$

Table 3. Rheological parameters with implications for processing at 190°C.*

Materials	E_a shift kcal/mol	E_a Eq[6] kcal/mol	E_a Eq[6] kcal/mol	n	MS cN
ESI-1	15.0	16.7	14.9	0.11	29.19
ESI-2	14.2	16.2	14.4	0.20	4.56
ESI-3	15.4	17.2	15.3	0.35	-
ESI-4	8.7	9.7	8.7	0.11	18.42

*Values are the activation energy determined from data by the shift factor approach, E_a shift,activation energy determined from Eq [6], E_a Eq [6], activation energy from a modified form of Eq [6], E_a Eq[6] mod., the power-low index, n, and the melt strenght, MS.

In Eq 5, K is a constant and T_o is the reference temperature (in this case, 190°C). The activation energies are shown in Table 3, and plotted in Figure 5. The E_a's are fairly constant at 14.2-15.4 kcal/mol at roughly 75 wt% styrene level, and decrease to 8.7 kcal/mol for a styrene level of 30 wt%. The high styrene level E_a's are comparable to those for highly branched low density polyethylene,[5] while the low styrene levels are similar to those reported for long chain branched polymers produced using INSITE™ Technology.[12] The measured activation energies for ESI agree well with those predicted using the relationship between the molar volume at the boiling point of the two substituents X and Y on the second carbon atom of a polyhydrocarbon, V_{X+Y}, and the activation energy:[11]

$$\log E_a = 0.784 + 0.0060 V_{X+Y} \tag{6}$$

in which $V_{X+Y} = 7.4$ ml for polyethylene and $V_{X+Y} = 96.0$ for polystyrene. By substituting for V_{X+Y} in Eq 6:

$$V_{X+Y} = 96(wt\% \, styrene) + 7.4(wt\% \, ethylene) \tag{7}$$

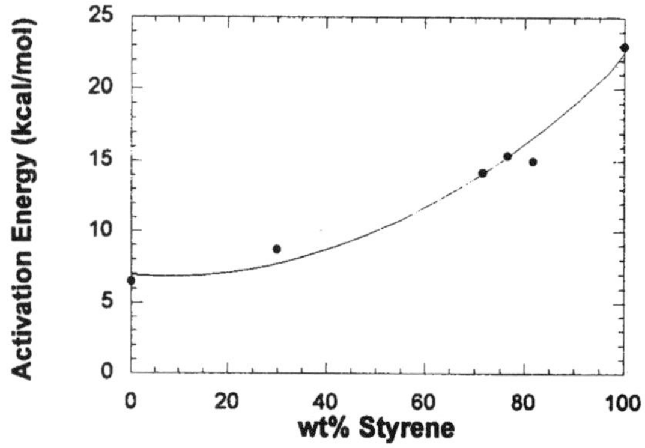

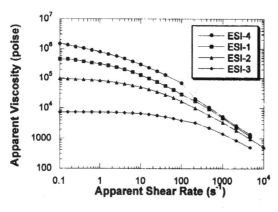

Figure 5. Activation energies of ethylene/styrene interpolymers including literature values for polyethylene and polystyrene.[11]

Figure 6. Complex viscosities and capillary viscosities at 190°C of four ethylene/styrene interpolymers.

E_a's within 2 kcal/mol of those measured were calculated as shown in Table 3. With a modified intercept of 0.739 and slope of 5.94×10^{-5} in Eq 6, activation energies were predicted within 0.2 kcal/mol of those measured as shown in Table 3. This variation in activation energy with styrene content is important to consider when processing ESI, since the percent change in viscosity with temperature will vary as a function of the styrene level.

CAPILLARY RHEOLOGY

The capillary data are overlaid with the complex viscosity data in Figure 6. The Cox-Merz rule[5] was obeyed well for all samples except ESI-4, which is a very low melt index sample (I_2=0.03). Power-law indices[5] were determined from the data ranging from 464-4,642 s^{-1} and are shown in Table 3. A polymer is more shear sensitive, i.e., the viscosity decreases more rapidly with increasing shear rate, if the polymer has a lower power-law index. The more shear sensitive ESI are at higher molecular weights at equivalent styrene level; no significant conclusion can be made from these data on the effect of styrene content on n. ESI's processability are similar to that of polystyrene (n=0.22) and improved over that of either high density polyethylene (n=0.56) and low density polyethylene (n=0.46).[13]

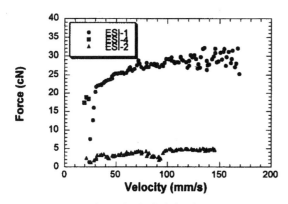

Figure 7. Melt strengths at 190°C of ethylene/styrene interpolymers described in Tables 1-3.

RHEOLOGICAL MODELING

The effects of molecular weight and styrene content on melt rheology were discussed above based upon the four interpolymers listed in Table 1. Other interpolymers have been prepared extending these ranges of molecular weights and styrene levels. Based upon these samples, a preliminary rheological model has been developed in order to allow predictions of rheological behavior depending upon important molecular parameters, such as styrene level and molecular weight, as well as the process variables of shear rate and temperature. The model form is similar to that discussed previously for INSITE™ Technology polymers,[4] in which a nonlinear Cross model is used to predict individual viscosity curves, and the parameters of the model are related to the molecular properties. The temperature dependence is accounted for through an Arrhenius relationship between the zero shear viscosity and temperature. Preliminary analysis of the data led to the conclusion that the Cross parameters were primarily functions of molecular weight and styrene level, while the E_a was primarily a function of only the styrene level.

MELT STRENGTH

The melt strengths,[14,15] an approximate measure of elongational viscosity, of the ESI samples are shown in Table 3 and Figure 7. The melt strengths are in general excellent (18-29 cN), being comparable to a fractional melt index low density polyethylene and polystyrene but with improved elongation. The lower molecular weight, higher styrene sample, ESI-2, had a melt strength more similar to that of a typical linear low density polyethylene.

SUMMARY

Melt rheological properties of ethylene/styrene interpolymers were presented. These polymers were shown to have 1) good thermal stability at processing temperatures, 2) viscosities which are a strong function of styrene level and molecular weight, 3) excellent processability comparable to that of polystyrene, and 4) excellent melt strength behavior comparable to that of low density polyethylene and polystyrene.

REFERENCES

1.   European Patent Application, **EP 416 815**, 1991.
2.   J. C. Stevens, *Proc. MetCon'93*, 157, Houston, 1993.
3.   Y. W. Cheung and M. J. Guest, *Proc. SPE ANTEC*, 1634 (1996).
4.   T. A. Plumley *et al.*, *Proc. SPE ANTEC*, 1221 (1994).
5.   J. M. Dealy and K. F. Wissbrun, **Melt Rheology and Its Role in Plastics Processing**, Van Nostrand, 1990.

6.   W. W. Graessley in **Physical Properties of Polymers**, 2nd Edition, *ACS*, 1993.
7.   S. Lai and G. W. Knight, *Proc. SPE ANTEC'93*, 1188 (1993).
8.   J. F. Vega *et al.*, *Macromolecules*, **29**, 960 (1996).
9.   B. J. Meister in **Encyclopedia of Polymer Science and Engineering**, *John Wiley*, 16, 138 (1986).
10.   F. Chambon, *Proc. SPE ANTEC'95*, 1157 (1995).
11.   H. Schott, *J. Appl. Polym. Sci.*, **6(23)** 29 (1962).
12.   Y. S. Kim *et al.*, *J. Appl. Polym. Sci.*, **59**, 125 (1996).
13.   Z. Tadmor and C. G. Gogos in **Principles of Polymer Processing**, *John Wiley*, New York, 1979.
14.   M. H. Wagner *et al.*, *Polym. Eng. Sci.*, **36**, 925 (1996).
15.   A. Ghijels *et al.*, *Int. Polym. Proc.*, **IX(3)** 252 (1994).

Characterizing the Melt Relaxation Behavior of Metallocene Polyethylenes

S. H. Wasserman

Union Carbide Corporation, Bound Brook, P. O. Box 670, NJ 08805, USA

ABSTRACT

Polyethylene has a broad relaxation spectrum that makes it ideal for blown film and blow molding. In polymer processing, melt and thermal phenomena are superimposed to generate semicrystalline morphology that delineates solid-state properties. A polymer melt is exposed to deformations varying in both time scale and strength. In many processes, melt stress simultaneously grows and decays so characterizing time-dependent stress relaxation is important. Stress-strain behavior in polymer processing is determined by the relaxation spectrum. Commercial polyethylenes have had narrow to broad molecular weight distributions; metallocene catalysis can give polyethylene with quite narrow molecular weight distributions. This affects their relaxation spectra, the breadth of which is quantified by the Relaxation Spectrum Index, RSI. This dimensionless index is a sensitive, reliable indicator of long-range melt state order. The RSI illustrates unique rheology of two metallocene polyethylene families (SSC-1 and SSC-2) used in polyethylene manufacture by the UNIPOL process. Their melt rheology is compared to competitive metallocene polyethylenes and other commercial resins. The melt behavior of SSC-1 and SSC-2 contribute to their processability, clarity, and toughness advantages compared to LLDPE and HP-LDPE.

INTRODUCTION

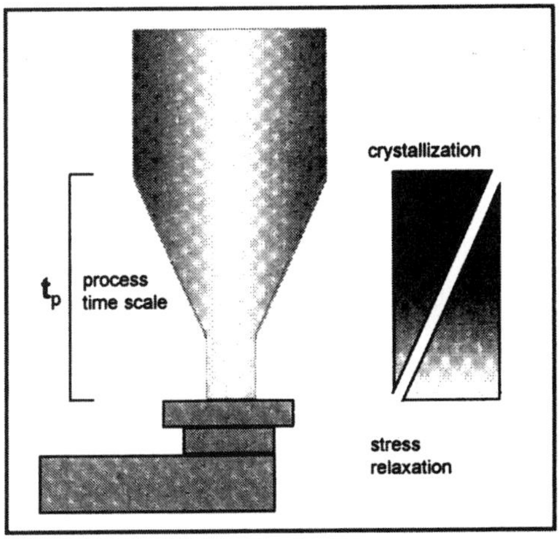

Commercial polymers depend on their ability to store and relax excess energy imparted on them by various deformation processes. In blown film processing, for example, polyethylene, PE, undergoes shear and extensional deformations during extrusion and residual shear stress is then relaxed during the subsequent crystallization process. As illustrated in Figure 1, a processing time parameter, t_p, can be defined by the time required for a film element to pass from the extruder exit to the frost line (which approximates the end of the crystallization process). According to molecular theories such as reptation, double reptation, etc., stress relaxation on a molecular scale is, in turn, specified by a characteristic relaxation time.[1] The ratio of this relaxation time t to the process time, called the Deborah number:[2]

Figure 1. Blown film example of process time scale.

$$De \equiv \tau / t_p \tag{1}$$

is an important parameter that can be used to characterize structure-property-processing relationships. The Deborah number can be calculated for any polymer melt and fabrication process. The level of residual shear stress that is able to relax during the melt state process will dictate the chain orientation in the finished article and therefore contribute to its solid state morphology and physical properties. The higher the Deborah number, the higher the stress that remain unrelaxed, leaving the longest chains aligned in the flow direction during crystallization.

In polymer melts, stress relaxation is actually composed of a series of relaxations covering times scales that vary in order of magnitude. Dynamic oscillatory shear experiments generate storage and loss moduli data which can be used to calculate, among other things, the discrete relaxation spectrum. This relaxation time distribution, RTD, is best calculated by a simultaneous fit of both storage and loss moduli to a series of Maxwell-type relaxation modes according to:[3]

$$G'(\omega) = \sum_{i=1}^{N} g_i \frac{(\omega\lambda_i)^2}{1+(\omega\lambda_i)^2} \tag{2a}$$

$$G''(\omega) = \sum_{i=1}^{N} g_i \frac{\omega\lambda_i}{1+(\omega\lambda_i)^2} \tag{2b}$$

Each of the N individual relaxation modes in the above relations are described by a distinct strength, g_i, and relaxation time, λ_i. Increasing the breadth of the molecular weight distribution, MWD, or level of branching in the polymer system is reflected by an enhanced RTD, particularly at high relaxation times. In order to differentiate polymers with different or changing relaxation spectra, we have introduced a quantity known as the Relaxation Spectrum Index, RSI. If the first and second moments of the RTD are:

$$\lambda_I = \sum_i g_i / \sum_i g_i / \lambda_i \qquad \lambda_{II} = \sum_i g_i \lambda_i / \sum_i g_i \tag{3}$$

the RSI is defined as their ratio:

$$RSI = \lambda_{II} / \lambda_I \tag{4}$$

Eqs 3 and 4 are analogous to those used to calculate the polydispersity index, PDI, which describes the breadth of the MWD from size exclusion chromatography, SEC. The RSI has proven to be a sensitive indicator of differences in molecular structure, especially where those differences affect the breadth of the relaxation spectrum. The RSI is a dimensionless quantity, but varies with T and MW since we choose to run our experiments over a fixed frequency range (i.e., 0.1-100 sec^{-1}).

Low deformation experiments such as dynamic oscillatory shear are considered to be the best for detecting differences in features of molecular structure such as PDI and levels of long chain branching, LCB. In earlier work, the RSI was used to follow thermo-oxidative degradation during extrusion of un-stabilized LLDPE and to quantify its effect.[4] Here, we use the RSI to illustrate unique rheology of two metallocene PE families (SSC-1 and SSC-2) developed by to take advantage of the UNIPOL process

Table 1. Sample identification and characterization

Sample	Class	MI[a]	ρ^b	RSI	nRSI[c]
1	SSC-1	0.5	0.938	5.7	3.7
2	SSC-1	0.7	0.937	3.6	2.9
3	SSC-1	0.8	0.920	6.4	5.7
4	SSC-1	0.9	0.920	4.1	3.9
5	SSC-1	1.0	0.920	2.8	2.8
6	SSC-1	1.2	0.924	2.7	3.1
7	SSC-1	1.5	0.924	2.2	3.0
8	SSC-1	1.7	0.929	2.2	3.0
9	SSC-1	2.0	0.926	2.4	3.7
10	SSC-1	3.2	0.916	1.7	4.0
11	SSC-2	3.0	0.920	63.9	26.3
12	SSC-2	0.9	0.921	36.9	34.1
13	SSC-2	1.5	0.917	21.6	29.1
14	SSC-2	1.6	0.906	24.0	33.9
15	SSC-2	1.8	0.917	20.5	31.7
16	SSC-2	1.9	0.928	26.9	43.1
17	SSC-2	2.0	0.923	21.4	35.6
18	SSC-2	2.4	0.906	19.8	37.8
19	SSC-2	2.8	0.924	18.5	39.6
20	SSC-2	2.9	0.903	16.5	36.2
21	LLDPE	0.5	0.923	5.3	3.5
22	LLDPE	1.0	0.918	4.8	4.8
23	LLDPE	2.0	0.922	2.8	4.3
24	HDPE	0.4	0.953	30.4	16.2
25	HDPE	0.8	0.950	19.8	16.8
26	HDPE	8.2	0.963	2.4	11.4
27	HP-LDPE	0.1	0.920	126.5	23.2
28	HP-LDPE	0.2	0.921	44.5	13.6
29	HP-LDPE	2.0	0.923	13.3	21.4
30	mLLDPE-A	0.7	0.917	1.5	3.2
31	mLLDPE-A	1.0	0.917	2.1	2.1
32	mLLDPE-A	3.4	0.917	2.4	2.0
33	mLLDPE-B	1.0	0.915	7.4	7.4
34	mLLDPE-B	2.5	0.935	5.0	9.8
35	mLLDPE-B	3.2	0.905	3.2	7.5

[a]MI units g/10 min at 190°C; [b]units g/ml; [c]nRSI is defined as RSIMIα where is 0.61 for SSC-1, LLDPE, and mLLDPE-A and 0.74 for others

metallocene PE families (SSC-1 and SSC-2) developed by to take advantage of the UNIPOL process for PE manufacture. Relaxation spectra and RSI values have also been calculated for other metallocene PEs, along with commercial LLDPE and HDPE made with Ziegler-Natta and Cr-based catalysts, and HP-LDPE.

EXPERIMENTAL

DYNAMIC OSCILLATORY SHEAR

Dynamic oscillatory shear experiments were conducted with a commercial Weissenberg Rheogoniometer manufactured by TA Instruments. Frequency sweep experiments at 190°C and 2% strain in parallel plate mode were run under nitrogen from 0.1 to 100 s^{-1}. Sample are typically 1200-1600 mm thick and 4 cm in diameter with care taken to ensure that the samples completely fill the gap between the upper and lower platens. Discrete relaxation spectra were calculated with the commercially available IRIS™ software package. The number of relaxation modes calculated for the samples reported here was typically 4-8. The relaxation times in the final spectrum are not necessarily evenly distributed on a logarithmic scale since the number of modes and their distribution are adjusted by the software to optimize the simultaneous solution of Eqs 2a and 2b.

SAMPLES

Table 1 includes descriptions of all PE samples used in the discussion that follows. Samples 1-10 are representative of SSC-1 LLDPE. Samples 11-20 are example of SSC-2 LDPE. All SSC-1 and SSC-2 samples are ethylene hexene-1 copolymers made in the UNIPOL gas phase process. Samples 21-35 are comparative examples of commercial LLDPE, HP-LDPE, HDPE, and metallocene LLDPE, mLLDPE. The LLDPE and HDPE samples are UCC products made by traditional Ziegler-Natta and Cr catalysts, respectively, in the UNIPOL gas phase process; the HP-LDPE are made by UCC in a high pressure, tubular reactor process. Of the comparative mLLDPE examples, samples 30-32 (mLLDPE-A) are ethylene hexene-1 copolymers made in the gas phase; samples 33-35 (mLLDPE-B) are ethylene octene-1 copolymers made in a slurry process.

RESULTS & DISCUSSION

Figure 2 shows relaxation spectra calculated for typical LLDPE sample 23 and HP-LDPE sample 29 of comparable 2.0 MI. As shown, the spectrum for the HP-LDPE is broader than that for the LLDPE and includes a longest relaxation that is more than 10 times greater in time. The enhanced long relaxation for the long-chain branched HP-LDPE is consistent with reptation and related theories which predict significant delay in the stress relaxation process due to branches that are longer than the entanglement MW (approximately 104 carbon atoms in the case of PE3).

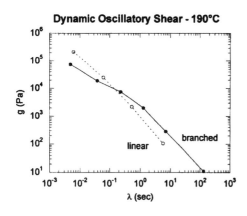

Figure 2. Relaxation spectra for linear and branched PE.

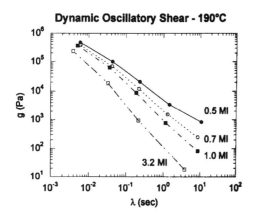

Figure 3. Relaxation spectra for SSC-1 LLDPE.

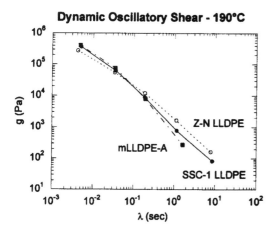

Figure 4. Relaxation spectra for SSC-1 LLDPE, Z-N LLDPE, and mLLDPE-A (all 1.0 MI).

Relaxation spectra for SSC-1 samples 1, 2, 5, and 10 are compared in Figure 3. As shown, the effect of decreasing MI (or increasing MW) is to shift individual relaxations to higher weights and times (effect is more substantial at higher times). Such a broadening of the spectrum also leads to an increase in the RSI, as shown in Table 1. Over a wide range in MI, the RSI-MI correlation closely follows a power law relation with the power law exponent ($\alpha < 0$) a characteristic parameter of the respective PE families. In fact, by multiplying the RSI by $MI^{-\alpha}$, the effect of MI variation can be removed from the RSI values. By this approach, we find that SSC-1 LDPE can be defined by the following expression:

$$2.5 \le RSI \cdot MI^{0.61} \le 6.5 \qquad [5]$$

where $RSI \cdot MI^{0.61}$ is a "normalized" RSI or nRSI. Comparative mLLDPE-A falls below the RSI-MI range given by Eq 5; the RSI-MI range for commercial LLDPE overlaps the upper range of Eq 5. The relaxation spectra for SSC-1 LLDPE are distinct from those for both mLLDPE-A and commercial LLDPE of comparable MI, as demonstrated in Figure 4 (at approximately 1.0 MI) for samples 5, 22, and 31. Such is the case for all MI considered.

Figure 5 compares the relaxation spectra for SSC-2 LDPE samples 11, 12, 17, and 20. As in the case of SSC-1, increasing MI leads to an enhancement in the relaxation spectrum and RSI although the power law exponent is noticeably higher in magnitude. In fact, for a given MI, the RSI values for SSC-2 LDPE are higher than any than any other PE homopolymer or ethylene α-olefin copolymer family. This unique RSI-MI behavior is defined by:

$$RSI \cdot MI^{0.74} > 2.6 \qquad [6]$$

For example, Figure 6 shows the relaxation spectra for SSC-2 LDPE, HDPE, and mLLDPE-B (samples 12, 25, and 33) of approximately 1.0 MI. Figure 7 compares the spectra for 2.0 MI examples of SSC-2 LDPE and HP-LDPE (samples 17 and 29). In both cases, the uniquely broad spectra for the

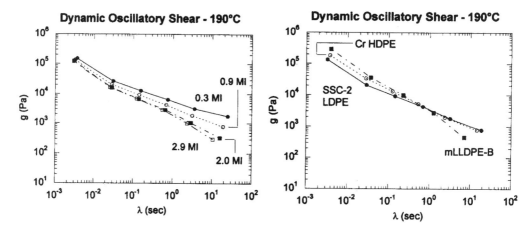

Figure 5. Relaxation spectra for SSC-2 LDPE.

Figure 6. Relaxation spectra for SSC-2 LDPE, HDPE, and mLLDPE (all 1.0 MI).

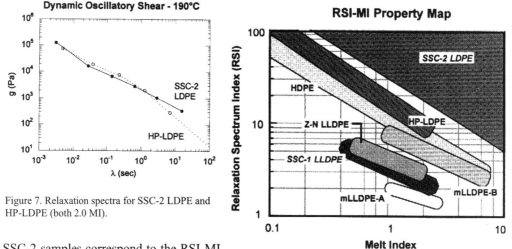

Figure 7. Relaxation spectra for SSC-2 LDPE and HP-LDPE (both 2.0 MI).

Figure 8. RSI-MI property map for families of PE.

SSC-2 samples correspond to the RSI-MI behavior given by Eq 6 (see Table 1). Again, the comparison is qualitatively the same for all MI studied.

Based on the melt relaxation spectra and RSI values for PE families such as those discussed here, it is possible to construct an RSI-MI property map covering at least two decades in both RSI and MI, as shown in Figure 8. Such a property map is very helpful in characterizing the melt relaxation behavior of any PE sample or family. The same approach has allowed us to develop three distinct classes of met-

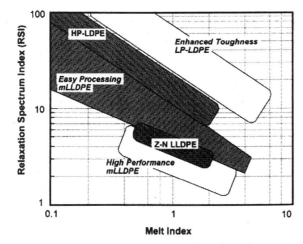

Figure 9. RSI-MI property map showing regimes of proprietary UCC metallocene PE.

allocene PE in terms of their RSI-MI behavior. Included among them are High Performance mLLDPE, Easy Processing mLLDPE, and Enhanced Toughness LP-LDPE, as illustrated in Figure 9. Technology has been developed to produce PE in all three regimes with the gas phase UNIPOL process.

CONCLUSIONS

Dynamic oscillatory shear tests have been used to investigate the melt relaxation behavior of metallocene PE. The sensitivity of such measurements to features of molecular structures make them ideal for use in characterizing the rheology of entire PE families as defined by their MW, MWD, and LCB.

With melt rheology defined by stress relaxation over a distribution of time scales (i.e., the relaxation spectrum), an important parameter, called the RSI, is introduced to quantify the breadth of that distribution. In addition to its use to differentiate PE families and to define regimes of metallocene PE products, the RSI has also been used to predict processing behavior and fabricated article properties. For example, a high RSI value is indicative of improved extrudability. In film blowing, the RSI has been correlated with bubble stability and impact strength; in blow molding, relationships between the RSI and die swell, parison recovery, and bottle weight have also been delineated.

NOMENCLATURE

De	Deborah number
G_i	strength of mode in relaxation spectrum
LCB	long chain branch
N	number of modes in relaxation spectrum
NRSI	"normalized" relaxation spectrum index
PDI	polydispersity index
PE	polyethylene
RSI	relaxation spectrum index
RTD	relaxation time distribution
SEC	size exclusion chromatography
SSC-1	single-site catalyst, type 1
SSC-2	single-site catalyst, type 2
T_p	processing time parameter
α	power lax exponent
λ_i	relaxation time of mode in relaxation spectrum
λ_I	first moment of relaxation time distribution
λ_{II}	first moment of relaxation time distribution
τ	molecular relaxation time

ACKNOWLEDGMENTS

I acknowledge the fine experimental work done by our rheology laboratory technicians and thank many of my colleagues at UCC for important contributions to the interpretation of this work.

REFERENCES

1.   J. E. Mark, A. Eisenberg, W. W. Graessley, L. Mandelkern, E. T. Samulski, J. L. Koenig, and G. D. Wignall in **Physical Properties of Polymers**, 2nd Ed., *ACS*, Washington, D. C., 1993.
2.   J. M. Dealy and K. F. Wissbrun in **Melt Rheology and its Role in Plastics Processing**, Van Nostrand Reinhold, New York, 1990.
3.   J. D. Ferry in **Viscoelastic Properties of Polymers**, 3rd Ed., *John Wiley & Sons*, New York, 1980.
4.   S. H. Wasserman and G. N. Foster, *Proc. XIIth Intl. Congress on Rheology*, Quebec City, Canada, p. 48 (1996).

Studies on the Thermal Stability and Processability of Syndiotactic Polystyrene

Ch.-M. Chen, H.-R. Lee, Ch.-J. Wu, H.-M. Chen,
Engineering Plastics Division
Union Chemical Laboratories, 321 Kuang-Fu Road, Sec. 2 HsinChu, Taiwan

ABSTRACT

This study examines the effects of antioxidants and processing conditions on the thermal stability and processability of syndiotactic polystyrene, sPS. Evaluation results indicate that, at a higher temperature, the antioxidant compositions AO-01 and AO-02 raise the oxidation induction temperature of sPS and reduce the degradation. Meanwhile, AO-01 decreases the discoloration of sPS caused by a high shear rate. Also, a long processing time decreases the molecular weight of sPS and increases the melt flow index. Consequently, such a phenomenon affects the processability.

INTRODUCTION

Syndiotactic polystyrene, sPS, was first synthesized by Idemitsu Kosan Co., Ltd. with a metallocene type catalyst in 1985.[1] Idemitsu Petrochemical Co., Ltd. and Dow Chemical are the two major developers of sPS applications.

SPS is a semicrystalline polymer owing to its syndiotactic configuration. The crystallinity offers sPS more characteristics than a conventional atactic polystyrene. SPS has attractive properties such as low density, easy molding, and good resistance to chemicals and heat. Moreover, its dielectric property is superior to that of other plastics except fluoroplastics. These properties provide sPS with the opportunity of being a material for electronic parts of SMT (surface mounting technology) grade. SPS not only competes with LCP and PPS in producing electronic parts, but also holds advantages over other engineering plastics in lieu of its low monomer price. Besides electronic parts, sPS is a promising material for automobile parts, electrical housing, and mechanical parts.

The processing temperature of sPS is extremely high due to its high melting temperature (ca. 270°C). The mechanical shearing and the high temperature affect the sPS stability during processing. The degradation of atactic polystyrene has been thoroughly discussed,[2] whereas sPS has only received limited attention. According to Gacther and Muller's proposed degradation mechanism[3] of polymers, adding antioxidants prevents or inhibits the oxidation reactions.

This study establishes an evaluation process for thermal stability. The relation between the thermal stability and processability of sPS is also discussed in terms of different antioxidant compositions and processing variables.

SPS (M_w = 481000) was purified with methyl ethyl ketone and vacuumed dried. The purified sPS was then compounded with different compositions of antioxidants using a Brabender PL 2100. The antioxidants were Mark® AO-60 (Adeka Argus Chem. Co. Ltd.), Mark® PEP-36, and Irganox® 1076 (Ciba Geigy). The processing temperature was 295°C. Table 1 lists various combinations of the screw speed and the processing time. The oxidative induction temperatures, OIT, of the sPS compounds were measured by a differential scanning calorimeter (DSC, Perkin Elmer 7 series). The heating rate was 10°C/min. and the nitrogen flow rate was 20 ml/min. The time needed to cause degradation, TD, was measured by a thermogravimetric analyzer (TGA, Perkin Elmer 7 series) at 300°C under a gas flow rate of 20 ml/min. The TD was then taken as the time to cause a weight loss of 1.5 wt% when the flow gas was switched to air from nitrogen. The melt index, MI, was measured by an MI meter (Toyoseiki Seisakusho Ltd. Type 3056) following ASTM D1238 with a load of 2160g. The yellowness index, YI, was measured by a colorimeter (Juki JP7100F) following ASTM D1925. The molecular weight was measured by gel permeation chromatography (GPC, Waters GPC 150CV).

Table 1. Recipes and processing Parameters

Sample	sPS	PEP-36	AO-60	Irganox 1076	Screw speed rpm	Proc. time min
C025	100	-	-	-	30	4
C035	100	-	-	-	30	9
C026	100	-	-	-	30	12
C033	100	-	-	-	60	4
C027	100	-	-	-	90	4
C028	100	0.3	0.3	-	30	4
C029	100	0.3	0.3	-	30	12
C030	100	0.3	0.3	-	60	4
C034	100	-	0.6	-	30	4
C032	100	-	-	0.6	30	4

Note: The processing temperature was 295°C.

RESULT AND DISCUSSION

Table 1 presents the recipes and processing variables of the sPS compounds. Table 2 summarizes the experimental data. The OIT was taken as the temperature where the DSC curve under air began to deviate from that under nitrogen due to the oxidation. Figure 1 displays the DSC curves of different compounds. The variations of OITs were clearly observed in this figure. This variation indicates that the OIT is an accurate indicator of the antioxidants' influences. The TD in an isothermal state (300°C) was chosen as another indicator of thermal stability. Figure 2 depicts the TGA curves. These indexes are references for the thermal stability during processing.

Table 2. Characterization of Various Blends after Processing

Sample	M_n 10^3	M_w 10^3	Polydisp- ersity	MI g/10 min	YI	OIT °C	TD min
C025	132	403	3.04	4.1	35.4	274	12.7
C035	69	269	3.92	11.1	31.0	263	5.2
C026	122	361	2.94	4.8	33.6	271	8.8
C033	131	402	3.07	3.6	gray	272	9.6
C027	139	428	3.09	2.5	dark gray	270	14.7
C028	137	414	3.02	3.7	38.0	313	>28
C029	98	302	3.08	11.0	38.2	309	>28
C030	134	405	3.02	4.2	32.1	311	>28
C034	126	415	3.30	3.9	28.1	310	>28
C032	134	407	3.04	3.5	32.5	293	18.2

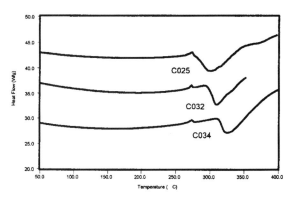

Figure 1. Comparison of OITs between various antioxidant compositions.

According to the OIT analysis, adding antioxidants promotes the sPS stability. By comparing the OIT of sample C028, C034, C032 and C025, it revealed that AO-60 and the mixture of AO-60 and PEP36 provided higher OITs (36 and 39°C) than Irganox 1076 (19°C) owing to their superior thermal stability under processing at higher temperatures. The influence of shear force and processing time on the thermal stability of sPS can be observed from the TD. The thermal stability of sPS without antioxidant (C035, C026, C025, C033 and C027) can be correlated to its molecular weight after processing. The lower the molecular weight implies the shorter the TD, i.e., the easier the degradation. By adding AO-60 or the mixture (PEP-36/AO-60), the thermal stability of sPS (C034, C029) was improved to some extent regardless the

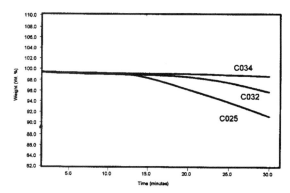

Figure 2. Comparison of TDs between various antioxidant compositions.

molecular weight of sPS. This finding confirms that AO-60 and the mixed antioxidant (PEP-36/AO-60) are effective to sPS.

Referring to the GPC data in Table 2, the M_n and M_w of sPS decreased as the processing time increased. The molecular dispersion became wider as well (C025 vs. C035). This phenomenon could be interpreted by the chain scissions. However, the M_n and M_w were recovered as the processing time lasted even longer (C026). This recovery was owing to the coupling of free radicals derived from the chain scissions. For C029, the likelihood of coupling decreased since the antioxidant eliminated the free radicals. Therefore, the M_n and M_w of C029 were lower than those of C026. Furthermore, the molecular weight dispersion of C034 was wider than that of C028. This finding suggests that the mixed antioxidant (PEP-36/AO-60) provides better stability during processing.

The melting index, MI, yields a similar result as GPC, i.e. the smaller the M_n and M_w imply a higher MI. It indicates that a long processing time affects the process stability. The low MI of C027 was due to degradation as well.

The Yellowness Index, YI, shows no direct relationships between processing time and YI values. At a high screw rate, an increasing of shear force caused a high local temperature and degradation, thereby making the sample dark gray. This phenomenon could be avoid by adding the mixed antioxidant (C033 vs. C030). At a low screw rate, the addition of antioxidant made the sample a yellowish color (C028, C029 vs. C025, C026). Meanwhile, C034 and C032 did not become yellow during the formation of oxidized products at a high temperature.

CONCLUSIONS

- The antioxidant AO-60 and the mixture of AO-60 and PEP-36 not only significantly improved the OIT of sPS, but also inhibited degradation at a high temperature (300°C). They are superior to Irganox 1076 which is commonly used.
- The mixed antioxidant provided better processing stability, while AO-60 lowered the YI value.
- Increasing the processing time caused sPS chain scissions, subsequently caused the molecular weight to decrease and MI to increase. As the processing time became even longer, the coupling reaction occurred and the molecular weight was recovered.
- A high speed shear force caused degradation of sPS. The degradation was prevented by adding the mixed antioxidant.

ACKNOWLEDGMENT

The author would like to thank the Ministry of Economic Affairs of the R.O.C. for financially supporting this work under Project No. 85-EC-2-A-17-0171. Ms. Stacy Huang is appreciated for performing thermal analysis and MI testing.

REFERENCES

1. P. Ravanetti and M. Zino, *Thermo-chimica Acta*, **207**, 53 (1992).
2. V. Kovacevic, M. Bravar, and D. Hace, *Angew. Makromol. Chemie*, **137**, 175 (1985).
3. R. Gacther and H. Muller in **Plastic Additives Handbook**, 2nd Ed., *Hanser*, New York, 1983.

The Performance of Primary and Secondary Antioxidants in Polyolefins Produced with Metallocene Catalysts. Part 1. Preliminary Studies Comparing m-Syndiotactic and Isotactic Polypropylenes

G. J. Klender, R. A. Hendriks, J. Semen, and K. P. Becnel
Albemarle Corporation, Albemarle Technical Center, Baton Rouge, LA 70820, USA

INTRODUCTION

The development of metallocene catalyzed polyolefins has resulted in the introduction of materials that have unique properties which were not obtained with Ziegler-Natta catalyst systems. Polyolefins with uniform insertion of comonomers and with different stereoregularity have been produced on a commercial scale by single site catalyst technologies. These polymers can have densities and crystal morphologies which may affect their solid state properties and behavior. Furthermore, the catalyst systems used to achieve these unique materials have proliferated in both the transition metal compound and co-catalyst used to produce the single site initiation of polymerization.

As a stabilizer producer, we have seen subtle effects with Ziegler-Natta systems which have been attributed to the interactions between catalysts and stabilizers.[1,2] The quenching of the acidity in these systems and the use of phosphites have had an effect, particularly, on the processing color of polyolefins. Also, the crystallinity of the polymer has been shown to play a part in the solubility and blooming characteristics of polyolefins.[3,4] Changes in crystallinity may also affect the long term heat aging of the polymer. The use of zirconium in place of titanium in these catalyst systems raises the question of the ability of these complexes to form colored metal phenolates.[1] Because of the variety of the single site systems, we expect that the responses of these catalysts to the stabilizer system may vary and be less predicable than the response of current Ziegler-Natta systems that are rather similar. The single site systems may respond more like the chromium-based systems for high density polyethylene which tend to be more variable particularly in regard to processing color.

The early introduction of a commercial syndiotactic polypropylene (sPP) by Fina Oil and Chemical[5] gave us an opportunity to look at a zirconium-based material in comparison to a titanium based isotactic polypropylene, iPP, produced by the same manufacturer. While one can argue that in such a study, we are comparing "apples to oranges," it does give us the opportunity to look at a zirconium catalyst system using testing protocols which we used to test the response of many Ziegler-Natta-based systems. Furthermore, since the slow nucleation of syndiotactic polypropylene requires the blending of some isotactic: polypropylene with syndiotactic polypropylene in order to get practical compounding and fabricating conditions, a blend of these materials can be studied in order to determine the effect of blending in regard to stabilization of the system. This paper presents some of our initial work on these systems.

EXPERIMENTAL
MATERIALS. COMPOUNDING AND TEST METHODS

Additives used in this work are identified by their chemical name in the glossary. These are commercial materials, obtained from the manufacturers and used without further purification. The phenolic antioxidant, AO-1, is a high molecular weight, thermally stable, high performance antioxidant that was used throughout this study. It was selected because of the large data base that we have on this antioxidant in all kinds of polyolefins. Its unique FDA clearance as an indirect food additive for all polymers does not limit its use in any new polymer that might be synthesized.

The syndiotactic polypropylene, sPP-1, used in this study was essentially a homopolymer with melt flow index of five and is based on a zirconium chloride metallocene catalyst system in a liquid full loop reactor.[5] The isotactic polypropylenes used in this study were homopolymers produced on a supported Ziegler-Natta catalyst system (third generation) by bulk processes. These materials are identified as iPP-1 (melt flow index of four) and iPP-2 (melt flow index of six).

Procedures for compounding and for multipass extrusion, as well as some polymer test methods, are described elsewhere.[6] The condition for multipass extrusion of isotactic polypropylene, TP-1, was a 180-266-266-266°C profile at 30 rpm which gave a polymer output rate of 12 grams per minute. For syndiotactic polypropylene, the temperature profile had to be lowered to 180-240-240-240°C at 30 rpm, TP-2. While a cold water bath is adequate cooling of the strand for cutting in the compounding and multipass extrusion of isotactic polypropylene, special procedures are needed for syndiotactic polypropylene because of the slow crystallization of this polymer. Syndiotactic polypropylene had to be cooled with 60°C water bath and taken up as a strand on a spool and allowed to cool for about one hour before it was strand-cut. The addition of 20% isotactic polypropylene to the syndiotactic polypropylene increased the crystallization rate to a point where it could be compounded and strand cut. One of us, KPB, devised a continuous procedure for compounding this 20/80 iPP/sPP blend in a twin screw extruder using the normal temperature profile, TP-3, of 180-240-240°C at 40 rpm. The strand from the die was fed into a 60°C hot water bath, then into an ice water bath and directly into the strand cutter.

Melt index measurements were made according to ASTM D1238, Condition L (230°C/2160 g) for polypropylene using a Kayeness Galaxy I Melt Indexer. Color measurements were made on compression molded samples that were 1.5 mm thick using a Hunterlab ColorQUEST Sphere. Yellowness index, YI-1, values are reported.

LONG TERM HEAT AGING

Long term heat aging experiments were performed in a forced convection laboratory oven, Blue M, which is designed to meet ASTM 0-2436. Samples are prepared from compositions that have been compounded on a twin screw extruder under nitrogen to minimize the consumption of the stabilizer during compounding. This material was then compression molded into 0.64 mm plaques that were cut into pieces and placed into glass petri dishes; five pieces for each composition. These were placed in the ovens at both 150°C and 120°C and observed for visible degradation. When three out of the five specimens showed degradation, failure was recorded. Oxygen induction time, OIT, measurements were made on some of the samples before and after aging at 195°C in air according to ASTM method D3895.

RESULTS AND DISCUSSION

EFFECT OF ACID NEUTRALIZERS DURING PROCESSING OF sPP

Titanium residues in Ziegler-Natta systems have been known to react with phenolic antioxidants to produce color.[1] As the catalysts have become more efficient and the amount of residue has diminished, the effect that these residues have on initial color has also diminished to the point where concentrations of titanium catalyst residues below 10 ppm in the polymer are not considered to be significant.[2] The acidity of the polymers has also diminished. In a corrosion test,

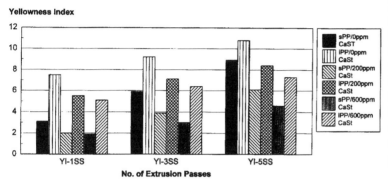

Figure 1. Effect of calcium stearate on processing color in sPP-1 & iPP-1. All samples contain 1000 ppm of AO-1. Multipass extrusion under condition TP-2.

iPP-1 shows a slight staining of a low carbon steel plate which is essentially eliminated with 200 ppm of the acid neutralizer, calcium stearate, CaSt; however, the acid neutralizer at higher concentrations will have a positive effect on color. A similar corrosion test run on sPP-1 also showed a slight staining of the low carbon steel plate which was also essentially eliminated by 200 ppm of calcium stearate. In Figure 1, a comparison of the processing color is made through five extrusions with various concentrations of calcium stearate. The yellowness index of IPP-1 shows high color which increases on multipass extrusion. The addition of 200 ppm of CaSt results in a reduction in the initial color by about 22% which is essentially maintained throughout the five pass extrusions. The addition of 600 ppm of CaSt results in about 33% reduction in color on fifth pass in comparison to fifth pass with no neutralizer at all. This behavior is typical of third and fourth generation Ziegler-Natta, supported catalyst systems, that is, acid neutralizers have a positive effect on color at concentrations greater than that necessary to eliminate detectable acidity. The processing color on first pass extrusion of sPP-1 was lower than iPP-1 but also showed significant color after five passes through the extruder. The addition of 200 ppm of CaSt also resulted in about a 35% decrease in color which remained consistent through five pass extrusions. The addition of 600 ppm of CaSt to sPP-1 also gave about a 49% reduction in color in comparison to no neutralizer at all; therefore, it appears that this zirconium-based sPP polymer is showing the same processing color responses as the titanium iPP polymer.

The melt flow rate of Ziegler-Natta polypropylenes is increased by the addition of calcium stearate. This may be, in part, a lubricity effect of the CaSt; however, inorganic neutralizers, such as synthetic dihydrotalcite (see glossary), have also shown an increase in melt flow with increasing concentration of the neutralizer. In Figure 2, the melt flow rate of iPP-1 increases about 70% during the

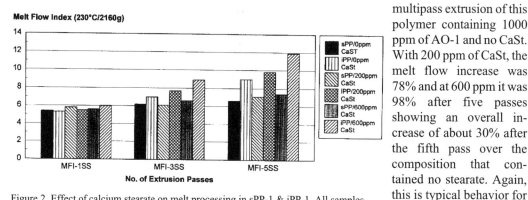

Figure 2. Effect of calcium stearate on melt processing in sPP-1 & iPP-1. All samples contain 1000 ppm of AO-1. Multipass extrusion under condition TP-2.

multipass extrusion of this polymer containing 1000 ppm of AO-1 and no CaSt. With 200 ppm of CaSt, the melt flow increase was 78% and at 600 ppm it was 98% after five passes showing an overall increase of about 30% after the fifth pass over the composition that contained no stearate. Again, this is typical behavior for isotactic polypropylene made by the Ziegler-Natta process.[2] The change in

melt index on five pass extrusions was only 35% for sPP-1 with no CaSt. The addition of 200 ppm of CaSt did not cause any increase in the change in melt flow on multipass extrusion. In fact, the change was only 21% and only 18% for the sample that contained 600 ppm of CaSt. This behavior is clearly different from the behavior in isotactic polypropylene but there is too little data to draw accurate conclusions. However, it is tempting to speculate that the narrower molecular weight distribution and lack of low molecular weight chains in the metallocene polymers are contributing to this effect.

PROCESSING STABILITY WITH PHOSPHITES

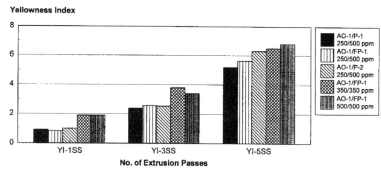

Figure 3. Effect of phosphites on processing color in sPP-1. All samples contain 1000 ppm of AO-1. Multipass extrusion under condition TP-2. All samples contain only 100 ppm of CaSt to maximize color.

Phosphites are used with phenolics in polypropylene to improve the melt stability of the polymer and to improve color. In well neutralized, supported Ziegler-Natta catalyzed polypropylenes, the effects of phosphites on color are generally small.[2] In order to accentuate the processing color, only 100 ppm of CaSt is added as a neutralizer in compounding the polypropyl-

ene. In Figure 3, a low level of AO-1 was used in combination with three different phosphorus stabilizers (P-1, P-2, FP-1) at a 1:2 ratio in this partially neutralized sPP-1. There was very little difference in color after five passes through the extruder between the phosphorus stabilizers but the relative

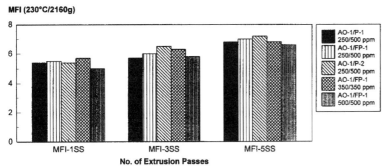

Figure 4. Effect of phosphites on melt processing of sPP-1. All samples contain 1000 ppm of AO-1. Multipass extrusion under condition TP-2. All samples contain only 100 ppm of CaSt to maximize color.

order of least to most color was P-1 < FP-1 < P-2. The better color stability of an aliphatic type phosphite, P-1, followed by the fluorophosphonite, FP-1, and finally the aromatic phosphite, P-2, is the general pattern found in Ziegler-Natta polypropylene compounds.[6] Also, in Figure 3, two concentrations of AO1/FP-1 at a 1:1 ratio were examined. Both showed a higher color after the first pass in comparison to the 1:2 ratio composition, as well as, a little higher color after the fifth pass. This effect of the ratio of the phenolic/phosphite regarding color has been also observed in Ziegler-Natta systems.

In Figure 4, the melt flow behavior of these five AO-1/phosphite compositions on multipass extrusions are shown. Differences in melt flow are very small but in the expected increasing melt index order of P-1 < FP-1 <P-2, which has been found in polypropylenes made by Ziegler-Natta catalyst systems, is also seen here. The 1:1 AO-1/FP-1 blends showed similar processing stabilities as the 1:2 blends. Larger differences might be observed if a higher processing temperature could be used; however, the melt strength of a strand of sPP was insufficient to allow take-up of the polymer at higher stock temperatures.

COMPOUNDING sPP WITH 20% iPP: EFFECTS ON PROCESS STABILITY

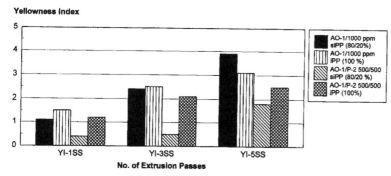

Figure 5. Comparison of processing color of iPP-2 & iPP-2/sPP-1 blend. All samples contain 600 ppm of CaSt. Multipass extrusion under condition TP-1. 20% of iPP-2 was blended with 8-% of sPP-1 before compounding.

Blends of syndiotactic and isotactic polypropylenes and their effects on their crystallization have been studied.[7,8] The blends separate in the liquid phase and on cooling, do not depress their melting points. Consequently, isotactic polypropylene can crystallize first and be a nucleating agent for sPP. However, the rate of crystallization will depend on crystallization temperatures and the rate of supercooling. This will result in a complex crystal morphology in the solid state which is strongly dependent on its thermal history. Since sPP crystallizes faster at a lower supercooling rate, we were able to compound, cool (at a slow rate) and strand cut a blend of 20% iPP with 80% sPP in a continuous manner so that our regular multipass extrusion procedure could be performed on the blend. The effect of stabilizers could be measured and compared to iPP at processing temperatures that have been typical for Ziegler-Natta catalysts.

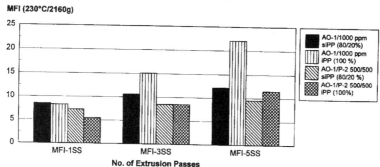

In Figure 5, the yellowness index changes on multipass extrusion of a blend of 20% of iPP-2 with 80% of sPP-1 is compared to 100% iPP by itself. Both formulations contain 1000 ppm of AO-1 and 600 ppm of CaSt. In the first and third passes, the iPP/sPP blend appears to be more stable than the iPP polymer while after five passes, the blend is less stable. This is an anomaly. A blend of sPP-1

Figure 6. Comparison of processing stability of iPP-2 & iPP-2/sPP-1 blend. All samples contain 600 ppm of CaSt. Multipass extrusion under condition TP-1. 20% of iPP-2 was blended with 8-% of sPP-1 before compounding.

and iPP-2, with a stabilizer system consisting of AO-1/P-2 at 500/500 ppm, has the better color throughout the five pass extrusions than the iPP-2 alone.

A melt stability comparison of these compositions is made in Figure 6. The melt stability of AO-1 at 1000 ppm and the combination of AO-1/P2 at 500/500 ppm is better after five passes in the IPP-2/sPP-1 blend than in iPP-2 alone. This reflects the same results as in the comparison of AO-1 in sPP-1 and iPP-1 and shows the sPP-1 to be the more stable polymer in processing.

OVEN AGING OF sPP/iPP BLENDS

The DSC thermogram of the 20/80% blend of iPP-2 and sPP-1 shows the two distinct melting endotherms for the syndiotactic and isotactic crystalline regions of the polymer. The percent crystallinity was calculated based on values in the literature.[9] The blend showed iPP crystallinity of 32.3% compared to 51.4% in 100% iPP-2 and sPP crystallinity in the blend of 22.0% compared to a value of 21.6% in 100% sPP-1. This shows that the isotactic polypropylene in the blend did not develop the same crystallinity that the iPP-2 polymer developed when compounded alone. The slower crystallization rate of the syndiotactic polypropylene allowed it to completely crystallize in the blend under the conditions that were used to cool the blend (hot water bath followed by a cold water bath). Mulhaupt[7] shows that a 20/80 blend of IPP/sPP can result in a dispersed phase of spherulites of isotactic polypropylene in a matrix of needle like syndiotactic polypropylene but that this structure can be distorted greatly by the thermal history. This suggests that the oven aging characteristics of these blends might be affected by their thermal history.

Table 1. Oven Aging of iPP and iPP/sPP.*

Composition AOX/CaSt/Phos Thioester	AOX/CaSt/Phos Conc. ppm	Time to Failure Hrs iPP/sPP**	Time to Failure Hrs iPP
AO-1	1000	48	192
AO-1	1000/1000	108	216
AO-1/CaSt/Fp-1	1000/600/1000	108	192
AO-1/CaSt/DSTDP	600/600/2500	636	1200

*blend at 150°C 0.64 mm samples in circulating air oven
**Blend of 20% iPP-2 and 80% sPP-1

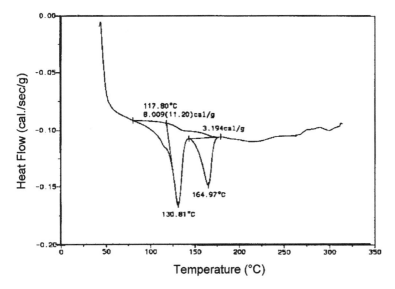

Figure 7. DSC of blend of 80% sPP-1 and 20% of iPP-2. Temperature rise at 10°C/min under nitrogen at 60 cc/min.

Polymers are normally screened in our laboratory for long term heat aging, LTHA, at 150°C and 120°C. While 150°C oven aging is not an applicable test for this sPP polymer because of its lower crystalline melting point, we performed the test anyway in order to see the degradation behavior of the melted polymer. While the iPP/sPP blend melted, it retained its shape and degraded by changing color rapidly in several hours. The iPP, as expected, maintained its shape and degraded by deterioration in the normal manner. Table 1 shows that the iPP had greater oven aging stability than the iPP/sPP blend at 150°C as expected. The phosphorus stabilizer, FP-1, was not effective in oven aging but the DSTDP was effective both in the partially melted iPP/sPP blend and in the pure iPP polymer.

Samples that are being oven-aged at 120°C are still being aged and have not failed. The solid iPP/sPP blends have not discolored. In order to estimate the amount of the stabilizer that had been used up, a comparison was made between an unaged and aged sample by oxygen induction time, OIT. The OIT of AO-1 has been found to be linear for a large range of concentrations in iPP[9] and we have confirmed the linearity of the response of AO-1 to concentration by statistical experimental design.[10] After 1344 hours of exposure, the OIT of composition of 20/80% iPP-2/sPP-1 containing 1000 ppm of AO-1 and 1000 ppm of CaSt had lost 20% of its OIT, while a similar composition based on iPP-2 alone had lost 35% of its OIT. This suggests that a reverse order of stability on oven aging will be found in the solid state for this iPP/sPP blend than was seen at 150°C. Therefore it appears that the sPP blend has better oven aging resistance. However, it is too early to make a conclusive statement on this. It will be necessary to look at the effect of different heat histories to see if the crystallization changes that have been observed in IPP/sPP blends will have an effect on oven aging stability.

CONCLUSIONS

A syndiotactic polypropylene based on a high activity zirconocene chloride catalyst showed similar responses to the phenolic, AO-1, as an isotactic polypropylene based on a high activity titanium chloride catalyst, particularly in regard to processing color. The metallocene-based sPP appeared to be more stable and discolors less during processing. Addition of 20% of the isotactic Ziegler-Natta-based polypropylene to the sPP gave a product that was processed more readily and one that was more stable to chain scission and to discoloration than the iPP alone. OIT data on oven aged samples at 120°C suggests that the iPP/sPP blend is less susceptible to oxidation in the solid state than the iPP polymer but much more data is needed to confirm this finding. Overall, the study provides no surprises with expected low processing color and good processing stability with these stabilizers because of the very low catalyst levels in these polymers. However, one must be careful not to generalize these results because of the very limited data.

ACKNOWLEDGMENT

The authors express their appreciation to Fina Oil and Chemical for their support and for providing the materials that made this study possible. We also acknowledge the efforts of Ms. Jodie Legg for her contribution in obtaining and interpreting the thermal data.

GLOSSARY

AO-1	1,3,5-trimethyl-2,4,6(3,5-di-tert-butyl-4-hydroxybenzyl)benzene
AOX	phenolic antioxidant
Phos	Phosphorous containing stabilizers
P-1	Bis(2,4-di-tert-butylphenyl)pentaerythritol diphosphite
P-2	Tris(2,4-di-tert-butylphenyl)phosphite
FP-1	2,2'-Ethylidenebis(4,6-di-tert-butylphenyl)fluorophosphonite
CaSt	Calcium stearate
Hydrotalcite	$Mg_{4.5}Al_2(OH)_3CO_{13}*3.5H_2O$
DSTDP	Distearyldithiodiproprionate

REFERENCES

1.   G. J. Klender, R. D. Glass, W. Kolodchin, and R. A. Schell, *Proc. SPE ANTEC'85*, **43**, 989 (1985).
2.   G. J. Klender *et al.*, *SPE Polyolefins V RETEC*, 225 (1987).
3.   G. Ligner *et al.*, *SPE Polyolefins VIII RETEC*, 385 (1993).

4.   R. Spatafore, K. Schultz, T. Thompson, and L. T. Pearson, *Polymer Bulletin*, **28**, 467 (1992).

5.   E. S. Shamshoum *et al.*, *Proc. 3rd International Business Forum on Specialty Polyolefins, SPO'93*, 207 (1993).

6.   M. K. Juneau, L. P. J. Burton, G. J. Mender, and W. Kolodchin, *SPE Polyolefins VI RETEC*, 347 (1989).

7.   R. Thomann, J. Kressler, S. Setz, C. Wang and R. Mulhaupt, *Polymer,* **37(13)** 2627 (1996).

8.   R. Thomann, J. Kressler, B. Rudolf, and R. Mulhaupt, *Polymer*, **37(13)** 2635 (1996).

9.   G. R. Hawley *et al.*, *Proc. 3rd International Business Forum on Specialty Polyolefins, SPO'93*, 91 (1993).

10.   T. Schwartz, G. Steiner, and J. Koppelmann, *J. App. Polym. Sci.*, **37**, 335 (1989).

11.   J. T. Books, *unpublished Albemarle data.*

Metallocene Plastomer Modification of Clear Polypropylene for Impact Enhancement

Th. C. Yu and Donna S. Davis
Exxon Chemical Company Baytown, TX 77522, USA

INTRODUCTION

Recently, nucleating agents such as dimethyldibenzylidine sorbitol were introduced to produce clear polypropylenes. Rigid containers with high clarity can now be produced via the injection stretch blow molding process similar to the process for making polyethylene terephthalate, PET, bottles. In the first step of this process, a preform with an integral neck was injection molded. It was then indexed to the conditioning and stretch blow station where the preform was conditioned to reach a uniform temperature. This was accomplished, for example, by allowing the preform to spin at a constant revolution in a constant temperature chamber for a preset time. At the end of the tempering period, a stretch rod was lowered into preform and pressurized air was introduced through orifices on the rod to expand the preform to fill the mold cavity. The finished parts together with the closed mold were indexed to a cooling station, and the molded part was ejected after a predetermined cooling time. A color concentrate such as ClearTint by Milliken Chemical can also be used to produce clear colored containers. Known applications for this types of oriented rigid container are packaging of medicinal tables, baby food jars, baby bottles, and many others where a combination of clarity, stiffness, and drop resistance are essential. Advantages over PET containers made by the same process are a 10% lower weight, and elimination of resin pre-drying.[1] Commercial injection stretch blow molding machines for PET resins such as those made by Aoki, Nissei ASB, and Johnson Controls, can be used to run clear polypropylene resin as well with only minor adjustments. Due to difference in shrinkage of PET and polypropylene, PET tooling may not perform satisfactorily for making clear parts.

Many types of impact modifiers were evaluated in order to improve the drop impact resistance of clear polypropylene without success. Although the impact resistance of the clear polypropylene can be improved, its clarity suffers. Recently we found that metallocene plastomers with a density of 0.900 were effective impact modifiers and when combined with a polypropylene resin would impart very little opacity.[2] Plastomers are a new family of ethylene α-olefin copolymers with a density range from 0.910 to 0.865. These plastomers were produced with a metallocene catalyst[3] which inserted comonomers uniformly along the ethylene backbone, such that at relatively low comonomer incorporation level, the copolymer exhibited both plastic and elastomeric behavior. In this study, we attempted to determine the best plastomer for impact modification of both medium and high melt flow polypropylenes, based on considerations of melt index, MI, and comonomer type. The traditional Ziegler-Natta

Table 1. Plastomers Evaluated

Grade Name	Melt Index dg/min	Density g/cm³	Comonomer Type	Catalyst
EXACT™ 3035	3.5	0.9	Butene	Metallocene
EXACT™ 4042	1.1	0.899	Butene	Metallocene
Affinity PL 1880	1	0.902	Octene	Metallocene
Tafmer A 4085	3.6	0.88	Butene	Ziegler-Natta
Flexomer 1085 NT	0.8	0.884	Butene	Ziegler-Natta

catalyst could be used also to produce plastomers. The performance of two such produced plastomers was compared with those of metallocene plastomers.

EXPERIMENTAL

Several different types of plastomers were selected for this study, and all had a density of approximately 0.900. At this density the reflective indices of polypropylene and plastomer matched well. To evaluate the effect of comonomer type, an ethylene-butene plastomer (EXACT™ 4042) and an ethylene-octene plastomer (Affinity PL 1880) both at about 1 MI were evaluated. The effect of melt index was evaluated by comparing a 1 MI ethylene-butene plastomer (EXACT 4042) with a 3.5 MI ethylene-butene plastomer (EXACT 3035). Two other non-metallocene produced ethylene-butene plastomers were also included in this study. They were a broad molecular weight distribution (Flexomer 1085 NT), and narrow molecular weight distribution (Tafmer A 4085) ethylene-butene copolymers. Details of these plastomers are given in Table 1. The clear polypropylene selected was a 12 melt flow rate, MFR, Escorene PD9374 and a 35 MFR Escorene ID 9465. Both are random polypropylene copolymers with a small amount of ethylene incorporated in the backbone. Since most of the applications for clear polypropylene molded parts required only drop impact resistance at freezer temperature (0°C), the level of modification was therefore selected at 10 and 20 wt% only (see Table 2).

Table 2. Impact Modification of Medium and High Flow Clear Polypropylenes

Sample	wt %	MFR
Clear polypropylene	80, 90	-
Plastomer	20, 10	-
*Escorrene PD 9377	-	12
Escorene PD 9465	-	35

*All blends were prepared using a 30mm ZSK twin screw extruder

A 30 min ZSK twin screw extruder was used to melt blend candidate plastomers into polypropylene base resins. Sample pellets were injection molded into ASTM test specimens as well as 40 mils thickness haze plaques. Direct injection molding of a dry blend of plastomer and polypropylene to produce preforms was feasible, if a mixing screw was used in the injection molding machine. This could easily be accomplished by adding an intensive mixing segment just before the check valve of the injection screw. By adjusting the screw speed and back pressure timing screw retraction, adequate plastomer dispersion could be achieved.

Drop impact resistance was measured using a falling weight impact tester, Ceast Fractovis equipped with a 1/2" hemispherical test dart , and a 3" test anvil. The test specimen used was 3-1/2" and 1/8" thickness injection molded discs. Test speed was set at 4 m/sec so that the kinetic energy loss during testing was less than 10% of the total input energy, and test dart penetrated the test specimen with very little velocity slow down.

RESULTS AND DISCUSSIONS
CLARITY OF MEDIUM MELT FLOW (12 MFR) POLYPROPYLENE

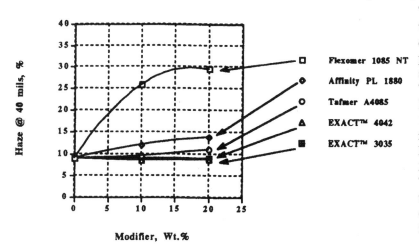

Figure 1. Clarity of medium flow (12 MFR) polypropylene blends.

The clarity of plastomer modified medium melt flow polypropylenes is shown in Figure 1. The broad molecular weight Flexmor 1085 NT showed rapid reduction in haze at both 10 and 20 wt% addition. This was because the Ziegler-Natta catalyst polymerized plastomer had uneven incorporation of comonomer. It not only contained a high comonomer tail, it also contained a segment which was essentially high density polyethylene. This comonomer lean crystalline segment was the cause of haze increase. Both the ethylene-butene EXACT plastomer modified polypropylenes showed essentially the same haze as unmodified polypropylene, and the ethylene-octene. Affinity plastomer showed higher haze increase especially at 20 wt%. Another Ziegler-Natta catalyst polymerized plastomer Tafmer A 4085 showed only slightly higher increase in haze than two EXACT plastomers. This was because the raw solution polymerized Tafmer was subjected to a final solvent washing step for the removal of both low amorphous and high density segments, so that the product was almost identical to a metallocene polymerized plastomer. We may therefore conclude that ethylene-butene plastomer such as EXACT 4042 or 3035 are best choices for impact modification of medium flow polypropylene.

CLARITY OF HIGH MELT FLOW (35 MFR) POLYPROPYLENE

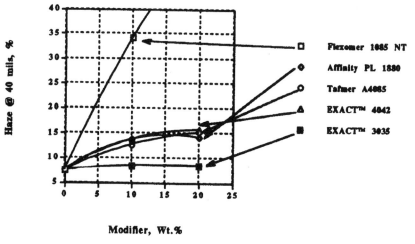

Figure 2. Clarity of high flow (35 MFR) polypropylene blends.

The haze versus blend ratio for the high melt flow polypropylene, PP, blends is shown in Figure 2. As expected, the broad molecular weight Flexomer 1085 showed very high increase in haze due to the presence of many crystalline segments inside the plastomer. It was surprising to find that only EXACT 3035 showed no haze increase over polypropylene base resin at both 10 and 20 wt% addition. Other plastomers, such as EXACT 4042, Affinity PL 1880, and Tafmer A4085 all caused significant haze levels. We may therefore conclude from Figures 1 and 2, that EXACT 3035 is the impact modifier of choice, which not only increases drop impact resistance but also causes virtually no haze increases in polypropylene, ranging in melt flow rates from 12 to 35 MFR.

MELT FLOW SUPPRESSION

Figure 3 shows another interesting observation. The addition of EXACT 3035 to the MFR polypropylene reduced the melt flow rate slightly. On the other hand, a much larger melt flow suppression was observed when the same plastomer is added to the high flow polypropylene.

STIFFNESS IMPACT BALANCE

As shown in Figure 4, the addition of EXACT 3035 to either polypropylene caused a linear reduction in stiffness. The magnitude of stiffness re-

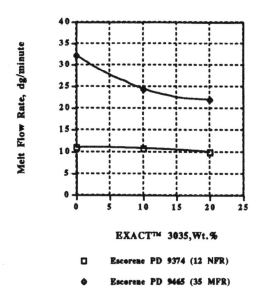

Figure 3. Melt flow rates of EXACT™ 3035 modified polypropylene blends.

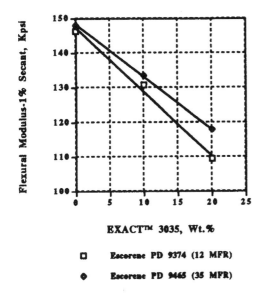

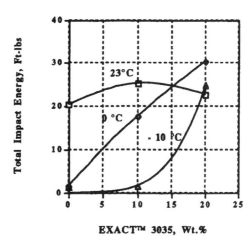

Figure 5. Impact enhancement of medium flow (12 MFR) polypropylene blends.

Figure 4. Stiffness of medium (12 MFR) and high flow (35 MFR) polypropylene blends.

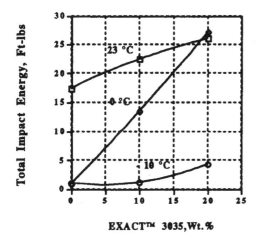

Figure 6. Impact enhancement of high flow (35 MFR) polypropylene blends.

duction can be estimated from Figure 4 to be 2,000 psi per wt% plastomer addition. The rigidity of the molded container, however, can be bolstered by incorporation of horizontal ridges or vertical panels into mold design.

Total impact energy absorbed by the medium flow polypropylene blends is shown in Figure 5. Addition of EXACT 3035 did not increase the impact resistance at room temperature at all. However, a significant increase in impact energy was observed when 10 wt% EXACT 3035 was added to the base resin. At 20 wt% of EXACT 3035, the material was found to be impact proof down to -10°C.

Figure 6 shows the impact improvement for the high melt flow polypropylene. In contrast to the medium flow polypropylene, addition of EXACT 3035 improved impact resistance at both the room and 0°C. Only 10 wt% EXACT 3035 was required to prevent drop failure at freezer temperature. Unlike the medium melt flow polypropylene, more than 20 wt% of EXACT 3035 was required to improve its

impact resistance to -10°C. This was however, seldom required for medical applications such as pipettes and antibiotic storage vials.

CONCLUSIONS

An ethylene-butene plastomer at 0.900 density and 3.5 melt index was found to be an effective impact modifier for both medium and high flow clear polypropylene. Incorporation of EXACT 3035 caused very little increase in haze level. About 10 wt% EXACT 3035 was required to provide drop impact resistance at freezer temperature.

ACKNOWLEDGMENT

The authors would like to thank Kelli Brightwell, Tony Flores, and the Polymer Application Laboratory in Baytown for their support of this study

REFERENCES

1.   S. E. Everett, *Proc. SPE ANTEC*, 1993.
2.   T. C. Yu, *Proc. SPE ANTEC*, 1994.
3.   C. S. Speed, B. C. Trudell, A. K. Metha, and F. C. Stehling, *SPE RETEC Polyolefins VII*, 1991.

Frictional Behavior of Polyethylenes with Respect to Density and Melting Characteristics

Y. S. Kim, C. I. Chung,
Rensselaer Polytechnic Institute, Troy, NY 12180, USA
S. Y. Lai
The Dow Chemical Co., Freeport, TX 77541
K. S. Hyun
The Dow Chemical, Midland, MI 48667, USA

ABSTRACT

The frictional behavior of 12 polyethylene samples with densities ranging from 0.963 down to 0.870 g/cc on a metal surface was studied. The samples with densities higher than 0.908 g/cc behaved as rigid plastics, sliding on the metal surface. The samples with densities lower than 0.905 g/cc behaved as elastomers, strongly adhering on the metal surface and tearing within the polymer. Melting occurred when the metal was heated to a temperature above the melting range of the sample. The frictional behavior of a polyethylene can be understood in terms of the density and melting temperature range of the sample.

INTRODUCTION

The frictional behavior of a polymer on a metal surface is an important material characteristic relevant to processing and product usage. In particular, the frictional behavior is the key factor in the solid conveying along extruder screws near the hopper until the feed material starts to melt. A high solid conveying rate results from a low frictional stress on the screw surface and a high frictional stress on the barrel surface. Thus, a highly polished screw surface is desired. Grooved barrel surfaces are used to aid the solid conveying in special cases.

Recent introduction of the new types of polyethylenes, PEs, produced by single-site

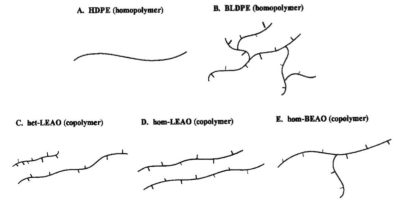

A. HDPE (homopolymer)   B. BLDPE (homopolymer)

C. het-LEAO (copolymer)   D. hom-LEAO (copolymer)   E. hom-BEAO (copolymer)

Figure 1. Molecular structures of different types of PE homopolymers and copolymers.

catalysts such as metallocene catalysts has greatly broadened the density range of commercial PEs.[1] Now, PEs with a density from about 0.970 down to 0.870 g/cc are available. A higher density of PE results from a higher crystallinity, and the rigidity or strength of PE increases as the density increases. Figure 1 shows the molecular structures of different types of PE homopolymers and copolymers. High density PE, HDPE, is a linear homopolymer of ethylene, E, with a very high crystallinity. Irregularities in molecular structure such as short chain branch, SCB, and long chain branch, LCB, cause crystal defects, reducing the crystallinity. The high pressure low density PE is also a homopolymer of E but with lots of SCB and LCB. It is called branched low density PE, BLDPE, because of its LCB. A polymer with SCB but without LCB is classified as a linear polymer. Alternately, the crystallinity of PE can be decreased by polymerizing E monomer with an α-olefin, AO, comonomer such as butene, hexene and octene. AO becomes SCBs along the molecules, reducing the crystallinity. Linear low density PE, LLDPE, is an ethylene/α-olefin copolymer, EAO, with a linear molecular structure. Since the AO comonomers in LLDPE are distributed heterogeneously at nonuniform content among the molecules, LLDPE was called a heterogeneous linear EAO, het-LEAO, in our previous paper[2] in order to allow a better scientific comparison between different types of EAOs. Low density PEs produced by single-site catalysts are also EAOs but with homogeneous distribution of the AO comonomers at uniform content among the molecules. There are two types of homogeneous EAOs, hom-EAOs, homogeneous linear EAO, hom-LEAO, and homogeneous branched EAO with a small amount of LCB, hom-BEAO. EAOs have higher viscosities due to their narrow molecular weight distribution than other PEs of the same melt index, and they are difficult to process. This study is a continuation of our project on the processability of EAOs.[2] The frictional behaviors of PE samples with a wide range of density, covering

Table 1. List of Samples

ID	Name	Producer	Monomers	Density g/cc	MI	DRI	*MR °C Start	*MR °C Peak	*MR °C End
A2	HDPE	Phillips	ethylene	0.963	3.50	N/A	90.8	138.2	145.9
D5	het-LEAO	Dow	ethylene/octene	0.935	1.00	N/A	61.8	125.8	133.8
B1	BLDPE	Exxon	ethylene	0.920	0.85	N/A	23.7	109.4	116.8
D2	het-LEAO	Exxon	ethylene/hexene	0.917	1.04	N/A	27.7	121.8	128.2
D4	het-LEAO	Dow	ethylene/octene	0.912	1.00	N/A	21.7	123.8	129.8
E1	hom-LEAO	Exxon	ethylene/butene	0.910	1.04	0.0	15.1	106.8	113.2
F1	hom-BEAO	Dow	ethylene/octene	0.908	1.04	1.1	15.1	103.8	114.2
F2	hom-BEAO	Dow	ethylene/octene	0.908	0.87	4.4	12.7	104.8	112.5
F3	hom-BEAO	Dow	ethylene/octene	0.908	0.69	14.0	17.7	104.8	112.8
D3	het-LEAO	Dow	ethylene/octene	0.905	0.79	N/A	10.0	121.2	129.5
F4	hom-BEAO	Dow	ethylene/octene	0.870	0.80	2.2	-25.5	55.4	80.8
F5	hom-BEAO	Dow	ethylene/octene	0.870	0.94	3.8	-24.3	61.4	84.8

*Melting Range

various types of commercial PEs, on a metal surface were studied at several temperature, pressure, and speed conditions. Their frictional behaviors are related to their densities and melting characteristics.

EXPERIMENTAL

12 PE samples with densities ranging from 0.963 down to 0.870 g/cc were studied, covering HDPE, BLDPE, het-LEAO, hom-LEAO and hom-BEAO. They are listed in Table 1 in the order of decreasing density, together with their supplier, density, type, melt index, MI, and melting characteristics. The density and, MI, values are the nominal values provided by the suppliers. The melting characteristics were measured using a differential scanning calorimeter, DSC, at 1°C/min heating rate under nitrogen purge.

  The frictional stresses of the samples were measured using an apparatus called Screw Simulator, SS, which was described in detail previously.[3] The SS basically consists of a large steel roll with a Colmonoy surface polished to about 1 μm finish and a freely swinging sample chamber. A sample placed in the sample chamber is pressed against the rotating roll surface. The roll temperature, T_b, sliding speed, U_b, and pressure, P, in experiments can be independently controlled. The frictional stress, τ_f, between the sample and the roll surface is detected by a transducer installed to resist the sample chamber and the signal is recorded on a chart paper. Visual observation of frictional behavior is possible with the SS. The underwater pelletized sample pellets with about 4 mm D at room temperature were placed into the sample chamber, and the gap between the sample chamber and the roll surface was set at the physical minimum of about 0.2 mm to avoid or minimize any leakage of the pellets through the gap. Measurements were made at 30 and 80°C under two speeds and two pressures at each temperature. Two speeds were 23.93 cm/s (9.42 in/s) and 47.88 cm/s (18.85 in/s), corresponding to about 51 and 103 RPM of an 8.89 cm (3.5 in) diameter extruder, respectively. Two pressures were 0.69 MPa (100 psi) and 1.725 MPa (250 psi).

RESULTS AND DISCUSSION

Possible frictional mechanisms of polymers on a metal surface were discussed previously.[3] A polymer sample pressed on a metal surface under a pressure will adhere on the metal surface. The adhesive strength, i.e., the adhesion, depends on the properties (chemical and physical) of the polymer and the metal surface as well as the metal surface

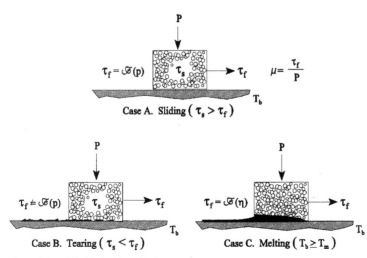

Figure 2. Possible frictional mechanisms.

temperature, the pressure, and the contact time. The contact time is controlled by the sliding speed. If the polymer is strong and the frictional stress is lower than the shear strength, τ_s, of the polymer but exceeds the adhesion, "sliding" will occur at the polymer-metal interface without tearing the polymer as shown in Figure 2A. In this case, the frictional stress increases with increasing pressure, and the frictional coefficient, μ, calculated by dividing the frictional stress by the pressure is meaningful. If the polymer is weak and the adhesion exceeds the shear strength of the polymer, sliding will not occur at the polymer-metal interface but "tearing" of the polymer will occur near the polymer-metal interface as sown in Figure 2B. In such case, the frictional stress reflects the shear strength of the polymer and it virtually does not depend on the pressure, making the frictional coefficient ($\mu = \tau_f /P$) meaningless and misleading. When the metal surface temperature, T_b, is higher than the melting point of the polymer, "melting" of the polymer will occur as shown in Figure 2C and the frictional stress depends on the viscosity of the polymer. An example of the recorded frictional stress is shown in Figure 3 for Sample B1, BLDPE, at 30°C. This recording is typical of the sliding mechanism, and a large oscillation of the frictional stress is observed. Since friction is a stick-slip type failure mechanism, such oscillation is expected and also poor reproducibility is expected.

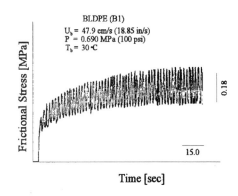

Figure 3. Chart recording of frictional stress.

The rigid pellets with a density above about 0.908 g/cc did not leak out of the sample chamber through the gap between the roll surface and the sample chamber during experiments. However, the soft elastomeric pellets with a density below about 0.905 g/cc leaked out, especially under a high pressure. When pellets leak, the measured frictional stress results from the combined effects of sliding or tearing stress, pressure and internal friction between the pellets. The leakage was so severe for small soft pellets with about 2.5 mm D that the measured data were useless. Furthermore, the SS has a freely swinging sample chamber which must be aligned perfectly on the center line of the roll in order to measure the frictional stress correctly. Any misalignment causes erroneous measurement of the frictional stress. The data presented in this paper should be considered in view of the large probable experimental error and the inconsistent nature of friction mechanisms.

The results of the friction tests are summarized in Table 2. Again, the samples are listed in the order of decreasing density. Both the frictional stress and the frictional coefficient are given, together with the observed frictional behavior. The frictional stress, at a given set of roll temperature, speed and pressure condition is plotted as a function of density in Figures 4-7. The data at two roll temperatures are shown in each figure. Since the frictional coefficient is meaningful only when the sample exhibits the sliding mechanism and it becomes meaningless when tearing or melting occurs, the frictional stress is used in these figures in order to compare all samples at all experimental conditions.

Referring to Table 2 and Figures 4-7, the frictional stress at 30°C and 0.69 MPa (100 psi) increases and the frictional mechanism changes from sliding to tearing as the density decreases. The rigid samples with a density above about 0.908 g/cc exhibited the sliding mechanism and developed lower

Table 2. Results of Friction Tests

A. $T_b = 30°C$; $U_b = 23.93$ cm/s; $P = 0.69$ MPa

ID	Density g/cc	Fractional Stress MPa	Fractional Coefficient	Fractional Behavior
A2	0.963	0.129	0.186	sliding
D5	0.935	0.248	0.361	sliding
B1	0.920	0.460	0.708	sliding
D2	0.917	0.441	0.642	sliding
D4	0.912	0.460	0.663	sliding
E1	0.910	0.547	0.795	sliding
F1	0.908	0.520	0.755	sliding
F2	0.908	0.524	0.762	sliding
F3	0.908	0.451	0.655	sliding
D3	0.905	0.902	1.299	grinding & tearing
F4	0.870	1.003	1.452	tearing
F5	0.870	0.864	1.252	tearing

Table 2. Results of Friction Tests

B. $T_b = 30°C$; $U_b = 23.93$ cm/s; $P = 1.724$ MPa

ID	Density g/cc	Fractional Stress MPa	Fractional Coefficient	Fractional Behavior
A2	0.963	0.294	0.170	sliding
D5	0.935	0.547	0.317	sliding
B1	0.920	1.182	0.685	sliding
D2	0.917	1.053	0.610	sliding grinding
D4	0.912	1.076	0.624	sliding
E1	0.910	0.938	0.544	sliding grinding
F1	0.908	1.118	0.647	sliding grinding
F2	0.908	1.126	0.653	sliding grinding
F3	0.908	1.177	0.680	sliding grinding
D3	0.905	1.683	0.972	tearing
F4	0.870	1.260	0.731	tearing
F5	0.870	1.205	0.699	tearing

Table 2. Results of Friction Tests

C. $T_b = 30°C$; $U_b = 47.88$ cm/s; P = 0.69 MPa

ID	Density g/cc	Fractional Stress MPa	Fractional Coefficient	Fractional Behavior
A2	0.963	0.138	0.200	sliding
D5	0.935	0.276	0.400	sliding
B1	0.920	0.644	0.936	sliding grinding
D2	0.917	0.445	0.653	sliding grinding
D4	0.912	0.455	0.659	sliding
E1	0.910	0.561	0.813	sliding grinding
F1	0.908	0.552	0.799	grinding & tearing
F2	0.908	0.565	0.819	sliding grinding
F3	0.908	0.478	0.695	sliding grinding
D3	0.905	0.745	1.079	tearing
F4	0.870	0.809	1.173	tearing
F5	0.870	0.763	1.106	tearing

Table 2. Results of Friction Tests

D. $T_b = 30°C$; $U_b = 47.88$ cm/s; P = 1.724 MPa

ID	Density g/cc	Fractional Stress MPa	Fractional Coefficient	Fractional Behavior
A2	0.963	0.281	0.163	sliding
D5	0.935	0.607	0.352	sliding
B1	0.920	1.159	0.668	grinding
D2	0.917	1.094	0.635	sliding grinding
D4	0.912	0.993	0.576	sliding grinding
E1	0.910	1.269	0.735	grinding & tearing
F1	0.908	1.159	0.672	grinding & tearing
F2	0.908	1.177	0.683	slight grinding
F3	0.908	1.159	0.672	sliding grinding
D3	0.905	1.444	0.837	tearing
F4	0.870	0.993	0.576	tearing
F5	0.870	1.049	0.608	tearing

Table 2. Results of Friction Tests

E. $T_b = 80°C$; $U_b = 23.93$ cm/s; $P = 0.69$ MPa

ID	Density g/cc	Fractional Stress MPa	Fractional Coefficient	Fractional Behavior
A2	0.963	0.138	0.201	sliding
D5	0.935	0.267	0.386	sliding
B1	0.920	0.538	0.782	grinding
D2	0.917	0.331	0.481	sliding
D4	0.912	0.405	0.586	sliding
E1	0.910	0.497	0.722	grinding
F1	0.908	0.428	0.621	slight grinding
F2	0.908	0.405	0.588	grinding
F3	0.908	0.395	0.575	grinding
D3	0.905	0.570	0.829	sliding
F4	0.870	0.373	0.540	melting
F5	0.870	0.373	0.540	melting

Table 2. Results of Friction Tests

F. $T_b = 80°C$; $U_b = 23.93$ cm/s; $P = 1.724$ MPa

ID	Density g/cc	Fractional Stress MPa	Fractional Coefficient	Fractional Behavior
A2	0.963	0.313	0.181	sliding
D5	0.935	0.616	0.358	sliding
B1	0.920	0.901	0.519	grinding
D2	0.917	0.607	0.352	sliding
D4	0.912	0.731	0.422	sliding
E1	0.910	0.938	0.544	grinding
F1	0.908	0.920	0.533	grinding & tearing
F2	0.908	0.855	0.496	grinding & tearing
F3	0.908	0.855	0.496	grinding
D3	0.905	0.902	0.522	tearing
F4	0.870	0.441	0.256	melting
F5	0.870	0.506	0.290	melting

Table 2. Results of Friction Tests

G. $T_b = 80°C$; $U_b = 47.88$ cm/s; $P = 0.69$ MPa

ID	Density g/cc	Fractional Stress MPa	Fractional Coefficient	Fractional Behavior
A2	0.963	0.143	0.207	sliding
D5	0.935	0.299	0.433	sliding
B1	0.920	0.441	0.642	grinding
D2	0.917	0.345	0.500	sliding
D4	0.912	0.372	0.639	sliding
E1	0.910	0.451	0.653	grinding
F1	0.908	0.524	0.759	grinding & tearing
F2	0.908	0.460	0.666	grinding & tearing
F3	0.908	0.460	0.666	grinding & tearing
D3	0.905	0.570	0.826	sliding
F4	0.870	0.368	0.533	melting
F5	0.870	0.331	0.480	melting

Table 2. Results of Friction Tests

H. $T_b = 80°C$; $U_b = 47.88$ cm/s; $P = 1.724$ MPa

ID	Density g/cc	Fractional Stress MPa	Fractional Coefficient	Fractional Behavior
A2	0.963	0.308	0.179	sliding
D5	0.935	0.699	0.406	sliding
B1	0.920	0.823	0.477	tearing
D2	0.917	0.685	0.398	sliding grinding
D4	0.912	0.828	0.480	sliding grinding
E1	0.910	0.805	0.467	tearing
F1	0.908	0.943	0.547	tearing
F2	0.908	0.869	0.504	tearing
F3	0.908	0.842	0.488	tearing
D3	0.905	0.902	0.523	tearing
F4	0.870	0.446	0.259	melting
F5	0.870	0.451	0.261	melting

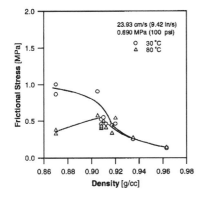

Figure 4. Frictional stress vs. density at 23.93 cm/s and 0.69 MPa.

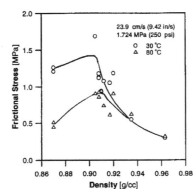

Figure 5. Frictional stress vs. density at 23.93 cm/s and 1.724 MPa.

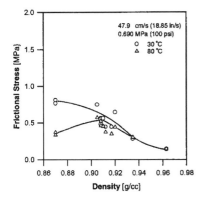

Figure 6. Frictional stress vs. density at 47.88 cm/s and 0.69 MPa.

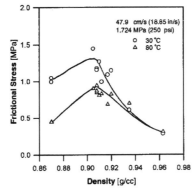

Figure 7. Frictional stress vs. density at 47.88 cm/s and 1.724 MPa.

frictional stresses. The sliding mechanism may involve a fast wear of the sample, producing large fragments of the polymer on the metal surface. Such mechanism will be called "grinding". The elastomeric samples with a density below about 0.905 g/cc exhibited the tearing mechanism and developed higher frictional stresses. When the pressure is increased, the frictional stress increases almost proportional to the pressure for the rigid samples but far less for the elastomeric samples. The frictional coefficients in Table 2 clearly prove such finding. The calculated frictional coefficient decreases only slightly for the rigid samples, but it decreases drastically for the elastomeric sam-

ples. When the speed is increased, the frictional stress increases slightly for the rigid samples but decreases slightly for the elastomeric samples.

The dependence of the frictional behavior of the samples on the roll temperature can be explained by the melting characteristics of the samples. Referring to Table 1, the melting range in general shifts to a lower temperature and becomes wider with a less pronounced peak as the density decreases. The melting characteristics, compared at a similar density, are controlled by the molecular composition and structure. Referring to Figure 1 and Table 1, het-LEAOs (D samples) melt at a higher temperature range than BLDPE, B1, or hom-EAOs (E and F samples) with a similar density. The DSC curves for Samples D3 and E1 are shown in Figure 8 as an example. Both samples melt over a wide range of tem-

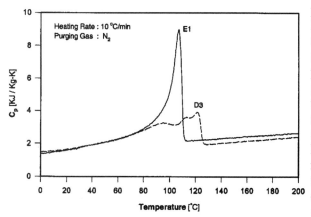

Figure 8. DSC thermograms of samples D3 and E1.

perature. The melting range of Sample D3 with 0.905 g/cc density starts at about 10°C and ends at about 130°C with a very broad peak at about 121°C. Sample E1 with a higher density of 0.910 g/cc than Sample D3 melts at a lower temperature range than Sample D3, starting at about 15°C and ending at about 113°C with a peak at about 107°C. Due to their higher melting temperature range, the temperature dependence of frictional behavior of het-LEAOs is comparable to those of hom-EAOs with a slightly higher density. If the roll temperature is below the starting point of the melting range of the sample, the frictional mechanism is sliding. If the roll temperature is above the ending point of the melting range, the frictional mechanism is melting. If the roll temperature is within the melting range making the polymer sufficiently soft near the roll surface, the frictional mechanism is tearing.

CONCLUSIONS

The density and melting characteristics of a PE are controlled by the composition and molecular structure. A PE with a higher density has a higher crystallinity and rigidity with a melting peak at a higher temperature. The frictional behavior of a PE can be understood by the density and melting characteristics. PEs with a density above about 0.908 g/cc behave as rigid polymers and exhibit the sliding mechanism, developing lower frictional stresses. PEs with a density below about 0.905 g/cc behave as soft elastomers and exhibit the tearing mechanism, developing higher frictional stresses. The new types of LDPEs, hom-EAOs, with lower densities produced by single-site catalysts develop high frictional stresses and melt at lower temperatures. These new hom-EAOs, compared at the same density, melt at a lower temperature range than het-LEAOs. They are expected to give a higher solid conveying capacity along a screw in extrusion than other LDPEs due to their higher frictional stresses. Thus, a screw with a less feeding depth and a lower compression ratio is recommended for these new LDPEs.

REFERENCES

1.   S. Y. Lai, T. A. Plumley, T. I. Butler, G. W. Knight, and C. I. Kao, *SPE-ANTEC*, **40**, 1814 (1994).
2.   Y. S. Kim, C. I. Chung, S. Y. Lai, and K. S. Hyun, *SPE-ANTEC*, **41**, 1122 (1995).
3.   C. I. Chung, W. Hennessey, and M. Tusim, *Polym. Eng. Sci.*, **17(1)** 9 (1977).

Crystallinity Dependence of Modulus and Yield Properties in Polyethylenes

J. Janzen and D. F. Register
Phillips Petroleum Co., USA

ABSTRACT

A large historical database of legitimately comparable deformation test results on various commercial and developmental polyethylenes has been carefully examined for information about how mechanical performance characteristics of these resins depend on molecular structure and semicrystalline morphology.

It was found that flexural moduli of specimens consistently prepared according to fixed protocols generally scale with density, irrespective of molecular weight, molecular weight distribution breadth, branching type and placement, etc., to within the statistical uncertainties in the data. A similar conclusion holds for values of strain and stress at yield; but the situation is less simple in the case of larger-strain failure properties.

INTRODUCTION

The question of the quantitative relationship between stiffness and crystalline content in polyethylenes was considered in detail in a pair of earlier publications[1,2] by one of us. The approach was to compare a large collection of modulus data with several composite micromechanics-based theoretical formulations available from the literature, in order to decide which of those formulations, if any, would agree satisfactorily with observations made on polyethylenes differing widely in density, molecular weight, and so forth. It was concluded that only the two-phase modulus mixing rule known[3] as the "self-consistent scheme", SCS, originally derived by Hill[4] and Budiansky,[5] and subsequently rederived by Berryman[6] via an independent approach, satisfactorily captures the density dependence of stiffness in unoriented semicrystalline polyethylenes (at ambient conditions). By far the largest part of the data considered were Young's moduli measured in flexural mode, but the studies did also incorporate the simultaneous coherent treatment of some data on shear and bulk moduli, sonic velocities, Poisson's ratios, etc., and did show that such characteristics can be well accounted for using only the selected mixing scheme along with plausible values for material constants of the crystalline and amorphous phases as inputs.

A question that could not be answered using the data assembled for the earlier studies was: Are apparent differences in modulus between resins of equal densities real effects attributable to geometric details of morphology, molecular architecture, and the like, or are they merely testing variations? This question remained because the data came from many sources, and inevitably entailed variations in measurement techniques, specimen preparations, etc. We have now amassed another sizable data set of flexural moduli and densities, all of which were determined at one location, with consistency in specimen preparation and measurement technique; it therefore seems appropriate to revisit the subject, and to attempt to refine and extend the earlier conclusions.

THEORETICAL BACKGROUND

A convenient working form shown previously[1] to represent the Young's modulus, E, obtained from the full solution to the self-consistent system of simultaneous equations is

$$E = \frac{E_a E_c + \lambda E(\varphi_a E_a + \varphi_c E_c)}{\lambda E + \varphi_c E_a + \varphi_a E_c} \qquad [1]$$

where λ is a function of the composite Poisson's ratio, ν, that varies so mildly that it is adequate to replace it by a constant with a value near to 3ν. The subscripted E's are effective (constant) isotropic Young's moduli for the amorphous and crystalline components, and the subscripted φ's are the corresponding volume fractions, which must sum to 1. If λ is constant, a rearrangement of Eq 1 that gives E explicitly is

$$E = \frac{h + \sqrt{h^2 + 4\lambda E_a E_c}}{2\lambda} \qquad [2]$$

where

$$h = E_a[\varphi_a(1+\lambda) - 1] + E_c[\varphi_c(1+\lambda) - 1] \qquad [2a]$$

Crystalline volume fractions are obtained from measured densities, ρ, by means of the usual dilatometric equation

$$\varphi_c = (\rho - \rho_a)/(\rho_c - \rho_a) \qquad [3]$$

here we take for the room-temperature crystalline and amorphous phase densities values of 1.010 and 0.852 g/cm^3, respectively, as discussed in the Appendix to (1).

Thus Eqs 2 and 3 compactly express a well grounded theoretical model for the density dependence of Young's modulus in terms of three semi-empirical parameters: E_a, E_c, and λ.

METHODS AND DATA SOURCES

We proceed by examining how well this model agrees with experimental data, allowing for plausible adjustments in values of the semi-empirical parameters. To do this in a rational manner called for application of a nonlinear least-squares optimization technique that is a generalization of the optimizations known for linear models as York[7] or major-axis regression; the key feature is to take into account experimental error known to be present in all the variables, not only a "dependent" one. The algorithm we use to do this will be detailed separately.[8]

Among several in-house databases available to the authors, one compilation contains density and mechanical test results obtained during the period 1984-1992 for all the distinct commercial grades of polyethylene manufactured by Phillips during that time. We excluded from consideration all grades compounded with fillers or colorants, etc., and were left with more than 60 distinct products for which enough measurements were present to allow their errors to be estimated. Resins made with both titanium halide-type and chromium oxide-type catalysts were represented. All measurements were made on compression-molded specimens prepared according to ASTM Standard Method D 1928. Densities were measured by density-gradient techniques according to D 1505. Flexural moduli and tensile yield data were obtained by methods almost indistinguishable from D 790 and D 638, respectively.

The average number of density determinations for each product was 40. The average number of flex modulus determinations for each product was 22, and for tensile yield stress it was 20; and each mechanical test "determination" is a mean for quintuplicate specimens. Thus our digested data are based on more than 16,000 individual test pieces.

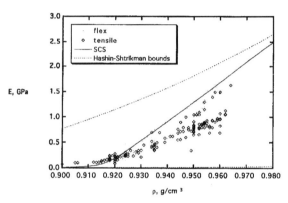

Figure 1. Summary of modulus-density results from Ref. 1.

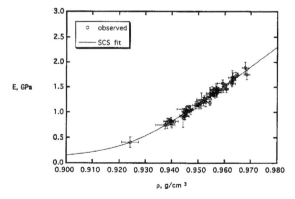

Figure 2. Polyethylene flexural modulus vs. density (product averages, room temperature).

RESULTS AND DISCUSSION

Figure 1 is a redrawn version of Figure 14 from Ref. 1; it illustrates the variability, at a given density, of results that are obtained by different laboratories using sometimes different techniques. Figure 2 shows our present flex modulus-density data, where each point coordinates are the averages for one product type. Thus the error bars shown represent standard variation not just from measurement error alone, but also include lot-to-lot and year-to-year variability in the manufactured product. Although these slight material variations are present in the data of Figure 2, it is clear that removing the interlaboratory testing variability, relative to Figure 1, brings about a great simplification of the picture, and in fact pushes any real effects on flexural modulus from factors other than density below the threshold of detectability by the present sort of measurement.

Scrutiny of the database also turned up a second property that appeared to have a simple dependence on density (or crystalline volume fraction): tensile elongation at yield. The available data, however were based on crosshead position, hence were not equivalent to true strains. We therefore turned attention to a much

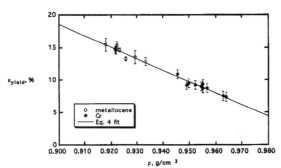

Figure 3. Yield strain vs. density, recent extensometer data.

smaller dataset obtained with a very recently acquired contact extensometer. Although the results are not nearly as voluminous as those in our commercial-product database, they span a respectably wide density range; here the resins represented were made by both Cr and metallocene catalysts, and differed appreciably in average molecular weights and breadths of distributions.

The extensometric yield strain results are plotted against densities in Figure 3, along with error bars indicating estimates of the standard errors of testing. (These are single-material determinations, not product averages.) These results are well represented by the simplest possible model, based only on the notion that the crystalline and amorphous fractions might contribute to the yield strain in proportion to their amounts, i.e.,

$$\varepsilon_{yield} = \varepsilon_a \varphi_a + \varepsilon_c \varphi_c \qquad [4]$$

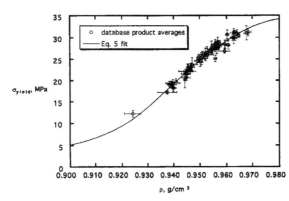

Figure 4. Yield stress vs. density.

where ε_a and ε_c are constants to be determined by curve fitting. The continuous curve in Figure 3 is the result of such fitting. The values obtained for ε_a and ε_c are such that the ratio $\varepsilon_c/\varepsilon_a$ does not differ significantly from the ratio of compliances, E_a/E_c.

We now turn back to the large database and consider product-average yield stress data, plotted in Figure 4. A working model for this property is simply constructed as follows: We write the yield stress as the product of the yield strain given by Eq 4 and a so-called "secant" modulus, which is the slope of the line from the origin to the yield point in the stress-strain curve. We then assume that this secant modulus is a constant fraction b of the SCS Young's modulus E given by Eq 2, and obtain the value of this fraction by curve fitting. Thus the curve through the data of Figure 4 represents

$$\sigma_{yield} = (\varepsilon_a \varphi_a + \varepsilon_c \varphi_c) bE \qquad [5]$$

where the constants ε_a and ε_c keep the same values as in Figure 3, and b is approximately 0.2. Again the property is remarkably well represented by the simple model depending only on density, with, in practical terms, almost no variability left over to be explained by other structural features.

CONCLUSIONS

Our conclusion from this database mining exercise is that for similarly prepared isotropic specimens of a wide range of semicrystalline commodity polyethylenes, the conventionally measured small-strain deformation properties are dominated almost entirely by the crystalline content as reflected in the density. Additional morphological and molecular structural characteristics become important in the larger-deformation properties, where we did not find such simple representations as the ones we have shown here.

ACKNOWLEDGMENT

The authors gratefully acknowledge the efforts of coworkers far too numerous to mention; without their having synthesized and tested the many thousands of resins upon which this study is based, the work would not have been possible. Particular thanks are due to Buddy Dillard, the original compiler of the body of data we utilized.

REFERENCES

1.   J. Janzen, *Polym. Eng. Sci.*, **32**, 1242 (1992).
2.   J. Janzen, *Polym. Eng. Sci.*, **32**, 1255 (1992).
3.   J. P. Watt, G. F. Davies, and R. J. O'Connell, *Rev. Geophys. Space Phys.*, **14**, 541 (1976).
4.   R. Hill, *J. Mech. Phys. Solids*, **13**, 213 (1965).
5.   B. Budiansky, *J. Mech. Phys. Solids*, **13**, 223 (1965).
6.   J. G. Berryman, *Appl. Phys. Lett.*, **35**, 856 (1979).
7.   D. York, *Can. J. Phys.*, **44**, 1079 (1966).
8.   D. F. Register and J. Janzen, *Non-linear least squares analysis of data with experimental uncertainties in all variables, to be published.*

Processing Characteristics of Metallocene Propylene Homopolymers

C. Y. Cheng and J. W. C. Kuo
Exxon Chemical Company, Baytown , TX, USA

ABSTRACT

Propylene polymers synthesized from metallocene catalyst systems have different rheological, thermal and morphological properties compared with conventional polypropylene made from the Ziegler-Natta catalyst systems. These properties cause the polymer to behave differently in a plasticating extruder and the subsequent fabrication processes. The first generation of metallocene propylene polymers has a narrow MWD and composition distribution as well as low crystallinity. The impact of these attributes in extrusion, fiber spinning and film casting is discussed.

INTRODUCTION

In recent years, polypropylene, PP, produced in metallocene catalyst systems has received great attention from polymer manufacturers and fabricators. Exxon Chemical announced the commercialization of metallocene-based isotactic propylene polymers, mPP, with the trade name ACHIEVE™ in October 1995. Since then, many other resin manufacturers have accelerated their development of metallocene-based propylene polymers. At the same time, fabricators have started commercial evaluation of this new family of polymers.

Among the key attributes found in the first generation of metallocene-based propylene polymers are narrow molecular weight distribution and composition distribution, lower crystallinity, and lower extractables.[1] These properties lead to significant differences in extrusion behavior and fabrication processes as well as end-product properties.

KEY PROCESSING RELATED PROPERTIES

RHEOLOGICAL PROPERTIES

The rheological properties of the resin are closely related to molecular weight, MW, and molecular weight distribution, MWD. The narrow MWD mPP exhibits lower shear viscosity in the low shear rate region and high viscosity at shear rates higher than 10 1/sec. The reduced viscosity in the low shear region is generally correlated with low elongational viscosity. This is advantageous for post extrusion processes where low extensional flow resistance is desirable. For example, fiber spinning processes require low elongational viscosity in order to draw to a finer fiber diameter. The low extensional viscosity is also favorable for spinning at a higher line speed.

Figure 1 compares the shear viscosity of a conventional resin with two metallocene resins over a shear rate range of 10 to 10000 1/sec at 230°C. The conventional resin is a narrow MWD grade

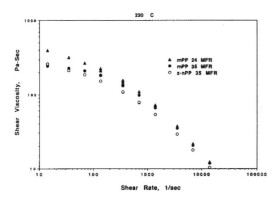

Figure 1. Shear viscosity comparison.

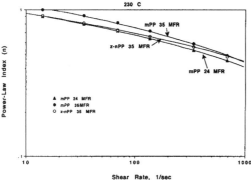

Figure 2. Power-law index comparison.

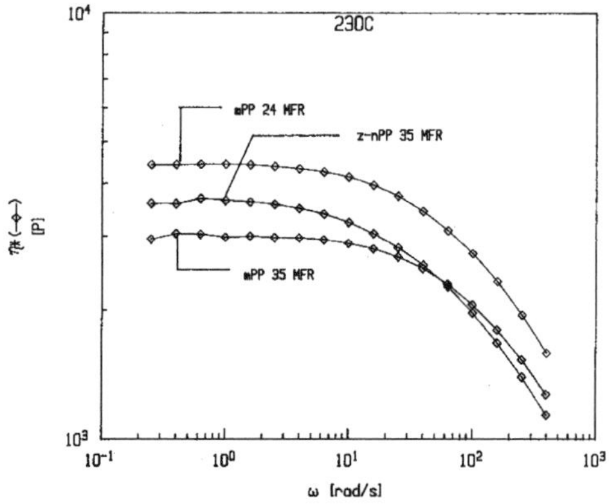

Figure 3. Complex viscosity comparison.

obtained by the visbreaking or controlled rheology, CR, process where a lower MFR resin is chemically broken down to a higher MFR to reduce MWD. The conventional Ziegler-Natta polypropylene, z-nPP, has a MFR of 35 while the two metallocene resins have MFR of 35 and 24. The z-nPP has a MWD of 2.3 while the metallocene propylene polymers has a MWD of 1.84.

Note in Figure 1 that in the high shear region, the mPP has a higher viscosity than the conventional resin. In the low shear rate region of below 20 1/sec shear rate, the reverse is true. This behavior can be further demonstrated by plotting the power-law index vs. the shear rate as shown in Figure 2. Note that the 35 mPP has a higher power-law index than z-nPP at shear rates up to 1000 1/sec.

Extending the low shear measurement using a dynamic cone-and-plate viscometer, the differences are more pronounced. Figure 3 compares the complex viscosity of the three resins at frequencies between 0.1 and 1000 rad/sec. Note that the viscosity of z-nPP resin eventually reaches the Newtonian plateau, but at a higher viscosity level and much lower frequency than mPP of the same MFR. The viscosity of the lower MFR mPP levels at a frequency higher than the z-nPP.

Although the elongational viscosity for low viscosity polymers at high extensional rate is very difficult to measure and quantify, the Newtonian plateau viscosity illustrated in Figure 3 implies that mPP has a lower elongational viscosity compared to the z-nPP of the same MFR. This attribute is advantageous in processes such as fiber spinning and cast film where polymer is subjected to extensional flow.

THERMAL PROPERTIES

Thermal properties of mPP are different from z-nPP due to the degree of crystallinity and type of crystalline structure. The occasional reversal of head-to-tail polymer chain growth (i.e., regio defects) hinders crystalline growth. The melting peak temperature is therefore, lower than conventional PP by approximately 8-10°C. The crystallization peak temperature for mPP is generally at 100°C to 101°C vs. the conventional resin at 105°C to 106°C.

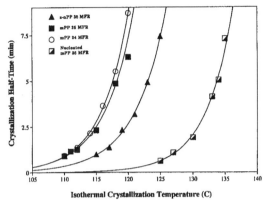

Figure 4. Isothermal crystallization half-time.

The difference in molecular structure also results in crystallization behavior differences. Figure 4 illustrates the isothermal crystallization half-times vs. temperature for mPP and conventional CR PP. At a constant temperature, the crystallization half time of mPP is approximately twice that of the conventional resin. However, since the crystallization temperature of the metallocene is approximately 5°C lower than the conventional resin, the 5°C difference should be taken into consideration when comparing isothermal crystallization. This would be a more accurate comparison since the driving force of crystallization is the temperature difference between the supercooling temperature and the crystallization temperature. In fact, Figure 4 shows the same crystallization half-time when the 5°C difference is taken into consideration.

Similar to conventional PP, an interesting crystallization behavior is demonstrated by the behavior of the metallocene resin in the presence of a nucleating agent. The crystallization half-time is vastly reduced as a result of providing nucleating sites.

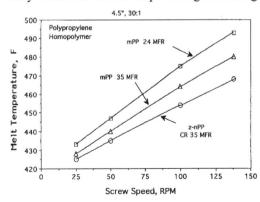

Figure 5. Melt temperature comparison.

EXTRUSION CHARACTERISTICS

In single screw extrusion, the shear rate is quite high. For example, the shear rate in the metering section of the screw can be well above 100 1/sec. The high shear region such as the melting film between the unmelted pellet and barrel, the clearance of the barrier flight as well as the Maddock mixing section, can have a shear rate well over 1000 1/sec.

Referring to Figure 1, the melt viscosity for the metallocene resin is higher than the conventional polymer of the same MFR. Therefore, one can expect the viscous shear heating in a plasticating extruder to be higher for the metallocene resin. The higher torque and extruder motor amperage load arise from the greater viscous shear heating.[2] The higher energy input into the polymer through viscous shear heating results in higher melt temperatures under similar barrel settings. Figure 5 shows the melt temperature differences for a 4.5" extruder processing mPP and z-nPP under the same processing condition.

Although the low shear sensitivity of the narrow MWD mPP results in higher extrusion torque and melt temperature, it does not pose a limitation in extrusion performance. Unlike metallocene polyethylene where most grades are in low MI (high MW), polypropylene grades are generally much lower in MW (higher MFR) and extrusion torque is well within the extrusion system capability.

FIBER SPINNING PROCESS
EXTRUSION

In fiber extrusion, the output rate is controlled by a melt pump speed. The extruder screw speed is regulated by the pump inlet pressure, which is generally maintained constant at 750 to 1500 psi. As the output rate increases by raising the melt pump speed, the screw speed is automatically increased to maintain a constant pump inlet pressure. While the pump speed is kept constant, the screw speed may fluctuate somewhat. A constant output rate is a must for fiber spinning in order to control the fiber size and maintain fiber diameter uniformity. The fiber size is controlled by the fiber take-up speed and output rate per capillary hole (e.g., gram/hole/min).

Since the capillary holes in the spinneret are only 0.4 to 0.6 mm in diameter, the shear rate of the capillary flow is relatively high. Therefore, the higher melt viscosity of the metallocene PP at high shear rate results in higher spinneret (die plate) pressure. Although the spinneret pressure of the metallocene is higher than conventional resin, it is still well within the acceptable pressure range. If a lower spinneret pressure is required, a larger capillary diameter spinneret or higher melt temperature can be used to reduce the spin pack pressure.

In general, the metallocene PP can be extruded at the same conditions as the conventional PP. However, the melt temperature will be somewhat higher than the conventional resin of the same MFR. A lower barrel temperature setting of 3°C to 5°C can be used to compensate for the higher viscous shear heating by reducing the external heat input at the melting section and increasing the cooling at the metering section of the screw.

FIBER SPINNING

The desired melt temperature is dictated by the fiber drawing and quenching requirements. In the conventional spinning process where the fiber speed is controlled by the fiber take-up velocity, a lower melt temperature increases the stress at the spin line and greater fiber orientation is obtained. For a spunbond process, the fiber draw force (or the attenuation air velocity) is constant. Therefore, a lower melt temperature increases extensional viscosity of the polymer, resulting in a larger diameter fiber. On the other hand, attempts to reduce fiber size by using a higher melt temperature may encounter fiber quenching difficulties, leading to fibers sticking together because of a longer quenching distance.

The single-sited metallocene catalyst produces very uniform molecular species. The MW of each polymer chain is very similar (and hence the narrow MWD product). Therefore, there is a very low MW fraction that volatilizes during extrusion. The emission of low molecular species during fiber

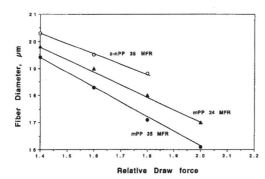

Figure 6. Fiber size comparison.

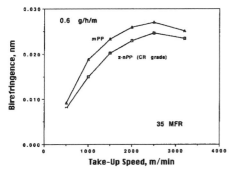

Figure 7. Birefringence comparison.

spinning is practically eliminated for mPP resin. This has a positive long-term impact on productivity (reduced down-time and improved safety) and environmental cleanliness.

FIBER PROPERTIES

The low elongational viscosity as measured by the low Newtonian plateau viscosity is advantageous for constant draw force spinning such as a spunbond process. Under a given draw force, the fiber diameter of mPP resin is smaller than the conventional resin of the same MFR extruding under the same melt temperature. The difference is greater as the draw force is increased. Figure 6 compares the fiber size of mPP and z-nPP.

Molecular orientation in the fiber spinning process is strongly dependent on the spin line stress near the crystallization point.[3] Being able to reach a finer fiber diameter increases the stress near the crystallization region and therefore, molecular orientation is enhanced. The narrow MWD further enhances the tendency of molecular alignment during extensional flow.

Figure 7 compares the molecular orientation of mPP and z-nPP of the same MFR as measured by birefringence. The fibers were spun at different speeds under constant capillary output rate. As expected, the differences are greater between the two polymers as the extensional rate increases. The higher molecular orientation leads to higher tenacity and lower elongation of the fiber as expected.

CAST FILM PROCESS

EXTRUSION

Cast film extrusion experiment was conducted on a 3.5", 30:1 L/D extruder using the same screw design and temperature settings. Resin samples of 35 MFR reported in the rheological and thermal property sections of this paper were selected to study the extrusion performance and drawdown capability.

At the same output rate of 300 lbs/hr, the screw speed for mPP was 108.7 rpm while the conventional PP was 115.7 rpm. That is, the metallocene PP has a higher specific output rate than the conventional CR resin. The same results have been reported earlier.[4] The most likely contributors to the higher specific output rate are lower melting point and greater shear viscous heating which promotes melting rate of the polymer.

As observed in fiber extrusion, the melt temperature was higher for mPP (460.5°F vs. 451°F than for z-nPP). This is because it has a narrower MWD and is less "shear sensitive," as discussed earlier. The extruder amperage is also higher due to higher screw torque. For a cast film process, the screw design is usually designed with a very high shear (shallower channel depth) to assure good melt homogeneity and film clarity. A lower shear screw can be used for mPP to reduce melt temperature.

The head pressure at the screen pack is higher for mPP. The higher pressure was due to higher viscosity of the melt at the shear rate of the downstream flow passages, including the die lip.

FILM CASTING

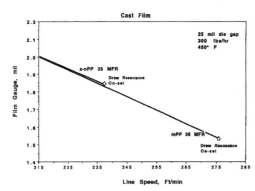

Figure 8. Drawdown capability comparison.

In the cast film process, it is important for the polymer to draw to light gauge and to process at a high line speed. However, as the drawdown ratio (die gap divided by film gauge) increases, the tendency of draw resonance increases. Being of a narrower MWD than conventional CR z-nPP, the metallocene PP has a better drawdown capability. Figure 8 compares the two 35 MFR polymers processed on a 42" (die width) cast film line. The die gap was set at 25 mil to test the drawdown characteristics. While maintaining constant output rate, z-nPP resin reaches maximum line speed of 235 ft/min at a film gauge of 1.85 mil when the draw resonance occurs. Under the same condition, mPP reaches the line speed of 275 ft/min and a corresponding film thickness of 1.58 mil before draw resonance occurs. The higher draw resonance speed of mPP is attributable to the low melt elasticity and low extensional viscosity of the polymer.

Comparing the film properties of the same gauge, mPP was found to have a higher stiffness than z-nPP (1% secant modulus of 135 kPsi vs. 118 kPsi). All other properties such as optical, tensile and elongation are the same. The higher stiffness of mPP is advantageous to many packaging applications.

The good drawdown capability and flexible processing window of mPP are major processing advantages for the cast film applications. Low volatiles resulting from narrow composition distribution are very desirable from an organoleptic perspective. The higher stiffness of mPP film can be very attractive to many applications.

CONCLUSION

The first generation of commercial metallocene propylene polymers are different from conventional PP due to the single-sited nature of the catalyst system. Although the polymer can be processed at the same conditions as the z-nPP of the comparable MFR, the mPP has a higher shear viscosity which leads to a higher melt temperature, pressure, and extrusion torque. A lower extruder temperature setting may be necessary to optimize the melt temperature. The lower Newtonian plateau viscosity makes the mPP easier to drawdown, producing finer fiber and thinner gauge film. The metallocene resin is also more susceptible to orientation resulting in stronger fiber and stiffer film.

REFERENCES

1. J. J. McAlpin, C. Y. Cheng, D. A. Plank, and G. A. Stahl, *INSIGHT'95*, October 22, 1995.
2. C. Y. Cheng, *SPE ANTEC*, 1990.
3. F. U. Lu and J. E. Spruiell, *J. Appl. Polym. Sci.*, **49**, 623 (1993).
4. C. Y. Cheng, *SPE ANTEC*, 1996.

Performance - Property of Novel Glass Fiber Reinforced Polypropylene Compounds and Their Applications

P. Davé, D. Chundury, G. Baumer, and L. Overley
Ferro Corporation, Filled & Reinforced Plastics Division
5001 O'Hara Drive, Evansville, IN 47711, USA

ABSTRACT

The objective of this work was to investigate the property enhancement of polypropylene compounds due to glass fiber reinforcement at various levels. General trends in properties, performance and application were carefully established and surveyed. Second generation metallocene- based glass fiber reinforced polypropylene compounds were also studied in terms of their *"property - performance"* characteristic and compared to the original series. Addition of nylon to glass fiber reinforced polypropylene compounds was investigated as a further possible avenue for creating a new family of polymer alloys. Comparison of properties, advantages, applications were examined.

INTRODUCTION

In recent years there has been an upsurge in the use of polyolefins to replace the more expensive engineering resin products in a wide number of applications.[1] They are being used in alloys, blends and compounds to produce an emerging class of new high strength, lower cost plastic products. These products offer a unique combination of high performance properties together with processing and fabrication advantages over traditional materials at a lower cost. The primary focus of this work will be to investigate the use of glass fiber reinforcement in polyolefins and see how well they *"bridge-the-gap"* between commodity plastics and the more expensive engineering resins.

Glass fiber reinforcement technology has been around for several decades.[2] One of the first consumers' products introduced utilizing this technology was a fishing rod which was followed by an experimental all-plastics car called the Scarab, and finally the introduction of the commercially successful Corvette sports car.[2] The scope of the present work will be limited to the use of chopped strands as a reinforcement.

It is well known in the plastics industry that fiber glass reinforced composites are produced on corotating twin screw extruders and represent a common and economical method of engineering enhanced performance into existing commercial polymers.[3] Grillo *et al.*[3] have done extensive work in investigating the effect of processing parameters like screw design, die design, and rheological characteristics[4] on the physical properties of fiber glass produced on corotating twin screw extruder. Mack has successfully extended this work on investigating compounding techniques for the incorporation of fillers like talc, mica and glass, at high loading levels to polyolefins on twin screw extruders.[5]

Nylon is a tough engineering resin and may be used to balance the strength-stiffness characteristics for polyolefins. Their hygroscopic nature is also diminished due to the presence of the polyolefin (being the second component). The recent technology of fiber glass reinforcement to nylon/PP polymer blends have been covered extensively and patented by Chundury.[7,8]

An advent of the new single site (metallocene-based) catalyst technology offers a unique class of olefin polymers with different molecular structures as compared to the traditional counterparts, which are produced using multisite Ziegler-Natta type catalyst systems.[6] The addition of this material as a modifier to our traditional glass fiber reinforced polyolefins may provide us with some unique properties and cost advantages which may not be easily attainable. A detailed survey of the property-performance of blends containing these unique modifiers with mineral fillers has been covered well by Chundury et al.[6] We plan to use these new materials in our study and go through a similar investigative approach.

EXPERIMENTAL APPROACH
MATERIALS

1. A polypropylene homopolymer (12 melt flow rate (MFR); 0.905 gm/cc density) resin was chosen for these studies.
2. The metallocene-based resin considered was of 1.1 MFR, 0.880 gm/cc density.
3. The nylon selected for this work was of a 6 type.
4. Glass fibers used in this study were 3/16" in length.
5. All custom compounds produced in this study contained stabilizer(s), pigments, and a proprietary coupling agent system.

COMPOUNDING

All glass fiber reinforced samples were produced on a laboratory type 40 mm corotating twin screw extruder machine at standard process zone temperatures between 450-500°F. The nylon-polypropylene alloys were processed at slightly higher zone temperatures. Extruder output rates, screw rpm's and other critical process parameters were kept unchanged during the course of the study. The extruded compound was stranded, water cooled and pelletized. Note: nylon pellets were dried before processing. The ratio of the nylon and PP was maintained at similar levels for all the proprietary formulations processed for the sake of consistency.

The pelletized samples were injection molded and tested (after curing) as per the standard ASTM procedures. The injection molding was carried out on a 90-ton machine at typical barrel temperatures.

RESULTS AND DISCUSSION
"FIRST GENERATION" GLASS FIBER REINFORCED PP'S

The physical property characteristics for the fiber glass reinforced polypropylene, PP, compounds are shown in Figures 1-5, respectively. The figures clearly demonstrate the effect on physical properties due to the incremental addition of glass fiber to the polymer matrix. We see an almost fourfold improvement in the flexural modulus due to glass fiber addition (refer to Figure 1). The homopolymer had a modulus of 260,000 psi; when 40% glass fibers were added to the polymer matrix, the modulus increased to more than 1,000,000 psi. A similar effect was seen in Figure 3 & 5. Figure 3 shows the

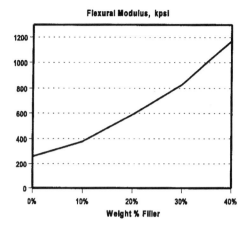

Figure 1. Effect of fiber glass on modulus.

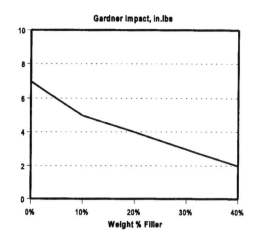

Figure 2. Effect of fiber glass on Gardner impact.

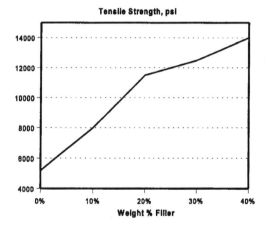

Figure 3. Effect of fiber glass on tensile strength.

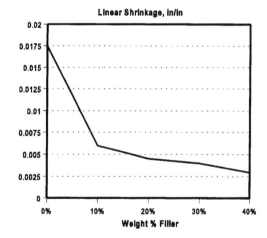

Figure 4. Effect of fiber glass on linear shrinkage.

tensile strength improvement from 5000 psi to 14,000 psi at 40% glass fiber addition. Figure 5 similarly shows an improvement in the heat deflection temperature from 150°F for the homopolymer to greater than 300°F for 40% glass fiber reinforced polymer system. Figures 2 and 4 show the effect of glass fiber addition on the Gardner impact and the linear shrinkage properties. Both these properties

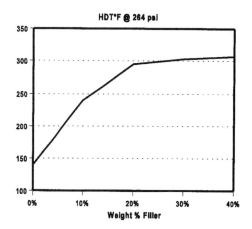

Figure 5. Effect of fiber glass on HDT.

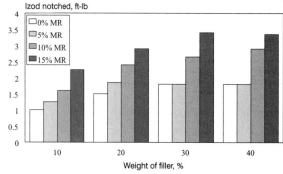

Figure 6. Effect of met resin, MR, on glass reinforced PP.

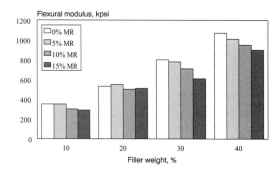

Figure 7. Effect of met resin, MR, on glass reinforced PP.

show lowering trends which was expected. By increasing the amount of glass fibers in the polymer matrix we are in fact reducing the polymer component in the formulation. The glass fibers are very brittle and rigid in nature.[2] This hence is responsible for the lowering of the falling dart property. On the other hand, by increasing the glass fiber addition we are reducing the linear shrinkage properties of the compound.

"SECOND GENERATION" GLASS FIBER REINFORCED PP CONTAINING METALLOCENE-BASED RESIN SYSTEM

Refer to Figures 6 to 9 to examine the physical property comparison due to the addition of 5, 10, and 15% of the metallocene-based resin, MR, to fiber glass reinforced PP's.

Figure 6 clearly demonstrated that by the addition of the MR to the polymer matrix we saw a substantial improvement in the Izod notched characteristic. This property seems to maintain itself well even at higher glass fiber percentages. This result was expected. The MR was contributing for the increase in the polymer chain movement resulting in an increase of the Izod numbers.

The remaining property for these materials (refer to Figure 7-9) shows a general reduction at higher addition of the MR. It may be noted that the property reduction is however not drastic; in fact the HDT remains almost the same even at 15% MR addition. Though there is some loss in the flexural and tensile properties there could be several applications where these may be acceptable and the improved Izod property may be the key selling point.

The MR may hence easily be used as an "impact modifier" to obtain a balance of properties and opening up the envelope of characteristics to attract its use in novel applications.

"THIRD GENERATION" NYLON/PP GLASS FIBER REINFORCED ALLOYS

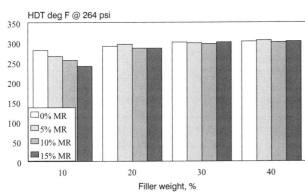

Figure 8. Effect of met resin, MR, on glass reinforced PP.

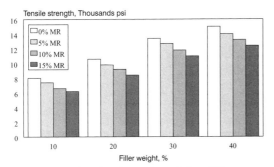

Figure 9. Effect of melt resin, MR, on glass reinforced PP.

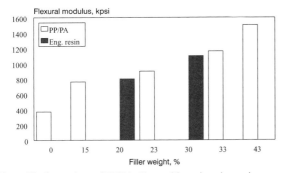

Figure 10. Comparison of PP/PA alloys with engineering resins.

Some of the major reasons for alloying the PP with nylon are: better dimensional stability, part cost effectiveness, and improved processability. Figure 10, 11, and 12 show some of the physical property trends obtained for the fiber glass reinforced nylon/PP alloy systems.

Figure 10 shows the improvement of flexural modulus by the addition of glass fibers to the nylon/PP alloy as compared to Figure 1 where we see that for plain PP with glass fibers. The flexural modulus for 23% glass fiber reinforced nylon/PP alloy was 900,000 psi when compared to 600,000 psi for 20% glass reinforced PP. Some of the improvement was derived due to the replacement of PP with a much tougher material whilst some of it may also be due to the compatibilizing effects of nylon with PP.[7,8] Figure 10 also compares these materials with a 20% and 30% fiber glass reinforced engineering resin, PPO-HIPS. Clearly both these products have comparative properties to the nylon/PP alloy system.

A similar improvement was seen when we compare Figure 11 to Figure 3 for tensile strength properties. By the addition of 43% glass fiber to the nylon/PP alloy the tensile strength was approx. 21,000 psi when compared to only 14,000 psi for PP containing 40% glass fiber reinforcement. This was almost a 50% improvement in strength. This improvement may also be seen due to the reason explained above. A similar comparison was also made with 20 and 30% glass fiber reinforced PPO-HIPS. The tensile strength in fact was superior for

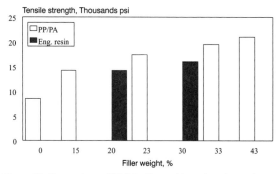

Figure 11. Comparison of PP/PA alloys with engineering resins.

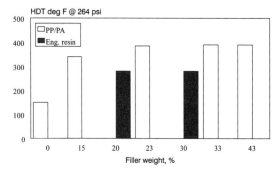

Figure 12. Comparison of PP/PA alloys with engineering resins.

the 23% and 33% glass reinforced nylon/PP alloys.

When the heat deflection temperatures of these novel alloys were compared to plain glass fiber reinforced PP's and the engineering resins we once again see a tremendous improvement. Refer to Figures 5 and 12 for more details.

The above property comparison clearly suggests that these new unique polymer alloy systems have a good "balance of properties" and are similar or superior in characteristics when compared to the engineering resin selected for this study.

CONCLUSIONS

1. A tremendous improvement in overall properties was seen due to the addition of glass fiber to the PP polymer matrix. There is also the advantage for cost savings and a convenient replacement for the more expensive engineering resins. These new materials may be used in the appliance, auto, floor care, liquid handling, lawn and garden industry.

2. The "second generation" glass fiber reinforced PP contains the metallocene-based resin which acts like an impact modifier. The performance-property characteristics of these new blends are very interesting and will be investigated further. These materials may be good where the application requires a high strength and stiffness characteristic with a good degree of flexibility.

3. The "third generation" novel glass fiber reinforced nylon/PP polymer alloys are much tougher and stronger than the two mentioned above. These new materials are very close to the engineering resin selected for this study. They may also be price competitive. Application of these materials may be similar to those mentioned in 1.

ACKNOWLEDGMENT

The authors wish to acknowledge the Filled & Reinforced Plastics Division of Ferro Corporation, for their encouragement and permission to publish this work. The authors also wish to thank Chuck Zhu & Ralph Henshaw for supervising all the processing and testing work carried out in the R&D facility.

REFERENCES

1.   A. D. McMaster and D. Chundury, *SPE ANTEC'94*, San Francisco, CA, 1994.

2.   J. F. Dockum, Jr., **Handbook of Reinforcement for Plastics**, Eds., J. V. Milewski and H. S. Katz, chaper 14, p. 233, 1987.

3.   J. Grillo, P. Anderson, and E. Papazoglou, *SPE ANTEC'92*, 1992.

4.   J. Grillo, E. Papazoglou, and S. Petrie, *Plastics Compounding*, September/October, 1993.

5.   M. H. Mack, *SPE ANTEC'90*, Dallas, TX, 1990.

6.   D. Chundury and Bill MacIver, *SPE ANTEC'95*, Boston, MA, 1995.

7.   D. Chundury, **U.S. Patent 5 278 231** (1994).

8.   D. Chundury, **U.S. Patent 5 317 059** (1994).

Injection Molding of Glass Fiber Reinforced Syndiotactic Polystyrene

Ch.-M. Hsiung, S. Zhang, and D. Bank
Louisiana Productivity Center, Chemical Engineering Department
University of SW Louisiana, Lafayette , LA 70504-4172, USA

ABSTRACT

Glass reinforcement has been applied to syndiotactic polystyrene, sPS, to enhance the dimensional stability, heat resistance, and electrical performance of this newly developed engineering resin. This research concentrated on the structure and property characteristics of the injection molded glass fiber reinforced sPS. Structural characteristics including crystallinity and the distribution of length, concentration, and orientation of glass fibers inside these injection molded parts were studied and correlated with processing conditions. Thermophysical behavior and mechanical properties of the glass fiber reinforced sPS were also studied.

INTRODUCTION

Syndiotactic polystyrene, sPS, is a newly developed engineering semicrystalline polymer that is based on metallocene technology. It has some attractive physical characteristics including high melting point (270°C), low specific gravity (1.045), excellent hydrocarbon resistance, a high degree of dimensional stability, enhanced mechanical performance at elevated temperature, and very good electrical properties.[1] This combination of properties opens a wide array of potential automotive applications including high and low temperature electrical/electronic components and fluid contact under hood devices. The structure-property characteristics of injection molded sPS parts have been studied by Hsiung and Cakmak.[2,3]

Thermoplastics filled with short glass fibers have been used in injection molding for many years in order to improve the properties such as strength, stiffness, dimension stability and heat distortion. While toughness, continuous service temperature, and processability are determined by the polymer matrix, the effects of reinforcement rely on the distribution of concentration, length, and orientation of the fibers.[4] There have been many studies on the orientation of glass fibers and its dependence upon injection-molding conditions. Kamal studied the fiber and matrix orientation in short glass fiber reinforced polypropylene samples. [4,5] Thermoelastic properties of short carbon fiber reinforced PEEK were correlated to the fiber orientation by Bozarth.[6] More recently, Lian[7] studied the fiber orientation by a nondestructive method in transparent injection molded short fiber reinforced resins.

In this work, we present our studies on the processing-structure-property relationship in the injection molding of short glass fiber reinforced sPS.

EXPERIMENTAL

Four different grades of glass fiber reinforced syndiotactic polystyrene, GFsPS, resin were supplied by the Dow Chemical Company. The glass fiber contents vary from 10 to 40% by weight. The weight average molecular weight of the sPS matrix is 400,000. Most experimental results shown in this paper were based on 30% glass fiber contents.

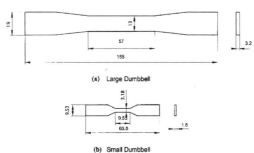

Figure 1. Dimensions (in mm) of large and small dumbbells parts prepared by injection molding.

End-gated ASTM standard large and small dumbbells (Figure 1a and b) were prepared by a Boy 15S injection molding machine with a constant injection pressure of 13.8 MPa and back pressure of 2.75 MPa. The melt temperature was set at 275°C and cooling time was held at 30 seconds. Three mold temperatures (40, 110, and 140°C) were used. The polymer was injected at three different speeds corresponding to volumetric flow rates of 2.1, 6.9, and 34.5 cm^3/s.

The crystallinity distribution within the molded parts was characterized by Differential Scanning Calorimetry.[3] The distribution of glass fiber concentration was measured by a Thermogravimetric Analyzer from TA Instruments. In order to study the distribution of the length of glass fibers, tiny specimens were first cut out at different depths from the surface then compression molded at 300°C to lay all the glass fibers flat. Then the molten thin disks of samples were quenched in ice water to vitrify the resin matrix so that the matrix became transparent. Finally, the distributions of length of glass fibers within these transparent thin disks were studied using a structural imaging system. The distribution of the orientation of glass fibers was characterized by a second order tensor $<a_{11}>$ originally suggested by Advani and Tucker.[8]

RESULTS

RHEOLOGICAL AND THERMAL DEGRADATION PROPERTIES

The shear rate viscosity relationship for GFsPS with four different glass fiber contents is shown

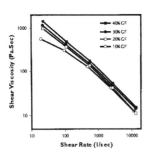

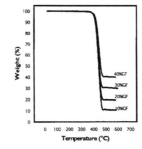

in Figure 2. Reinforcement of sPS with glass fiber increases the shear viscosity. At higher shear rates, GfsPS with different glass fiber contents show much more similar results. Results from thermogravimetric analysis of each sample indicate the level of glass reinforcement in each sample as well as the autoignition temperature which occurs in the range of 400°C.

Figure 2. Shear viscosity of GFsPS at 300°C.

Figure 3. TGA results of GFsPS with different GF contents.

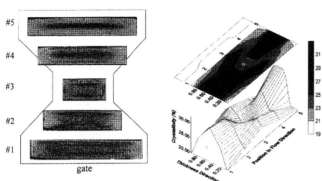

STRUCTURAL CHARACTERISTICS

Figure 4 shows the optical photomicrographs of the cross-sectional view of the small dumbbell sample cut perpendicularly to the flow direction at 5 different distances from the gate. Crystallinity distribution within the sample was measured by DSC scan of tiny specimens cut out at different locations and depths. Crystallinity was calculated through:

Figure 4. Optical pictures of injection molded GFsPS. Small dumbbell molded at 40°C and 6.9 cm³/s.

Figure 5. Distribution of crystallinity within injection molded GFsPS. Same conditions as in Figure 4.

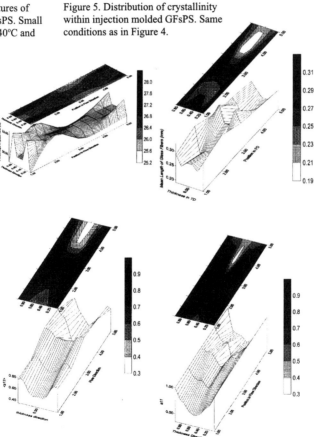

Figure 6. Distribution of glass fiber concentration. (a- upper left) - large dumbbell, (b- upper right) - small dumbbell. Distribution of glass fiber orientation factor. (c- lower left) - large dumbbell, (d- lower right) - small dumbbell. Dumbbells molded at 40°C and 6.9 cm³/s.

$$X = \frac{\Delta H_{exp}}{\Delta H^{o}}$$

where: $\Delta H_{exp} = \Delta H_{melting} - \Delta H_{cold\ crystallization}$, and $\Delta H = 53.2$ J/g is the heat of fusion of 100% crystalline sPS.[3] The opaque crystalline regions match quite well the corresponding crystallinity distribution

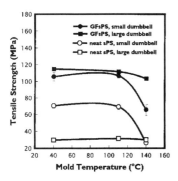 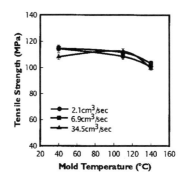

shown in Figure 5. The higher shearing effect caused by the neck geometry (location #3) changes the gapwise crystallinity profile around that region because of higher shear and correspondingly higher temperatures in this region.

Figure 7. Tensile strength of sPS and GFsPS molded at 6.9 cm³/s (left). Effects of injection speed on the tensile strength of GFsPS (large dumbbell) (right).

The morphological characteristics of glass fibers within small and large dumbbells are shown in Figure 6. Glass fibers tends to be more concentrated in the shear zone areas close to the gate in the large dumbbell sample (Figure 6a). In the small dumbbell sample, since the neck region is very narrow, the combined effects of high cooling rate and high shearing in that region trap most long glass fibers in the shear zone (Figure 6b). The distributions of the orientation of glass fibers in both large and small dumbbell show very similar trends: highly oriented in flow direction in most of the skin and shear zone areas,

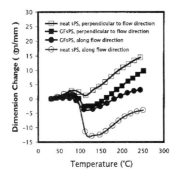 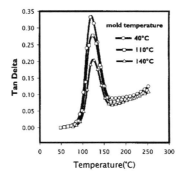

Figure 8. Thermal expansion behavior of sPS and GFsPS molded at 40°C (left). Damping characteristics of GFsPS of material molded at different temperatures. Large dumbbells, location #3, molded at 6.9 cm³/s, tested along flow direction.

close to random orientation at the core region before the neck (location #1), and random to transversely oriented at the core region after the neck (location #5). The shear zone area in the neck region shows the highest degree of orientation in the flow direction (Figures 6c and d). The overall orientation is higher in the small dumbbell compared to the large one.

MECHANICAL AND THERMOMECHANICAL PROPERTIES

The addition of glass fibers into the sPS resin can improve the mechanical strength as shown in Figure 7 (left). However, the best mechanical properties seem to be shown by samples molded at the lower mold temperatures. While the mold geometry plays an important role in determining the tensile strength of neat sPS, the GFsPS samples show little influence from the mold geometry. Changing injection speed also seems to show little effects on the tensile strengths of GFsPS samples (Figure 7 right).

The addition of glass fiber improves the dimensional stability of sPS as measured by coefficient of linear thermal expansion (Figure 8 left). It also decreases the anisotropic characteristics of the thermal expansion behavior of GFsPS when compared to the neat resin. The damping characteristics of GFsPS molded at different temperatures were studied by a Dynamic Mechanical Analyzer. The results show that samples molded at lower temperature have better damping characteristics due to the lower crystalline contents (Figure 8 right).

CONCLUSIONS

- The morphological characteristics (crystallinity and concentration, length, and orientation of glass fibers) of GFsPS samples show wide range of spatial variation within the molded parts. The mold geometry plays a crucial role in affecting the distributions of these morphological characteristics.
- The addition of glass fibers improve the mechanical strength of sPS.
- Injection molded GFsPS samples show better dimensional stability and less anisotropic characteristics in their thermal expansion behavior in comparison to neat sPS.

ACKNOWLEDGMENTS

This work was supported by Louisiana Education Quality Support Fund (Grant # LEQSF(1993-95)-RD-B-16 and LEQSF(1995-98)-RD-B-99). sPS material donation and financial support from the Dow Chemical Company is highly appreciated.

REFERENCES

1. J. H. Schut, *Plastics Technology*, **2**, 26 (1993).
2. C. M. Hsiung, J. Miao, Y. Ulcer, and M. Cakmak, *SPE Annual Technical Papers*, 1788, 1798 (1995).
3. Y. Ulcer, M. Cakmak, J. Miao, and C. M. Hsiung, *J. App. Polym. Sci.*, **60**, 669 (1996).
4. M. R. Kamal and P. Singh, *SPE Annual Technical Papers*, 240 (1989).
5. M. R. Kamal, Li Song, and P. Singh, *SPE Annual Technical Papers*, 133 (1986).
6. J. Bozarth, J. W. Gillespie Jr., and R. L. McCullough, *SPE Annual Technical Papers*, 568 (1986).
7. B. Lian, A. Nothe, J. Ladewig, and J. J. McGrath, *SPE Annual Technical Papers*, 608 (1995).
8. S. G. Advani and C. L.Tucker, *J. Rheology*, **31**, 751 (1987).

Injection Molding Optimization Procedures for Polyolefin Plastomers and Elastomers

S. Hoenig, Wendy Hoenig, and K. Parsely
Dow Plastics, USA

ABSTRACT

Polyolefin plastomers and elastomers for flexible durable goods are injection molded substantially different than the flexible PVC (f-PVC) or styrene block copolymers, SBC, they replace. Certain conditions, such as cold molds and fast injection, must be utilized to effectively produce parts.

The optimized injection molding conditions for polyolefin plastomers and elastomers, oil-modified polyolefin plastomers and elastomers, and oil- and filler-modified polyolefin plastomers and elastomers, were studied and are presented. The effects of eight injection molding variables on the physical properties of these unique polymers are given.

Using the optimized conditions outlined, a *step change procedure* was developed and is presented. This procedure allows for efficient switch-over from f-PVC, SBC or other resins to the polyolefin plastomers and elastomers wherein optimum molding conditions and optimum performance are readily achieved.

BACKGROUND

Polyolefin plastomers and elastomers are new to the flexible durable goods markets, offering polyolefin-like processing and properties to an industry burdened with temperature sensitive and difficult to mold plastics. The polymers can be produced in many processes, including solution, gas phase, and slurry, and generally involve the interpolymerization of ethylene and one or more α-olefin, such as octene, hexene, and butene. The catalyst systems used to produce these new polymers are single-site metallocene systems, preferably those of a constrained geometry, which result in polymers characterized by a very narrow molecular weight distribution, very narrow comonomer distribution, and long chain branches.

As polyolefins, these polymers can and should be processed at relatively high temperatures, shot into the mold as fast as possible, then cooled with the coldest water possible. This is in contrast to f-PVC, which must be processed relatively cold, injected slow, then cooled in a hot mold to allow filling. It is also in contrast to SBC which is processed cold, is injected slowly to prevent warping, and cooled down quickly.

Polyolefin plastomers and elastomers are semicrystalline polymers, containing both an amorphous phase and crystalline phase. The size and shape of these phases are highly dependent on the molding conditions to which they are subjected. Furthermore, the size and shape of these phases determine many of the final part properties.

STANDARD CONDITIONS

Standard conditions have been established for injection molding polyolefin plastomers and elastomers, which take into account their semicrystalline nature, their narrow molecular weight distribution, their very low density, and their narrow comonomer distribution. These conditions include:

- *Use the lowest possible mold temperature.* Use the coldest water available and circulate to achieve turbulent flow. Target: 10°C (50°F).

- *Aim for maximum part weight.* Pack out the part using first injection pressure, then hold time. Target: uniform surface.

- *Use the fastest possible injection speed.* Fast injection reduces viscosity, allowing for more efficient filling. Example: Less than 0.4-second injection time, 14 MPa (2000 psi) injection pressure, for a 250 g part. Filled systems, however, require slow injection times to achieve a good surface finish.

- *Keep barrel temperatures uniform (except for feed zone).* Strive for melt uniformity. Feed throat must remain below 120°C (250°F) to avoid bridging. Target: 175-290°C (350-550°F).

- *Raise melt temperatures to fill molds.* Metallocene catalyzed polyolefins will shear heat: melt temperature will be greater than set temperature. Target: High enough to flow easily, but cool enough to solidify quickly.

- *Use low back pressure for consistency.* Target: Back pressure should be maintained as low as possible.

EXPERIMENTAL

Polyolefin plastomers and elastomers used in this study were from Dow Plastics trade-named as Affinity™ Polyolefin Plastomers, POP, and Engage™ Polyolefin Elastomers, POE. Injection molding was performed on a 150 ton deMag injection molder equipped with an ASTM test mold, following conditions outlined in a designed experiment. The experiment was designed and the data analyzed using JMP statistical software produced by SAS. Trends in performance were deemed significant if the calculated molding parameter estimates, Prob.> |t|, was less than 0.05, thus giving 95% confidence the selected parameter had a significant influence on the performance property.

Three distinct studies were performed. In a broad scope screening, Study I, Affinity POP and Engage POE were molded while systematically changing eight molding parameters, which included melt temperature, injection speed, injection pressure, water temperature, hold time, mold closed time, shot size, and hold pressure. Analysis of the data from this study determined the four primary molding parameters which were used in subsequent Study II on oil-modified Engage POE/Affinity POP blend and subsequent Study III on oil- and filler-modified Engage POE/Affinity POP blends.

Performance properties tested were percent shrink, secant modulus, tensile, tensile impact, falling dart impact, shore A hardness, and part (as is) density, following ASTM or other standardized test methods.

Summarizing the optimum performance into a *step change procedure* was then done on the deMag injection molder and is presented.

RESULTS

STUDY I

Table 1. Molding parameters tested polyolefin plastomers and elastomers*

	High Setting	Low Setting
Melt Temp., °C	285	190
Water Temp., °F	70	50
Injection Speed, setting	63	10
Injection Pressure, setting	127	30
Hold Time, seconds	20	0
Mold Closed Time, seconds	60	0
Shot Size, setting	98	94
Hold Pressure, Setting	127	5

*30 MI, 0.870 g/cc density

Table 1 shows the Molding Parameters tested and the Range of Results for Affinity POP and Engage POE. It is shown that both the plastomer and elastomer gave significant performance variations depending on molding conditions. For example, shrinkage varied from 0.54% to 1.23% for the plastomer, and from 0.42% to 0.79% for the elastomer. Shore A hardness varied 5 units for the plastomer and 8 units for the elastomer. Such variability is expected for semicrystalline polymers and must be accounted for to achieve optimum performance properties.

Table 2 shows the relative fabrication effects on performance properties and the direction of the effect (increasing performance or decreasing performance) as the fabrication condition increased at 95% confidence for Affinity POP and Engage POE. In no case did the two classes of resins contrast,

Table 1A. Range of results polyolefin plastomers and elastomers.

	High Value 30 MI, 0.870 g/cc density	Low Value 30 MI, 0.870 g/cc density	High Value 30 MI, 0.885 g/cc density	Low Value 30 MI, 0885 g/cc density
Shrink, %	0.79	0.42	1.29	0.54
Secant Modulus, psi	2330	1680	6540	5820
Ultimate Tensile, psi	440	350	1010	910
Yield Strength, psi	280	245	565	540
Elongation, %	790	450	585	505
Shore A Hardness	68	60	87	82
Part Density, g/cc	0.874	0.871	0.890	0.889

Table 2. Molding effects on part performance properties polyolefin plastomers and elastomers

	Melt Temp	Water Temp	Hold Time	Mold Closed Time	Hold Pressure
Shrink, %	increases*	increases	decreases[a] decreases[b]	decreases[a] decreases[b]	
Secant Modulus	increases[a]				
Ultimate Tensile	decreases[a] decreases[b]				
Yield Strength		decreases[b]			decreases[a]
Elongation, %					
Shore A Hardness		decreases[b]		decreases[a]	
Part Density		decreases[a]			

[a]30 MI, 0.870 g/cc density; [b]30 MI, 0.885 g/cc density
As the fabrication condition increases, the performance property ...

both responded to the fabrication conditions similarly.

As melt temperature increased, the percent shrink, secant modulus, tensile yield, and percent elongation increased, while the ultimate tensile decreased. Melt temperature was by far the dominatant fabrication condition on part performance. As the water temperature increased, the percent shrink increased, while the secant modulus, yield, shore A hardness, and part density decreased. As mold closed time increased, the Shore A hardness and part density increased, while the percent shrink decreased. Increasing the hold time increased the shore A and decreased the percent shrink. The injection speed, injection pressure, shot size, and hold pressure all had minor, albeit scattered, effects on final part properties.

From this study, it was determined that melt temperature, water temperature (mold temperature), mold closed time (cooling time), and hold time were the most significant fabrication variables on the performance properties of polyolefin plastomers and elastomers.

STUDY II

Table 3 shows the molding parameters tested and the range of results for oil-modified polyolefin blends (Affinity POP/Engage POE/oil). It is shown that composition gave significant secant modulus performance variation depending on molding conditions. For example, secant modulus varied from 9000 psi to 5150 psi, Shore A hardness from 65.4 to 61.6, yet the part density did not change at 0.884 g/cc.

Table 4 shows the relative fabrication effects on performance properties and the direction of the effect (increasing performance or decreasing performance) as the fabrication condition increased at 95%

Table 3. Molding parameters tested oil modified polyolefin plastomers and elastomers*

	High Setting	Low Setting
Melt Temp, °F	500	350
Water Temp, °F	80	50
Injection Speed, Setting	2.5	0.4
Injection Pressure, Setting	127	30

*Engage POE/Affinity POP/Oil

Table 3A. Range of results oil modified polyolefin plastomers and elastomers.*

	High Value	Low Value
Shrink, %		
Secant Modulus, psi	2244	1820
Ultimate Tension, psi	445	368
Yield Strength, psi	284	227
Elongation, %	91	63
Shore A Hardness	65.4	61.6
Part Density, g/cc	0.885	0.884
Tensile Impact, ft lbs/in²	260	171

*Engage POE/Affinity POP/Oil

Confidence for oil-modified polyolefin blends (Affinity POP/Engage POE/oil). In this study, melt temperature, water temperature, injection speed, and injection pressure are highlighted. As melt temperature increased, the percent shrink and percent elongation increased, while the rigidity, tensile, hardness and tear strength decreased. As water temperature increased, the percent shrink increased and all other performance properties decreased. As the injection speed and injection pressure increased, very little, if any, effect on performance properties were noted.

STUDY III

Table 5 shows the molding parameters tested and the range of results for oil- and filler-modified polyolefin blends (Affinity POP/Engage POE/oil/calcium carbonate). Melt temperature, the most dominate fabrication variable seen in the other studies, could not be

Table 4   Molding effects on performance properties oil modified polyolefin plastomers and elastomers*

	Melt Temp	Water Temp	Injection Speed
Shrink, %	increases	increases	
Secant Modulus		decreases	
Ultimate Tensile	decreases	decreases	
Yield Strength		decreases	
Elongation, %			increases
Shore A Hardness	decreases	decreases	
Part Density		decreases	
Tensile Impact	decreases	decreases	

*Engage POE/Affinity POP/Oil
As the fabrication condition increases, the performance property....

Table 5. Molding parameters tested oil and filler modified polyolefin plastomers and elastomers*

	High Setting	Low Setting
Melt Temp, °F	350	350
Water Temp, °F	80	50
Injection Speed, setting	2.5	0.4
Injection Pressure, setting	127	30

*Engage POE/Affinity POP/oil/calcium carbonate

Table 5A. Range of results oil and filler modified polyolefin plastomers and elastomers*

	High Value	Low Value
Shrink,%		
Secant Modulus, psi	2415	2238
Ultimate Tensile, psi	527	465
Yield Strength, psi	261	152
Elongation, %	579	542
Shore A Hardness	66.2	64.6
Part Density, g/cc	1.048	1.049
Tensile Impact, ft lbs/in^2	341	264

*Engage POE/Affinity POP/oil/calcium carbonate

Table 6. Molding effects on performance properties oil and filler modified polyolefin plastomers and elastomers*

	Melt Temp	Injection Speed
Shrink, %	not tested	
Secant Modulus	not tested	decreases
Ultimate Tensile	not tested	
Yield Strength	not tested	
Elongation, %	not tested	
Shore A Hardness	not tested	
Part Density	not tested	
Tensile Impact	not tested	

*Engage POE/Affinity POP/oil/calcium carbonate

increased without significant burning, and so was left out of the design. It is shown that this composite gave minimal performance variation depending on molding conditions. Only tensile yield strength and tensile impact strength varied significantly.

Table 6 shows the relative fabrication effects on performance properties and the direction of the effect (increasing performance or decreasing performance) as the fabrication condition increased at 95% confidence for oil- and filler-modified polyolefin blends (Affinity POP/Engage POE/oil/calcium carbonate). Very little changes were seen in product properties as a function of molding conditions.

STEP CHANGE PROCEDURE

Optimum conditions for injection molding polyolefin plastomers and elastomers were then determined in a stepwise fashion on the deMag injection molder and are presented below:

1 Turn on the chilled water system. Target mold temperature should be as low as possible without mold sweating. Drain incumbent material and purge completely.
2 Change barrel temperatures to 165-200°C (330-392°F).
3 Reduce extruder back pressure to approximately 100 psi or less.
4 Decrease hold pressure as much as possible.
5 Increase injection pressure to maximum possible (decrease injection time to run as quickly as possible).
6 Increase the pack/hold time to fill the part out.
7 Adjust the shot size until part is just shorting.
8 Wait until all molding parameters have lined out (5-10 shots).
9 Finish filling the part by using hold pressure, or the pack/hold adjustment.
10 Decrease clamp time until part is ejected too hot, then add minimal clamp time to produce a good quality part.

SUMMARY

Polyolefin plastomers and elastomers can be optimally injection molded by following general guidelines for polyolefins with a few exceptions. Low back pressures improve surface quality and packing is the best adjustment for filling out a molded part. Understanding these processing nuances will allow molders to use these new polymers to compete in high value applications and obtain higher productivity from existing equipment.

Potential Film Applications of Metallocene-based Propylene Polymers from Exxpol® Catalysis

A. K. Mehta, M. C. Chen, and J. J. McAlpin

Exxon Chemical Company Baytown Polymers Center P. 0. Box 5200 Baytown, TX 77622-5200,
USA

INTRODUCTION

Metallocene-based isotactic propylene polymers have attracted considerable attention[1,2,3] in recent years. Exxon Chemical and catalyst development partner, Hoechst AG, are leaders in the field. They have independently developed and commercialized products (ACHIEVE™ Propylene Polymers for Exxon and HOSTACEN™ for Hoechst) based on their jointly developed catalyst technology (termed EXXPOL® technology by Exxon). Fibrous products and clarity injection moldings are examples of early end uses of the new product families. In this paper, attention will be focused on opportunities for the new materials in the film arena. While film properties of the metallocene resins will be shown to be excellent, their inherently narrow MWD and tacticity distribution, TD, will be seen to limit their performance in some film forming processes. However, the ability of the metallocene catalyst/product designer to meet the processing challenges of these film forming processes will be illustrated. Effects of MWD, composition distribution, CD, and tacticity distribution, TD, variations on resin performance in critical film applications will be discussed.

BACKGROUND

Table 1 compares EXXPOL resins with a Ziegler-Natta, Z-N, standard with respect to some typical properties. The Z-N standard chosen is based on second generation catalyst; it is widely accepted in the US film market. The designations MCN-1 and MCN-2 in Table 1 mean that the resins were produced using two different metallocene catalysts. While the tacticity of the resins made from the two catalysts is seen to differ somewhat (as evidenced by differences in melting point, stiffness), the striking difference is in molecular weight capability. MCN-1 can deliver a minimum melt flow rate, MFR, of only 10; for many film applications, lower MFR is needed and MCN-2 is seen to deliver that capability. Both catalysts produce resins with narrow MWD (NMWD) implying they are single-sited. The last column in Table 1 contains data for resins produced by catalysts which embody both metallocenes on a single support. The aim here is to broaden MWD and the data indicate control over a range from 2.5 to 5, depending on the metallocene mix and process conditions. This indicates the potential for designing film resins with rheology similar to the Z-N case, if that is desired. Homopolymers made using this mixed metallocene catalyst will be referred to as "BMWD resins" in the remainder of this paper.

Table 1. Resin properties comparison

Property	Z-N Standard	MCN-1	MCN-2	Mixed MCN
DSC* Peak Melting Point, °C	162	147	151	148-150
Heat Deflection Temp, °C	98	104	110	-
Stiffness (molded), MPa (k psi)	1380 (220)	1380 (200)	1483 (215)	-
Min Possible MFR, dg/min	≤0.1	10	≤0.1	1
MWD, M_w/M_n	≈4	2.0	2.2	2.5-5
Target Application	various	nonwovens	films	films

*DSC=Differential Scanning Calorimetry

DISCUSSION

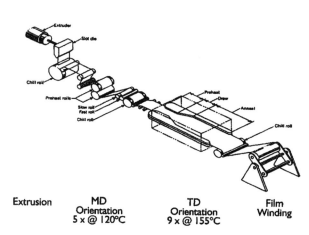

Extrusion   MD
             Orientation
             5 x @ 120°C

TD
Orientation
9 x @ 155°C

Film
Winding

Figure 1. BOPP via tenter orientation process schematic.

Polypropylene films are formed by two basic techniques, the chill roll casting process and the biaxially oriented film process, BOPP. By far the more important film forming process for polypropylene is the BOPP process. Figure 1 is a schematic of a typical BOPP line, operating via the tenter orientation setup. A thick polypropylene sheet of ~1200 μm is formed and stretched in perpendicular directions (sequentially or simultaneously) at a temperature close to the melting point of the resin. The resulting oriented film has excellent optics, high stiffness and strength and low water vapor transmission rate, WVTR. It finds a host of applications, primarily in packaging uses. General resin requirements for the OPP market include a low melting point to allow stretching at low temperatures and a broad MWD to broaden the range of temperatures over which the resin can be stretched (i.e., the stretching window). From the data of Table 1, it is clear that the single metallocene resins bring the low melting point but their MWD is likely to be found lacking. Thus one would expect the mixed metallocene BMWD resin of Table 1 to have better potential for this application.

The chill roll casting process deposits a very thin sheet of molten polymer onto a chilled, rotating roll from which the finished film is removed. The resulting films have high clarity and gloss and find application in packaging areas. requiring high quality optics and good sealability, among other attributes. The process often employs conventional resins with relatively narrow MWD; controlled rheology, CR, resins are sometimes used. Thus the first generation NMWD ACHIEVE products are finding application in this area.

GENERAL PURPOSE BOPP FILM POTENTIAL

Films in this category are used in a variety of packaging applications. Some key film requirements are outlined in Table 2. Given the fact that BOPP film lines tend to be large machines operating at high line speeds, it is not surprising that an important processing requirement is to minimize the occurrence of film breaks during operation. With regard to film properties, optics, moisture barrier and stiffness are among the key film attributes. As has been reported,[3] narrow MWD metallocene homopolymers can be fabricated to yield BOPP films with enhanced properties. However, particularly in the US market, competitive film manufacturing economics are crucial to the successful introduction of a new product, and the narrow MWD metallocene homopolymers are at a disadvantage in this respect. Figure 2 illustrates one aspect of their processing limitations. The processing window displayed in the figure is the range of stretching temperatures over which a particular resin yields a high quality film in a batch experiment at some standard stretching condition. If the temperature is too far below the melting point, the film is too strong to be stretched evenly and it breaks. As the

Table 2. Key BOPP film requirements

Processability
low number of film breaks/day at high(er) line speeds
good gauge uniformity
Film Properties
low haze
low shrinkage
high stiffness
good barrier (low WVTR)

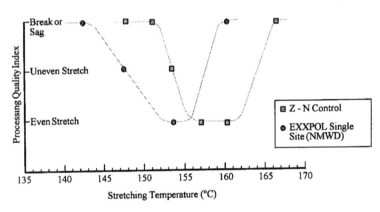

Samples stretched on batch, biaxial stretching apparatus
MD x TD Stretching Ratio = 6 x 6

Figure 2. Processing window during biaxial stretching.

temperature is raised, partial melting occurs and the film enters its processing window. At still higher temperatures, further melting occurs and the crystalline network is insufficient to support the film; it tends to sag before or during stretching leading to a poorly formed film or breakage. Single-site metallocene catalysts typically give resins with sharp melting points and narrow MWD. Thus one would expect their processing windows to be comparatively narrow. The data of Figure 2 bear out this expectation. It is seen that good film formation for the metallocene resin occurs over only a 5°C temperature range (152-157°C). By contrast the Z-N resin, designed (using a fourth generation catalyst) for BOPP application, forms film nicely over a range of about 10°C (156-164°C). While a tightly controlled BOPP process may be able to cope with a processing window of only 5°C, we shall see later that other limitations accompany the narrow processing range.

One of the chief advantages of metallocene catalysis over traditional Z-N catalysis is its ability to design multi-sited catalysts in which the performance of the various sites can be preselected. These designer catalysts should be able to produce resins meeting almost any reasonable set of performance criteria. If low stretching temperature with broadened BOPP processing window is the goal, the catalyst/product designer must broaden MWD, and TD or CD of the resins. From the data of Table 1, the mixed metallocene catalyst broadens MWD and TD. The melting range can be further influenced by CD control, through the addition of comonomer. The highly tailored resin produced by manipulating all these variables is referred to as "Tailored Molecular Distributions", TMD, resin in what follows. Table 3 illustrates the ranges in certain key parameters which can be obtained by these catalyst/process modifications. Figure 3 shows the extent of bimodality in molecular weight attainable with the present technology.

Table 3. Materials properties evolution of EXXPOL resins

Property	Single Site N MWD, TD	BMWD T MWD, TD	TMD* T MWD, TD, CD	Z - N
MFR, dg/min	4.8	2.0	1.0	2.6
M_w/M_n	1.9	2.7	3.6	3.9
Recoverable Compliance, Pa^{-1} 10^{-4}	1.1	3.6	5.2	3.9
T_m, °C	151	150	146.5	157
Xylene Solubles, wt%	≤0.7	≤0.7	0.7	3.1

MWD=Molecular Weight Distribution; TD=Tacticity Distribution; CD=Composition Distribution; B-Broad; N=Narrow; T=Tailored; *Tailored Molecular Distribution

Figure 4 displays the film processing windows of the modified resins and the Z-N standard. The BMWD homopolymer has a processing window about the same width as that of the standard product but shifted to lower temperature. The TMD resin results are even more remarkable. The processing window is even wider than that of the standard resin and good films can be made at temperatures up to

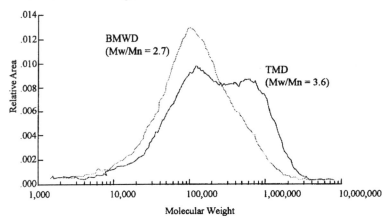

Figure 3. GPC molecular weight distribution of EXXPOL resins.

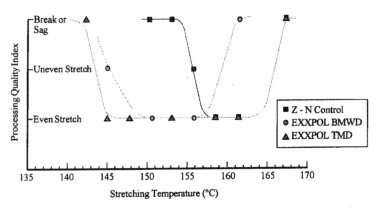

Figure 4. Processing window during biaxial stretching. Samples stretched on batch, biaxial stretching apparatus. MD x TD stretching ratio = 6 x 6.

15°C below that possible with the conventional product. This suggests the possibility of a resin with step-out processability, allowing higher line speed and improved manufacturing economics.

High BOPP line speeds require the film to be stretched at very high strain rates. The processing window experiments described above do not meet this criterion. To evaluate the ability of a particular resin to be stretched smoothly at very high strain rates, the following experiment was designed. A cast extruded sheet was stretched 7x in the machine direction at 110°C in the environmental chamber of an Instron tensile tester. Specimens were then cut perpendicular to the MD and these were stretched in the Instron tester at controlled temperatures covering the range 120 to 160°C. The speed and extent of the TD stretch were varied. The highest TD strain rate possible in the Instron equipment was 11,000%/min, a rate close to that achieved on a commercial BOPP line. In this test, the specimen was required to survive a TD stretch to 2,200% elongation at the high strain rate. Figure 5 highlights the Instron stretching conditions that provided the highest strain rate. The results of this experiment are shown in Table 4.

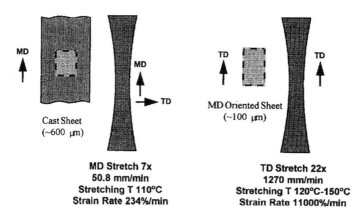

MD Stretch 7x
50.8 mm/min
Stretching T 110°C
Strain Rate 234%/min

TD Stretch 22x
1270 mm/min
Stretching T 120°C-150°C
Strain Rate 11000%/min

Figure 5. Instron elevated temperature sequential biaxial stretching.

The data of Table 4 lead one to several conclusions. First, the high speed stretching test poses a formidable challenge to even the best of today's Z-N based resins; the Z-N control is seen to survive the test only at temperatures within about 10°C of its melting point. Second, it is clear that the single-site EXXPOL® resin is deficient in high speed stretchability to the Z-N control. This confirms the expectation that its narrow melting range and narrow MWD would be negative influences under certain film fabricating conditions. Finally, one concludes that the systematic tailoring of molecular distributions employed in the TMD resin has produced a desirable result. This resin survives the Instron stretching test at temperatures nearly 20°C below its melting point and 30°C below the processing temperature of the Z-N control. Easier TD stretchability at lower stretching temperatures, without breaking, is one of the unique features of the TMD resin. Since film breaks during TD stretching are typically the weak link in BOPP film fabricated via the tenter frame process, it is anticipated that resins like the TMD resin will offer a significant processing advantage.

Table 4. Instron TD stretching at elevated temperatures*

Stretching Temp, °C	EXXPOL Single Site $T_m = 151°C$	EXXPOL TMD $T_m = 146.5°C$	Z - N $T_m = 157°C$
120	B	9.1 B (21x)	B
130	B	8.9	B
140	B	6.1	B
150	B	3.0	5.8 B(20x)
160	B	-	2.9

*Instron TD tensile strength, MPa
TD stretch to 22x; TD strain rate 11,000%/min; B=breakage

Table 5. Biaxially oriented film properties

Film Property	EXXPOL TMD MFR = 1.0 dg/min T_m = 146.5°C	Z - N MFR = 2.6 dg/min T_m = 157°C
Thickness, μm (mil)	18 (0.7)	15 (0.6)
Haze, %	0.3	0.3
Gloss, %	94	94
WVTR@ 37.8°C, 100% RH g/m^2/day per 25.4 μm	6.7	6.5
1% Sec Modulus, MPa (k psi)	2359 (342)	2729 (396)
Ultimate Tensile Strength MPa (K psi)	200 (29)	207 (30)
Ultimate Elongation, %	65	71

Films prepared on batch biaxial stretching apparatus; MD x TD stretching ratio = 6x6, optimum stretching temperature for EXXPOL TMD film is ~15°C lower than for Z-N control

This potential processing advantage will only be of value, of course, if the resulting films are competitive with today's BOPP standards. Table 5 compares films made (by batch stretching) from the TMD resin and from the Z-N control. The TMD film was produced at a stretching temperature about 15°C below that used to prepare the control sample. The data of Table 5 lead one to conclude that the performance levels of the two films are quite similar. With further resin tuning, any deficiencies in the TMD resin can probably be eliminated or even improved without sacrificing the processing advantage.

BOPP SHRINK FILM RESULTS

One use of biaxially oriented film of particular interest for this discussion is shrink film. Typically, the film is heated near its melting point to induce form-fitting shrinkage around the article to be packaged (compact disc film wrap is a typical example). In this high performance corner of the BOPP market, random copolymers, RCPs, with modest levels of ethylene are frequently used. In Table 6 general oriented film attributes and shrink behavior of 3 resins are compared. The resins are a conventional Z-N RCP, which is one of the standards of the shrink industry and two NMWD, narrow CD ethylene RCPs made with a metallocene catalyst. It is seen from the data of Table 6 that the metallocene resins provide control over a greater range of shrinkage and shrink tension levels. In many shrink applications, high shrink and shrink tension are desirable attributes and metallocene RCPs are seen to offer higher levels of both. In addition, the data of Table 6 show a favorable balance between the shrink level and tension and other film physical properties. At equivalent stiffness, WVTR and optics, the metallocene copolymers offer superior shrink capability. It is anticipated that the TMD technology described in the previous section will also apply to shrink films, yielding a superior balance of shrink performance and processing speed.

CAST FILM APPLICATION RESULTS

Table 6. Biaxially oriented shrink films comparison

Property	Z - N RCP 4.0 wt% C2	EXXPOL RCPs 1.1 wt% C2	EXXPOL RCPs 3.4 wt% C2
MFR, dg/min	3.8	3.9	4.0
T_m, °C	139	139	127
Hexane Extractables, %	3.0	0.4	1.4
Film Thickness, μm (mil)	17 (0.7)	15 (0.6)	27 (1.1)
Shrinkage @ 135°C, 180 s (%)	41	41	65
Shrink Tension @ 110°C, g force	86	125	215
WVTR @ 37.8°C, 100% RH g/m²/day per 25.4 μm	10.5	7.1	11.0
1% Sec Modulus, MPa (k psi)	1240 (180)	2096 (304)	1290 (187)
Haze, %	0.4	0.5	0.7

Films prepared on batch biaxial stretching apparatus; Z-N RCP and low C2 EXXPOL RCP stretched at 138°C; high C2 Exxpol RDP at 124°C

Early work with ACHIEVE® homopolymers in cast films has been encouraging. Films have been cast on state-of-the-art high speed equipment with no processing debits noted for the metallocene products. While the resulting films have shown good properties, the most interesting cast film results have come from RCP trials. As reported earlier,[1] hexene-1 copolymers offer a favorable balance of cast film properties. Table 7 compares film data for a Z-N ethylene RCP, a metallocene hexene-1 RCP and a 1:1 blend of the two resins. From the data of Table 7, it is seen that the metallocene hexene-1 RCP film compares very favorably to the Z-N ethylene RCP film. Perhaps more impressive, though, is the blend result. There appears to be clear synergy on blending, with film toughness of the blend film being higher than that of either of the component films and blend film modulus being lower than that of the components. In addition, the clarity of the blend film is about the same as that of the better of the components (the hexene-1 copolymer). One possible source of the blend synergy can be seen in Table 7, where data on crystallization rate of metallocene RCPs (ethylene and hexene-1) is displayed. The hexene-1 copolymer crystallizes faster than the ethylene RCP. It is speculated that in the blend system, the hexene-1 RCP is nucleating a favorable morphology. More work is in progress to understand and define the potential of such blend systems, but these early results offer hope for step-out products.

CONCLUSIONS

Table 7. Cast film properties comparison

Property	Z-N RCP 5.0 wt% C2	EXXPOL RCP 2.9 wt% C6	Z-N/EXXPOL 1:1 Blend	EXXPOL RCPs 1.9 wt C2	EXXPOL RCPs 1.6 wt% C6
MFR, dg/min	5.0	4.3	4.6	10	10
T_m, °C	133	128	129.5	130	131.5
Haze, %	5.2	0.6	0.5		
Dart Impact g/25.4 μm	59	51	392		
1% Sec Mod MPa (kpsi)	556 (81)	738 (107)	468 (68)		
Crystallization Half-Time*, min				4.9	2.6
Crystallization Half-Time**, min				19	8.2

Crystallization Half-Time is time for 50% of material to crystallize at given T_c
*T_c=104°C; **T_c=110°C
Film thickness ≈1.7 mil (43.2 μm);

Table 8 summarizes the conclusions of this work. The basic conclusion is there is great potential for EXXPOL® catalysis-based propylene polymers in film applications. In the very important general purpose BOPP film application, the flexibility in catalyst/product design possible with metallocene catalysts promises new materials offering step-out processing advantages. This new family of resins has the potential to move the BOPP industry to a new level of processing speed and manufacturing economics. In the BOPP shrink film application, metallocene-catalyzed RCPs show promise to deliver levels of shrink performance previously unattainable. In the cast film area, there appears to be substantial potential for metallocene-catalyzed propylene polymers, particularly RCPs of ethylene and of hexene-1. Blends of these copolymers with conventional Z-N-based RCPs also show promise. These favorable results reinforce the notion that metallocene catalysis will constitute the "new S-curve" for the polyolefins industry.

BIBLIOGRAPHY

1.   J. J. McAlpin, A. K. Mehta, D. A. Plank, and G. A. Stahl, *SPO '95*, September, 1995.
2.   M. J. Brekner, *Metallocenes'96*, Dusseldorf, March, 1996.
3.   D. Fischer, at al, Metallocenes'96, Dusseldorf, March, 1996.

Table 8. Potential BOPP and cast film application conclusion

- The ability of the metallocene catalysts designer to control MWD, TD, and CD will open the door to new families of resins with exceptional property/processability balances
- Early work on tailored molecular distribution, TMD, resins indicates potential for unprecedented BOPP processing performance and manufacturing economics
- Special BOPP applications such as shrink film will reach new levels of performance with metallocene-based polymers
- The novel copolymers possible with metallocene catalysis will bring value to cast films; blends of the new resins with conventional materials show promise, as well
- The work reported here further confirms the hypothesis that metallocene catalysis constitutes the "new S-Curve" in the polyolefins industry

Effect of Various Polyethylene Structures on Film Extrusion

C. M. Wong, H. H Shih, and C. J. Huang

Union Chemical Laboratory, Industrial Technology Research Institute
321 Kuang Fu Road, Sect 2, Hsinchu , Taiwan R.O.C.

ABSTRACT

Viscoelastic behaviors and film extrusion are concerned in this study. Several materials are tested. They are a low density polyethylene, a linear low density polyethylene, and two metallocene polyethylenes. Molecular weight, molecular weight distribution, storage modulus, and loss modulus have been measured to interpret the viscoelastic behaviors. The energy consumption and output rates are described at various screw revolutions. A wide range of drawdown ratios, blowup ratios with frost-line heights in film blowing is investigated in order to determine the region of bubble stability.

INTRODUCTION

Blown film extrusion and casting film extrusion are the common methods in film production. Especially, blown film extrusion is a major way to produce bags for low density polyethylene, LDPE, linear low density polyethylene, LLDPE, and high density polyethylene, HDPE. Long chain branches are believed to be an important structure in LDPE. The size of long branches increases with increasing the number-average molecular weight, M_n, of polyethylene, but the mean size of long branches relative to M_n decreases with increasing M_n.[1] Logarithmic plots of storage modulus, G', against loss modulus, G", are found to be strongly dependent on the molecular weight distribution, MWD, and the degree of side chain branching, but weakly sensitive to (or independent of) temperature and to weight-average molecular weight, M_w, of high molecular polymers.[2] The molecular weight distribution is an important effect on the output rate, extruder torque, and presure development in the extruder. Polypropylene, PP, LLDPE, and HDPE with narrow MWD have a higher output rate, extruder torque and pressure development in the extruder than those with broad MWD during extrusion.[3] The elongation flow behavior of PP, HDPE, and LDPE at different melt temperatures in blown film extrusion is studied to determine the elongational viscosity.[4] Analysis of the bubble deformation and heat transfer process involved in blown film extrusion has been carried out in reference.[5] The effects of die gap, die land length, blowup ratio, BUR, and drawdown ratio, DDR, on LLDPE and LDPE blown film properties and structure has been analyzed.[6] Die land length has no significant effect on film structure and properties. However, die gap, BUR, and DDR are able to influence film structure and properties such as tensile, tear, and impact properties.

References[7-10] are concerned with the relationship between bubble stability and processing conditions such as BUR and DDR, etc. A material containing long chain branches normally has a high melt strength. The high melt strength can result in a good bubble stability in film blowing. At equal melt index, MI, level, the melt strength of LDPE is about two times higher than that of LLDPE and HDPE.[7] The increase of processing temperature can lead to the decrease of bubble stability.[8] Blowup ratio and drawdown ratio are also important factors to affect the bubble stability.[9,10]

The present paper is devoted to the study of various polyethylenes in viscoelastic behaviors and blown film extrusion.

EXPERIMENTAL

MATERIALS AND METHODS OF INVESTIGATION

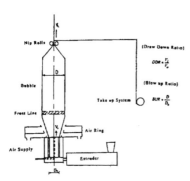

Figure 1. Schematic representation of blown film system.

A blown film system shown in Figure 1 includes an extruder (ϕ 40 mm, L/D=28) with a barrier screw, annular die (outer diameter: 60 mm, inner diameter: 56.4 mm), air ring, nip rolls, and take-up system. Polymer melt through annular die becomes a tube. The inside of the tube introduced by air leads to the inflation of the tube into a bubble. The ratio of take-up speed, V_L, by nip rolls to line speed, V_o, of polymer melt through die lip is drawdown ratio (DDR=V_L/V_o). The ratio of bubble diameter, D, to outer diameter of annular die, D_o, is blowup ratio (BUR=D/D_o).

The various thicknesses of blown film are produced by different drawdown ratios and blowup ratios. The bubble stability is the key factor to make a uniform film. Therefore, the bubble stability and instability have been studied under various drawdown ratios, blowup ratios, and frost-line heights. The experiments have been performed on various polyethylenes including LDPE (Asia Polymer), LLDPE (Exxon), Metallocene PE(1) (DOW), and Metallocene PE(2) (Exxon). Four materials having a similar MI and different molecular structures

Table 1. Characteristics of materials

	LDPE	LLDPE	MPE(1)	MPE(2)
Comonomer		Hexene	Octene	Hexene
Mole, %	-	3.5	4.6	2
Density, kg/m³	922	921	902	917
MI	1.1	0.8	1.0	1.0
T_m, °C	112	124	102	115
$M_n \times 10^3$	62	71.5	70	71
$M_w \times 10^3$	312	279	245	263
MWD	5.03	3.90	3.50	3.70

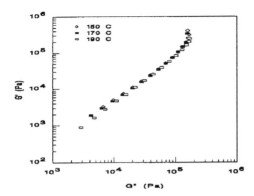

Figure 2. Storage modulus vs. loss modulus at a frequency 1 Hz and three different temperatures for MPE(1).

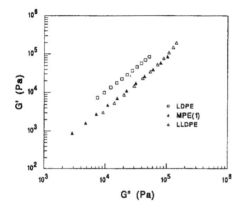

Figure 3. Storage modulus vs. loss modulus at frequency 1 Hz and temperature 190°C for three materials.

are used in film blowing. Melting temperatures, T_m, are obtained by differential scanning calorimetry, DSC. The amounts of comonomer are measured by nuclear magnetic resonance, NMR. M_n, M_w, and MWD are determined by gel permeation chromatography, GPC. G' and G" are investigated by cone-plate viscometer.

RESULTS AND DISCUSSION

Table 1 gives the characteristics of four materials. The four materials have a similar MI which is around 1.0. MPE(1) has the lowest melting point and the narrowest MWD. LDPE has the broadest MWD. MWD of MPE(2) is narrower than that of LLDPE. Long chain branches exist in LDPE and MPE(1). The length of short chain branches is uniform in LLDPE, MPE(1), and MPE(2). MPE(1) and MPE(2) have a uniform distribution of short chain branches. Log G' versus log G" at three different temperatures for MPE(1) is plotted in Figure 2. The experimental results show that log G' vs. log G" for MPE(1) is independent of temperatures. Figure 3 shows log G' vs. log G" for three materials at a frequency (1 Hz) and a temperature (190°C). As a G" is considered, G' for LDPE has the highest value, but G' for LLDPE and MPE(1) is very similar. Generally, the results partially correspond to the conclusion of reference,[2] log G' vs. log G" is strongly dependent on MWD. However, the results show log G' vs. log G" is not only strongly dependent on MWD, but also on molecular structures such as long chain branches.

Figures 4 to 6 present the dependence of output rate, Q and energy consumption, A, on screw revolutions for four materials. The materials go through the same screw and annular die, and are run at the same processing temperature. The output rate of four materials at various screw revolutions is in Figure 4. Four materials have a similar output rate at low screw revolutions. When screw revolution increases, MPE(1) has the highest output rate and MPE(2) has the lowest output rate. The output rate LLDPE is slightly larger than that of MPE(2) at the same screw revolutions. General speaking, narrow MWD material has higher output rate than broad MWD material in the study. The experimental results match the observation in reference.[3] However, the output rate of LLDPE and MPE(2) contrary to the result of narrow MWD material having higher output rate is possibly caused by the different distribution of short chain branches in material.

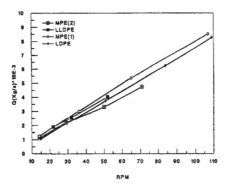

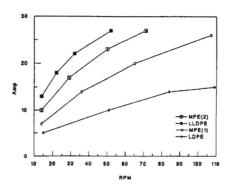

Figure 4. Dependence of output rates on screw revolutions for four materials.

Figure 5. Dependence of energy consumption on screw revolutions for four materials.

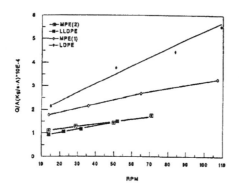

Figure 6. Dependence of output rate per energy consumption on screw revolutions for four materials.

Figure 5 gives the comparison of energy consumption at various screw revolutions for four materials. X-axis is ampere and Y-axis is screw revolutions per minute in Figure 5. The maximum ampere is 30A for the processing system. The order of energy consumption at a screw revolution for four materials is: LLDPE>MPE(2)>MPE(1)>LDPE. The material with long chain branches has lower energy consumption than the one without. Figure 6 shows the experimental results of output rate per energy consumption, Q/A, at various screw revolutions. LDPE appears the highest value in Q/A indicating that LDPE needs less energy during extrusion than other three. LLDPE and MPE(2) have almost the same values in Q/A. Experimental data show that LLDPE and MPE(2) are extruded at less than 70 screw revolution due to the limit of energy consumption, but LDPE and MPE(1) can be extruded at more than 100 screw revolution. It seems that long chain branches strikingly affect flow behaviors of material in extrusion.

Figures 7 to 10 describe the bubble conditions, stable, metastable, or unstable, in film blowing at various drawdown ratios and blowup ratios with a frost-line height, FLH, for four materials. Figures 7 is for LDPE at a FLH, 0.2 and Figures 8 is for LLDPE at a FLH, 0.2. Results in Figures 7 and 8 show that LDPE used in film blowing has a wide region of bubble stability, but LLDPE doesn't have such situations. Generally, LLDPE is more difficult to be processed in film blowing than LDPE. A possible way to obtain a stable bubble for LLDPE is to control frost-line heights below a certain value. The region of bubble stability for MPE(1) at a FLH(0.2m) and for MPE(2) at a FLH(0.13m) is presented in Figures 9 and 10, respectively. The narrow region of bubble stability tell us that both metallocene poly-

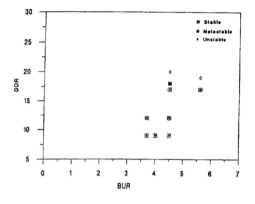

Figure 7. The bubble conditions at various drawdown ratios and blowup ratios with frost-line height, 0.2 m, for LDPE.

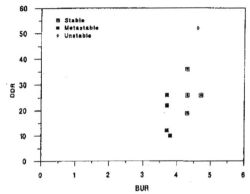

Figure 8. The bubble conditions at various drawdown ratios and blowup ratios with frost-line height, 0.2 m, for LLDPE.

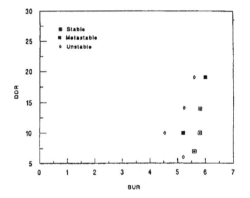

Figure 9. The bubble conditions at various drawdown ratios and blowup ratios with frost-line height, 0.2 m, for MPE(1).

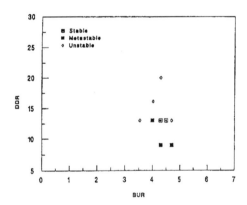

Figure 10. The bubble conditions at various drawdown ratios and blowup ratios with frost-line height, 0.13 m, for MPE(2).

ethylenes are difficult to be processed in film blowing. The range of FLH at which stable bubbles are found for four materials summarize in Table 2 as LDPE>LLDPE>MPE(1)>MPE(2). The wider range of FLH at which stable bubbles are observed represents that material has better processability.

CONCLUSIONS

LDPE, LLDPE, MPE(1) and MPE(2) have been studied in film extrusion. Four materials have different molecular structures and molecular weight distributions. Log G' vs. log G" associated with the four materials not only strongly depends on MWD, but also molecular structures such as long chain branches. LLDPE is the highest energy consumption during extrusion and LDPE is the less. The range of FLH at which bubble stability exists can be used to evaluate the proscessability of the material in film blowing. Metallocene polyethylenes are generally more difficult to be processed in film extrusion than LDPE and LLDPE.

REFERENCES

1.   D. C. Bugada and A. Rudin, *J. Appl. Polym. Sci.*, **33**, 87 (1987).
2.   L. O. Han and M. S. Jhon, *J. Appl. Polym. Sci.*, **32**, 3809 (1986).
3.   R. E. Christensen and C. Y. Cheng, *Plastics Engineering*, June, 31 (1991).
4.   C. D. Han and J. Y. Park, *J. Appl. Polym. Sci.*, **19**, 3257(1975).
5.   C. D. Han and J. Y. Park, *J. Appl. Polym. Sci.*, **19**, 3277(1975).
6.   R. M. Patel, T. I. Butler, K. L. Walton, and G. W. Knight, *Poly. Eng. Sci.*, **34** (19) 1506 (1994).
7.   A. Ghijsels, J. J. S. M. Ente, and J. Raadsen, *Intern. Polymer Processing*, **4** (1990).
8.   C. D. Han and J. Y. Park, *J. Appl. Polym. Sci.*, **19**, 3291 (1975).
9.   T. Kanai and J. L. White, *Polym. Eng. Sci.*, **28** (15) 1185 (1984).
10.   J. J. Cain and M. M. Denn, *Polym. Eng. Sci.*, **28** (23) 1527 (1988).

Table 2. The range of frost-line height having stable bubbles in film blowing for four materials

	Range of FLH having bubble stability min-max, m	ΔFLH, m
LDPE	$0.1 \approx 0.35$	0.25
LLDPE	$0.1 \approx 0.25$	0.15
MPE(1)	$0.15 \approx 0.25$	0.10
MPE(2)	$0.07 \approx 0.14$	0.07

An Innovative Approach to Understand Metallocene Based Polyolefins and Thermoplastic Resins for Film Applications

N.S. Ramesh and Nelson Malwitz
Sealed Air Corporation, Danbury, CT 06810, USA

ABSTRACT

In this paper, a simple rheological technique based on the entrance pressure drop method is used to predict the behavior of new resins under film blowing conditions to make heat shrink films. The influence of irradiation on extensional viscosity and bubble stability will be discussed.

IMPORTANCE OF THIS STUDY

Metallocene-based polyolefins are new polymers that can be tailored to achieve better physical properties. There is a great interest in industry to study metallocene-based resins for various end use applications. The purpose of this study is to understand the application of thermoplastic resins including metallocene in the polyethylenes film blowing process. Since hundreds of resins are available in the market, the industry needs a simple but powerful experimental technique successfully predict the suitability of these materials for stable processing in film blowing production. Film blowing involves biaxial extensional flow, where the blown film is stretched in the machine and cross-machine directions. The more stretch-stiffening the polymer melt, the more stable the bubble during the film blowing process. The more elastic is the deformation, the less susceptible it is to expand uncontrollably in the radial direction and cause process instability. The stretching or extensional viscosity of a polymer plays an important role in controlling the bubble stability during the film blowing process.

The measurement of biaxial viscosity requires either extensive equipment, great skill, or both. Hence, it is difficult to find any published data on biaxial extension of polyethylenes. On the other hand, converging flow experiments generate uniaxial extension, which can be used to compare resin performance qualitatively. The described experimental technique is simple, inexpensive and capable of screening the conventional or metallocene-based polyolefins for various polymer processing applications. This will allow polymer engineers to evaluate the new resins. Previous studies in the literature, discuss the complicated role of extensional viscosity on bubble stability. There is a need to develop a simple but reliable technique to understand the fundamentals of rheology in the extrusion of blown films. The purpose of this study is to accomplish such goal.

PROBLEM STATEMENT

The main objective is to determine if extensional rheology technique used here is effective for screening new thermoplastic and metallocene-based polyethylenes for the blown film process to prepare heat shrink films. The rheology of irradiated resins and its effect on bubble stability using a Haake rheometer was also investigated.

FILM BLOWING PROCESS AND VARIOUS RESINS

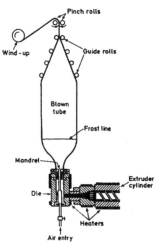

Figure 1. Schematic diagram of blown-film process.

Tubular film extrusion is one of the most important polymer processing operations used in industry. A schematic diagram of tubular film extrusion is shown in Figure 1. Polymer orientation and crystallization occur simultaneously in both machine and transverse directions due to air cooling from outside of the bubble. Conventionally, low density polyethylene, LDPE, and high density polyethylene, HDPE, are used for manufacturing blown films. LDPE exhibits good bubble stability due to its short chain branching, SCB, characteristics. SCB offers higher melt strength, higher extensional viscosity at higher stress and thus better bubble stability. HDPE is a popular choice for stiffer films although it is more difficult to achieve uniform properties in both machine and transverse direction due to its extensional viscosity characteristics. Ionomer resins have unique characteristics in offering good melt strength under cooling below its melting point lending itself for good bubble stability. On the other hand, the new metallocene-based polyethylenes, mPEs, exhibit lower melt strength and lower extensional viscosity which make them more challenging to run using the conventional equipment and process conditions. An attempt has been made here to experimentally characterize the extensional viscosity of these resins and then correlate their behavior to processing characteristics to predict the stability of blown film in shrink film applications.

EXPERIMENTAL WORK

MATERIALS

Seven resins were selected to study the effect of irradiation on extensional viscosity. This is a key factor in determining the stability of the bubble in our film blowing process.

EVA's
1. Dupont Elvax 3120, 1.2 MFI, 12% VA, 0.93 density
2. Dupont Elvax 3135, 0.25 MFI, 7.5% VA, 0.94 density
3. Dupont Elvax 3175, 6.00 MFI, 18% VA, 0.95 density

Metallocene Polyethylenes - VLDPE
4. DSM Teamex 1000, 2.2 MFI, 0.903 density
5. Dow Affinity PF1140, 1.6 MFI, 0.895 density
6. Dow Engage EG 8150, 0.5 MFI, 0.868 density

Ionomer
7. Dupont Ionomer Resin

The first six resins were irradiated to 1.0 MRad dosage levels. The ionomer resin was not irradiated. The extensional rheology experiment using the entrance pressure drop method was performed on virgin and 1.0 MRad irradiated resins using a Haake capillary rheometer fitted with a zero length capillary die having a 45° angle converging conical entrance zone. The extensional viscosity data of selected resins were compared to that of ionomer resin which appears to give good bubble stability for heat shrink film preparation.

EXTENSIONAL VISCOSITY MEASUREMENTS

The seven resins and irradiated samples were evaluated using capillary rheometry. Shear viscosity was determined by conventional capillary methods. End loss correction was applied to establish true shear viscosity as a function of true shear rate.

Elongational viscosity was determined by the converging cone method. Cogswell[1] has derived an expression using entrance pressure drop data. As melt flows from the reservoir of a capillary rheometer into a conical shape zero length die or orifice die, the streamlines converge and thus produce a strong extensional flow. The key assumption is that the extensional strain is so large that the stored elastic energy due to extension is approximately equal everywhere to some maximum value and that the flow is predominantly viscous. The viscous resistance to flow is described by power-law model. The equations which are often used and widely recognized are:

Extensional stress: $\sigma_E = \dfrac{3}{8}(n+1)P_o$

Extensional strain rate $\dot{\varepsilon} = \dfrac{4\dot{\gamma}_A^2 \eta_A}{3(n+1)P_o}$

Extensional viscosity $\eta_E = \dfrac{9(n+1)^2 P_o^2}{32\eta_A \dot{\gamma}_A^2}$

where P_o is the orifice pressure drop; η_A is the shear viscosity at fully developed flow at a shear rate $\dot{\gamma}_A$; n is the power law index.

A typical strain rate encountered in the process is in the range of 1 to 3 s^{-1}. Hence the rheological data in this same range was gathered to deduce useful conclusions.

THERMAL CHARACTERIZATION

The samples were evaluated for melting point, percent crystallinity, and melt flow index, MFI, to study irradiation effects. Perkin-Elmer DSC 7 machine was used to perform the DSC analysis.

RESULTS AND DISCUSSION
DSC RESULTS

The changes in MFI and thermal characteristics by DSC are summarized in Table 1. DSC values from the second heating were used to eliminate any manufacturing effects. The second heating effects are pertinent in a production film process. They can be taken as representing the true thermal characteristics of a resin erasing any morphology history built in to a resin by the manufacturing pelletization step.

Table 1. MI and thermal comparison of virgin and irradiated resins

Resin	MRads	Resin Density	MI	MI 1/Mi x Incr	DSC MP °C	$X_{100\%}$=289.9 J/g % crystallinity
Elvax 3120	-	0.93	1.20		97.1	19.0
Elvax 3120	1.0		0.22	5.5	96.3	19.9
Elvax 3135	-	0.94	0.34		94.0	21.5
Elvax 3135	1.0		0.09	3.8	93.0	21.8
Elvax 3175	-	0.95	6.30		70.7	4.0
Elvax 3175	1.0		0.68	9.3	69.7	3.4
DSMT 100	-	0.903	2.60		122.3	16.8
DSMT 100	1.0		1.34	1.9	123.2	17.2
Dow PF 1140	-	0.895	1.80		93.0	18.0
Dow PF 1140	1.0		1.10	1.6	93.4	16.0
Dow EG 8150	-	0.868	0.52		54.7	1.3
Dow EG 8150	1.0		0.18	2.9	44.8	1.1

Conclusions
- The thermal properties, such as melting point, the degree of crystallinity, change very little on irradiate resin as can be seen from Table 1.
- There is a profound decrease in MFI with irradiation at 1 MRad especially for the EVA resin.

EXTENSIONAL VISCOSITY RESULTS-BUBBLE STABILITY

Figures 2 and 3 show that irradiation does not seem to affect shear viscosity at working shear rates. In other words, the crosslinking cannot be detected by shear viscosity techniques meaning that the weight average molecular weight of samples remained same. However, a significant difference in extensional viscosity can be observed indicating that the molecular weight distribution has broadened. Figures 4 to 9 show the results of extensional viscosity data.

Conclusions

Upon analyzing the experimental results, the following conclusions have been made:
- Inspection of Figure 4 indicates that ionomer resin exhibits tension thickening (viscosity increase) when it is stretched at an extensional strain rate of 2 to 3 s^{-1}. For a good bubble stability this behavior is needed to achieve increasing resistance with the stretch of the blown film. An increased slope, followed by a flat elongational rheology curve characteristic is best

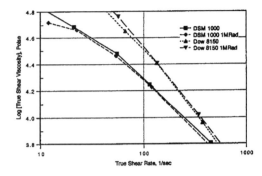

Figure 2. Effect of irradiation on shear viscosity. DSM Teamex 1000, 2.2 MFI, 0.903 density. DOW 8150 Engage elastomer, 1.6 MFI, 0.895 density. T = 130°C.

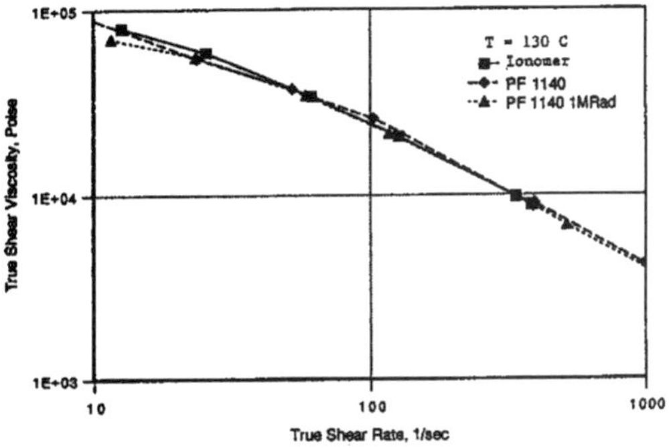

Figure 3. Effect of irradiation on shear viscosity.

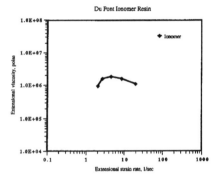

Figure 4. Rheology of ionomer resin at 150°C. DuPont ionomer resin.

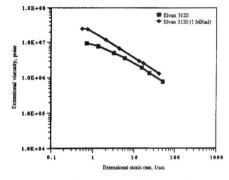

Figure 5. Effect of irradiance on DuPont Elvax 3120 at 130°C. 1.2 MFI, 7.5% VA, 0.93 density.

for achieving bubble stability. Indeed, in our experiments we found that ionomer resin gave excellent bubble stability during the film blowing process.

- Figures 5 to 9 show the rheology curves for EVA and mPE resins. They exhibit tension-thinning behavior (viscosity decreases with an increase in stretch rate) which is not favorable for achieving good bubble stability in preparing heat shrink films. Experimental trials showed that the above resins were difficult to process due to reduced melt strength at the desired stretching rates.

- It is also observed that the slope of extensional thinning curves for EVAs and mPEs gets steeper with increasing elongational strain rate indicating the lower melt strength. So, the irradiation is not expected to help stabilize the bubble with the new resins.

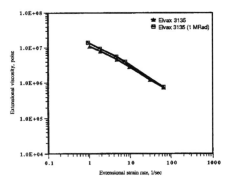

Figure 6. Effect of irradiation on DuPont Elvax 3135 at 130°C. 0.25 MFI, 12% VA, 0.94 density.

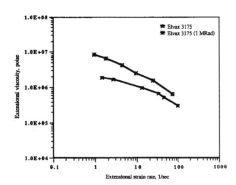

Figure 7. Effect of irradiation on DuPont Elvax 3175 resin at 130°C. 6.0 MFI, 18% VA, 0.95 density.

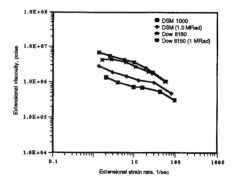

Figure 8. Effect of irradiation on very low density polyethylene (VLDPEs) at 130oC. DSM Teamex 1000, 2.2 MFI, 0.903 density. Dow 8150 Engage, 1.6 MFI, 0.895 density.

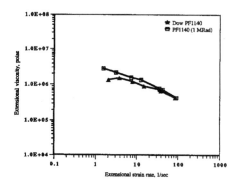

Figure 9. Effect of irradiation on Dow PF1140 VLDPE resin at 130°C. 1.6 MFI, 0.895 density.

- Although only a few metallocene resins have been evaluated, this experimental technique seems to correlate well with film processing for a new generation of metallocene-based polyolefins. More experimental work is required to understand and utilize metallocene-based polyolefins.

ACKNOWLEDGMENT

The authors wish to thank Leslie Marti for typing this manuscript and Sealed Air Corporation for allowing us to publish it.

Extrusion Behavior of Exxpol Metallocene Polypropylene

C. Y. Cheng
Exxon Chemical Company, Baytown, TX

ABSTRACT

Metallocene based polypropylene has many unique properties that are unattainable with current catalyst systems. Some of these attributes are narrow molecular weight and composition distribution, low extractable, low melt elasticity, and no vis-breaking required to achieve the narrow molecular weight distribution. The lower melting point and lower heat of fusion compared to conventional polypropylene, PP, are favorable to the extrusion process. Experimental results show that the metallocene PP can be extruded at the same conditions and equipment as the conventional PP with the same or higher output rate, depending on the temperature profile and screw design used. The new family of PP is chemically more stable during the extrusion process, resulting in less molecular breakdown and MFR changes following each process step.

With the continuing development of the metallocene catalyst system and the product family, fabricators will have more choices of resin attributes to meet the challenges in the market place.

INTRODUCTION

The metallocene catalyst technology development was initiated in the late '70s by academic researchers. These investigators discovered that certain metallocene species in combination with an oligomeric aluminum compound were interesting because the metallocene sites are all identical. Therefore, this new family of catalyst is also called "single-site" catalysts. This is in sharp contrast to the typical heterogeneous catalyst systems which are "multi-site" and have been used commercially for several decades. The distinct single-site characteristics of metallocene catalyst systems lead to the discovery of very interesting polymer products.

Extensive developments were launched by major polymer producers worldwide in the late '80s to commercialize metallocene based polymers. Ethylene polymers from a single-site catalyst were commercialized by Exxon Chemical Company in 1991. The new family of polyethylene extends the resin attributes well beyond the capability of the conventional multi-site catalyst. The commercialization of metallocene based propylene polymers, with the trade name of Achieve™, was also announced by Exxon Chemical Company in October 1995. The availability of these new resins has provided a new challenge and opportunity for the plastic industry.

KEY ATTRIBUTES OF METALLOCENE POLYPROPYLENE

This new generation of polymers has many unique properties that are unachievable with current conventional catalyst systems.[1] These properties affect both end use applications and fabrication processes. Table 1 summarizes the key attributes of the ACHIEVE metallocene propylene polymer.

Table 1. Comparison of metallocene propylene polymers vs. conventional polypropylene

Attributes	Achieve propylene polymers	Conventional polypropylene
Stereoregularity Melting point	147-158°C	160-165°C
Tacticity/compositional distribution Extractables	0.7%	3.3%
Molecular weight capability MFR range	0.01-5000	0.01-1000

Molecular weight distribution
    Typical
    Possible
	2.0
	2.0-4*
		3.5-6.0
		2.5**

*homopolymer, **CR grade

THERMAL PROPERTIES

The current generation of metallocene propylene homopolymer has a melting point of 147-158°C compared to 160-165°C for conventional polypropylene. This is a result of occasional reversals of the propylene molecules as they attach to the growing polymer chain. A reversal of the head-to-tail prevents the growing of the crystalline structure and the melting point is suppressed. The melting point of a particular ACHIEVE metallocene polypropylene depends on the catalyst system used.

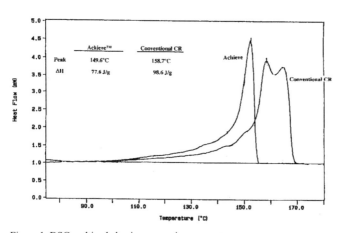

Figure 1. DSC melting behavior comparison.

The low melting point can have an advantage in various aspects of the extrusion process. For example, the melt temperature of the extrudate can be lowered to reduce the energy of extrusion and the subsequent cooling process. This would be beneficial for processes such as blow molding and thermoforming where low melt temperature provides an improved melt strength. Lower melting temperature may also make it easier to laminate with lower melting point substrates such as PP copolymers, polyethylene, etc.

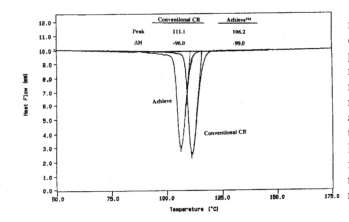

Figure 2. DSC crystallization behavior.

Figures 1 and 2 compare the melting and crystallization behavior differences of an ACHIEVE homopolymer sample and a conventional narrow MWD PP of comparable molecular weight. The metallocene resin shows a lower heat of fusion and crystallization. The combination of lower melting point and lower heat of melting are beneficial for the extrusion process. The lower melting temperature leads to earlier melting and therefore the initiation of viscous shear heating occurs sooner than conventional PP. The lower heat of fusion also improves melting as less energy is required to change the polymer from the solid state to the molten state. The crystallization temperature of the metallocene sample is only 5°C below the conventional grade although the melting point difference is approximately 10°C.

NARROW MOLECULAR WEIGHT DISTRIBUTION, MWD

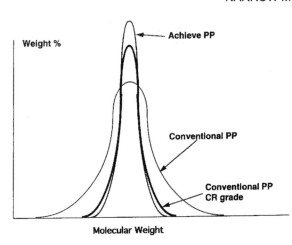

Figure 3. Comparison of molecular weight distribution of conventional vs. metallocene PP.

Polypropylene polymers produced by the current metallocene catalyst have a molecular weight distribution, MWD, as measured by M_w/M_n, of 2.0. This is in contrast to a broad range of MWD of 3 to 6 from conventional catalysts. The broad MWD of the polymer produced by the conventional catalyst is a result of multiple active sites of the catalyst solid. Figure 3 depicts the molecular weight distribution of PP with and without post-reactor modification from conventional catalyst, and the metallocene produced polypropylene of the same average molecular weight. For applications where broad MWD is required, the catalyst system currently being used to produce ACHIEVE resin can be tailored to produce specific MWD and composition distribution to meet the process and product needs.

LOW MELT ELASTICITY AND ELONGATIONAL VISCOSITY

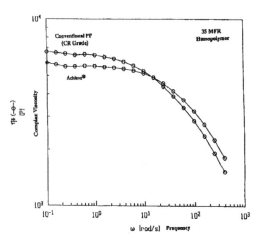

Figure 4. Viscosity comparison.

Due to narrow MWD, the metallocene based propylene polymer has a low melt elasticity and elongational viscosity. The low melt elasticity reduces die swell at the die exit and the melt can be drawn down more easily to form thinner film, finer fibers, etc. The low elongational viscosity reduces spin line stress, and the polymer can be extended easier.

Figure 4 compares the complex viscosity of a typical 35 MFR polypropylene and the metallocene ACHIEVE resins of the same MFR. Although the MFR is similar, the conventional resin shows a higher shear thinning effect due to broader MWD. Note that where the two viscosity curves cross each other is near the shear rate of MFR measurement.

NO CR (VIS-BREAKING) REQUIRED

Currently, PP made from conventional catalyst has broad MWD and must be thermally or chemically broken down in the post-reactor extrusion to obtain narrow MWD polymer grade for such applications as films and fibers. When peroxide initiated breakdown of the PP is employed, it is generally referred to as Controlled Rheology (or "CR" process).

Since the ACHIEVE products are more uniform in MW and have a narrower MWD compared to a typical CR product, it does not require post-reactor modification. In fact, the granules from the reactor (after the addition of stabilizer) can be fed directly into the extrusion system without pelletizing. This eliminates the extrusion step and preserves the molecular attributes.

LOW VOLATILE AND EXTRACTABLE

The CR process also produces small amounts of low and very low molecular weight species. Being a non-CR product, the ACHIEVE polymers have minimal low MW volatile components or fumes at the die exit. This can have significant environmental benefits, and at the same time reduces equipment downtime for equipment cleaning and housekeeping. The low extractable is especially attractive for food packaging applications.

EXTRUSION EXPERIMENTS

EXTRUDERS

Extrusion experiments were conducted on three different sizes of extruders over a broad range of screw speeds. Pressure profile is also monitored to get a better understanding of melting mechanism. The extruder size and the screw types used are listed below.

Extruder	L/D	Screw Type (Single Stage)
3/4"	25:1	Conventional
2.5"	24:1	Barrier with mixing section
4.5"	30:1	Barrier with mixing section

The 3/4" extruder utilized a conventional single metering screw without a mixing section. Since barrier screws have been used and analyzed extensively[2,3] barrier screws were used for the larger extruders. The 2.5" screw was a general purpose PE barrier screw with Maddock mixing section. The 4.5" extruder employed a high shear barrier screw designed for low viscosity materials since most of the resins evaluated have relatively low melt viscosity. In all experiments, the same restriction (adapter and die) was used to simulate the actual extrusion process.

RESINS

The following PP homopolymer resins were used in the experiment:

Conventional PP	**AchieveTM**
35 MFR broad MWD	35 MFR
35 MFR narrow (CR grade)	25 MFR
25 MFR narrow MWD (CR grade)	

All resins have the same additive packages and are in pellet form.

RESULTS AND DISCUSSION

EXTRUSION OUTPUT RATE

Having narrow molecular weight distribution, the metallocene polypropylene has improved extrusion characteristics. Earlier studies[4] have indicated that, in general, the narrower the MWD, the higher the achievable extrusion output rate. The result is an extrusion process that is less susceptible to surging.

The output rate for the 3/4" extruder shows that the output increases linearly with screw speed. That is, the output per RPM is constant throughout the screw speed tested (up to 130 RPM). This is not surprising for a small extruder where the external heating is a major source of energy for melting. The ACHIEVE resin shows a slightly higher output than conventional resin of CR and non-CR grade.

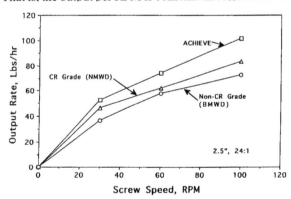

Figure 5. Output rate comparison. 35 MFR.

For the 2.5" extruder, the output difference is more significant. This may be due to the fact that a lower barrel temperature was used (380°F, 380°F, 450°F, and 450°F from feed to metering section). It has been found[5] that extrusion is favored by a reversed temperature profile due to its high melting temperature and heat of fusion. An ascending temperature profile is used to observe the differences between the two families of polypropylene. Figure 5 shows the output differences of ACHIEVE versus conventional PP. With the particular screw design and the temperature profile used, the ACHIEVE resin shows a higher output than the conventional PP. Note that the output rate does not increase in direct proportion with screw speed and the specific output rate (output per RPM) decreases significantly with screw speed. This may be the combined effect of the type of screw design and the temperature profile.[5]

For the 4.5" extruder, a reversed temperature profile was used (from 450°F at the feed to 420°F at the end of the metering section). With this temperature profile, the output rate of ACHIEVE resins of 35 and 25 MFR are consistently higher than the conventional resin. However, the difference between the two families of resin is smaller because of the high external heating. The output rate also increased linearly with the screw speed.

These results indicate that the metallocene based polypropylene can be processed at the same conditions as the conventional resin with the same or higher output rate. Such characteristics makes it easier for the processor to convert the resin usage from conventional to the metallocene based resin. Since the extrusion rate is controlled by the screw design, the output differences between the two resins will depend on the type of the screw used. The extrusion performance may be further enhanced by optimizing the temperature profile for each resin to account for melting behavior differences.

MELT TEMPERATURE AND EXTRUSION TORQUE

While less shear sensitive due to narrower MWD, the melt temperature of the ACHIEVE resin is expected to be higher than conventional PP with broader MWD. This is the case for the 4.5" and 3/4" extrusion experiments where the output rate of the ACHIEVE resin is only moderately higher than the conventional resin. When the specific output rate (output per RPM) of the ACHIEVE resin is higher, the melt temperature will be lower due to lower specific energy input. This is in fact observed on the 2.5" extruder as shown in Figure 6.

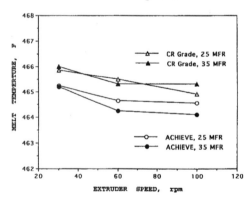

Figure 6. Melt temperature comparison. 2.5", 24:1.

For polypropylene, the melt viscosity of the most commercial resins is relatively low and therefore narrow MWD does not pose extrusion difficulties in terms of extruder torque or excessive melt temperature. For example, a typical filament yarn grade, film grade, and fiber grades have a MFR of 3-4, 4-12, and 12-42, respectively. This is in contrast to most polyethylene grades that have MI of 2 or less. Therefore, the melt temperature of an ACHIEVE resin can be controlled through proper screw design and barrel temperature settings.

The extruder torque is reflected in the extruder drive motor amperage draw. The narrow MWD ACHIEVE resin is expected to have a higher torque than conventional resin. This is less an issue with PP due to lower melt viscosity, and the extruder torque is well below most extruder capacities.

EFFECTS OF MULTIPLE EXTRUSION

For polypropylene resin, thermal and oxidative breakdown of the polymer during each extrusion step is a major concern. For example, it is not unusual for MFR to increase from 8 MFR to 11 MFR for film extrusion and from 35 to 45 for fiber spinning, depending on the resin stabilizer package and extrusion

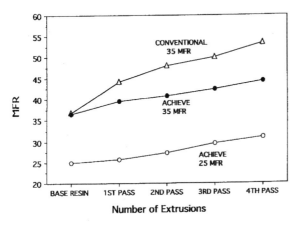

Figure 7. MFR changes due to multiple extrusions.

conditions. Being less shear sensitive and lacking high MW species present in conventional PP, the MFR of ACHIEVE resin is much more stable following each extrusion step. Figure 7 compares MFR changes resulting from successive extrusion. The metallocene resin shows significantly less MFR or average molecular weight changes after repeated extrusion. This can have a significant impact on the finished product physical properties and recycling process. The minimal molecular breakdown preserves the MFR and the properties of the extruded product.

In terms of recycling when ACHIEVE resin is used, the recycled material will have a closer MFR to the virgin material and therefore, the rheological properties of the two will be much closer. Hence, a larger amount of recycled material can be used, preserving process stability and product quality.

CONCLUSIONS

The ACHIEVE propylene polymer produced from the metallocene catalyst technology has unique attributes that can be used to design new products and processes. Among the key attributes are narrow molecular weight, uniform composition distribution, lack of low molecular weight fractions, and excellent organoleptic properties.

The extrusion characteristics of ACHIEVE resins are not very much different from the conventional polypropylene resin. However, with proper screw design and extruder temperature profile, the low melting point and low heat of fusion of the ACHIEVE polymer has the advantage of a higher output rate without excessive extruder torque and melt temperature. Hence, the ACHIEVE resin can be processed at the same condition as the conventional resin without sacrificing extrusion rate or stability.

The development of metallocene based propylene polymer is still in its infant stage. As the new catalyst technology develops, the processor will have broader choices of resin attributes to meet their market needs.

REFERENCES

1. J. J. McAlpin, C. Y. Cheng, D. A. Plank, and G. A. Stahl, *INSIGHT '95*, October 22, 1995.
2. C. Y. Cheng, *Plastics Engineering*, November, 1978.
3. K. Amellal and B. Elbivli, *45th ANTEC*, 1987.
4. C. Y. Cheng and R. E. Christensen, *Plastic Engineering*, **47(6)** 1990.
5. C. Y. Cheng, *SPE ANTEC*, 1990.

The Effects of Long Chain Branching on the Processing-Structure-Property Behavior of Polyethylene Blown Film Resins

A. M. Sukhadia

Phillips Petroleum Company, 94-G PRC, Bartlesville, OK 74004, USA

ABSTRACT

Long chain branching, LCB, in polyethylenes, PE, has been recognized as a very important structural parameter for over 30 years. In film blowing, it has been well established that higher levels of LCB improve the bubble stability of the resin. This work was undertaken to fully explore the effects of LCB on the processing-structure-property, P-S-P, behavior of PE blown film resins. LCB introduced by different methods such as peroxide addition, finishing/stabilization and polymerization changes was explored in this work. The effects of LCB on rheology, blown film processability and film properties are discussed in detail.

INTRODUCTION

Long chain branching, LCB, in polymers has been recognized as an important molecular structural parameter for over 30 years. Earlier work has shown that even small amounts of LCB can profoundly affect the rheological properties of these materials.[1-5] LCB in PE can be introduced in many different ways such as peroxide addition,[6] radiation treatment,[7] finishing/stabilization,[8] and through an appropriate choice of polymerization conditions.[9] Since the modes by which LCB can be introduced in the resin are many and small amounts of LCB affect resin properties significantly, it becomes critical to understand how the different melt and solid state properties are affected.

It is generally recognized in the literature that higher levels of LCB in film blowing resins give improved bubble stability.[10-13] This is thought to be due to the extensional strain hardening characteristics of melts with LCB. While the improved bubble stability with higher LCB is perhaps well established and accepted, several important questions about the role of LCB in film blowing resins are believed to be unknown. For example, is there a difference in the influence of LCB on resin behavior and properties from LCB due to polymerization effects versus LCB due to finishing effects? What is the effect of LCB on film properties viz. impact and tear strengths? Is there a density dependence, i.e., does LCB affect high density resins differently from low density resins? Besides an improvement in bubble stability, does LCB affect other processing behavior such as melt fracture or drawdown? This work was undertaken in order to try and answer some of these questions in order to gain a more comprehensive understanding of the effects of LCB on the processing-structure-property (P-S-P) behavior of polyethylene blown film resins.

EXPERIMENTAL
RHEOLOGY

Melt Index, MI, was measured as per ASTM D-1238, Condition F (190°C, 2.16 kg). High Load MI, HLMI, was measured as per ASTM D-1238, Condition E (190°C, 21.6 kg).

The rheological data was obtained on a Rheometrics Mechanical Spectrometer (model RMS-800) using a parallel plate geometry. Temperature-frequency sweeps were performed between 170-250°C in 20°C increments. This data was reduced to a single master curve at 190°C using the well known Williams-Landel-Ferry, WLF, time-temperature superposition method,[14] through a proprietary algorithm to estimate the flow activation energy. This master curve was then fitted with the three parameter Carreau-Yasuda, CY, empirical model to obtain the CY parameters viz. zero shear viscosity, η_0, characteristic maximum relaxation time, τ_{relax}, and A parameter. Details of the significance and interpretation of these three parameters may be found elsewhere.[15,16]

BLOWN FILM EVALUATIONS

The film blowing experiments were conducted on a laboratory scale blown film line. The line consists of a 38 mm (1.5 inch) diameter single screw Davis Standard extruder (L/D=24; 2.2:1 compression ratio) fitted with a barrier screw with a Maddock mixing section at the end. The die used was 50 mm (2 inch) in diameter from Sano and was fitted with a single lip air ring. The die gap was 0.89 mm (0.035 inch). The film blowing was done at typical high density polyethylene, HDPE, conditions as follows: 4:1 blow up ratio, BUR, 7:1 frost line height, FLH, to die diameter ratio and 220°C extruder and die set temperatures. Films of 0.0254 mm (1 mil) gauge were made in each case. The film properties were obtained by processing at 6.8 kg/hr (15 lb/hr) output rate using a screw setting of 25 RPM. These conditions were chosen since the film properties so obtained scale directly with those from larger commercial scale film blowing operations.[17] The processability characteristics viz. melt fracture and drawdown were evaluated at a higher output rate of 18.2 kg/hr (40 lb/hr) using a screw RPM setting of 75 RPM. This was done to provide more severe film blowing conditions in terms of die shear rates and shear stresses. Melt temperatures were about 225°C at the 25 RPM conditions and 237°C at the 75 RPM conditions.

BLOWN FILM PROPERTIES

The film properties were measured as follows:
  1. Dart impact strength - ASTM D-1709 (method A)
  2. Machine, MD, and Transverse, TD, direction Elmendorf Tear strengths - ASTM D-1922

BLOWN FILM RESINS

Five experimental resins were used in this study. The first is a high molecular weight, high density polyethylene, HMW-HDPE, type resin designated as P1 having the following characteristics: 7 HLMI, 0.03 MI, 0.950 g/cc density. The second is a low density linear, LLDPE, type resin having a medium molecular weight and broad molecular weight distribution with the following characteristics: 24 HLMI, 0.24 MI, 0.923 g/cc density. Three more resins designated P3, P4, and P5 which are all HMW-HDPE type resins were also used in this study and they are all nominally 11 HLMI, 0.08 MI and 0.950 g/cc density. Other rheological and molecular weight data is presented in the body of the paper and thus will not be repeated here.

There are considerable difficulties involved in the accurate measurement of LCB levels through conventional methods such as light scattering, intrinsic viscosity, gel permeation chromatography, GPC, and nuclear magnetic resonance, NMR, especially at low LCB levels of the order of 0.1 LCB/1000 C atoms.[1] The LCB levels in the chrome catalyst HDPE resins used here are estimated to be of the order of 0.01 - 0.1 LCB/1000 C based on the earlier studies.[1] LCB levels in this study were therefore assessed on a relative rather than absolute basis through the use of the rheological flow activation energy, E_a, as demonstrated in earlier studies.[1,18] An increase in the flow activation energy at largely constant weight average molecular weight, M_w, and molecular weight distribution, MWD, was attributed to be due to an increase in the LCB level in the polymer.

RESULTS

One method that researchers have used to introduce small but controlled amounts of LCB in polyethylene resins has been through peroxide addition.[6,19] The peroxide used in this study was 2,5-dicumyl peroxide, DCP, and the experiment is described below in further detail.

Two lots of P1 type resin differing in the weight average molecular weight, M_w, were used for the study. Resin P1-A had a 6 high load melt index, HLMI, while resin P1-B had a 9 HLMI. The samples were prepared as follows. Fluff from resins A & B was finished using a proprietary stabilization package and no peroxide. Three more samples were made with Resin B fluff containing 20 ppm, 40 ppm, and 60 ppm peroxide in addition to the stabilization additives. Note that a processing aid (0.07 wt% FX-9613 fluoropolymer from 3M Co.) was used with all the P1 resins.

Table 1. Effects of peroxide induced long chain branching on P1-A and P1-B rheology and molecular weight data

Resin	6 HLMI (P1-A)	9 HLMI (P1-B)	P1B+20 ppm	P1-B+40 ppm	P1B+60 ppm
Eta(0), Pas	9.89E+5	7.53E+6	2.12E+6	4.33E+6	1.23E+7
Tau Relax, s	4.5	3.2	11.5	29.2	103.7
A	0.194	0.192	0.170	0.166	0.152
E_a, kJ/mol	32.3	32.6	34.6	35.9	38.8
M_w	291000	275000	272000	288000	289000
M_n	14.3	14.0	12.0	13.3	11.9
M_w/M_n	20.4	19.6	22.8	21.6	24.2

P1 PEROXIDE STUDY

1. Gel Permeation Chromatography, GPC: The molecular weight data viz. weight average molecular weight, M_w, number average molecular weight, M_n, and molecular weight distribution (MWD or M_w/M_n) for these 5 resins is shown in the bottom half of Table 1. Resin A is slightly higher in M_w than Resin B. After peroxide addition, the three samples with 20, 40, and 60 ppm peroxide show a slight increase in the M_w from that of the parent resin P1-B. It is worthwhile to note that the molecular weight data of Resin A and Resin B + 40 ppm peroxide are largely similar.

2. Rheology: The quantitative rheological data in terms of the CY parameters is shown in the top half of Table 1. The addition of peroxide results in a systematic increase in η_0, τ_{relax} and flow activation energy, E_a, but a decrease in the A parameter. Particular attention is drawn to the flow activation energy data where the E_a increases from about 32 kJ/mol for the control P1 resins (A & B) to about 39 kJ/mol for the 60 ppm peroxide sample. The typical E_a values for polyethylene resins with no LCB are reported in the literature to be about 26-29 kJ/mol.[18,20] The two control resins with E_a values of 32 kJ/mol are, therefore, believed to contain some LCB. However, it is the relative increase in the level of LCB with peroxide addition that is being suggested as the cause of the rheological changes being observed.

3. Blown Film Processability and Properties: The blown film processability and property information is summarized in Table 2. The P1 control resins showed no melt fracture as expected due to the presence of the fluoropolymer. However, as LCB is introduced in the resins, the film begins to exhibit sharkskin melt fracture, SSMF.

Table 2. Effects of long chain branching on the blown film properties and processability of P1 resins

Resin description	Dart impact, g	Tear, MD, g	Tear, TD, g	Comments
P1-A, 6 HLMI	277	26	256	0.07% FE; no melt fracture
P1-B, 9 HLMI	246	23	286	0.07% FE; no melt fracture
B + 20 ppm peroxide	177	30	46	0.07% FE; very slight melt fracture
B + 40 ppm peroxide	93	25	26	0.07% FE; slight melt fracture
B + 60 ppm peroxide	53	20	13	0.07% FE; moderate melt fracture

Film properties are for 1.3 mil thick films

It was mentioned earlier in the introduction that LCB improves bubble stability in the blown film resins. However, the P1-B + 60 ppm peroxide could not be drawn down at standard processing conditions to the normal 1 mil thickness due to the bubble breaking or tearing off in the stalk. The lowest gauge possible with this sample was 1.3 mils. To keep all the five samples on a comparable basis for film properties, the remaining P1 samples were also made at 1.3 mils.

The film property results show a significant reduction in the dart impact strengths as well as the TD tear strength of the P1 films with increasing LCB. No effect, however, on the MD tear strength is observed. The biggest effect of the LCB appears to be on the TD tear strength which reduced from 286 g to 46 g with the addition of just 20 ppm peroxide to P1-B as shown in Table 2. These property reductions are very significant and not intuitively obvious in light of the GPC and rheology data.

LCB VIA FINISHING/STABILIZATION

The process of converting reactor fluff (or powder) to pellets is often referred to as either finishing and/or stabilization. This is typically accomplished by dry blending the requisite amount of stabilization additives with the fluff prior to pelletization.[21] If the additives and/or the levels used are insufficient, the possibility of forming LCB is increased.[22,23]

One lot of fluff of a LLDPE type resin, P2, was finished at the same extrusion conditions using seven different proprietary stabilization packages. The samples are designated as X1 - X7 in the following discussions.

1. GPC: The GPC traces for the seven samples after finishing showed no significant differences in the molecular weight characteristics viz. M_w, M_n and M_w/M_n. For the sake of completeness, however, it is sufficient to state here that values of M_w = 227,370, M_n = 12,530 and M_w/M_n = 18 were observed on average for these seven samples.

2. Rheology: The rheology of these seven samples in terms of the CY parameters is shown in Table 3. The flow activation energies, E_a, and relaxation times for samples -X4, X6, and X7 are the highest in this set, although E_a for sample - X4 is not different from the other four samples. Based on both the relaxation time and flow activation energy data, it is postulated that samples -X4, X6, and X7 contain somewhat higher LCB levels relative to the others. The LCB are formed during finishing due to inadequate stabilization from the additive packages.

Table 3. Effects of long chain branching via finishing/stabilization on the rheology of P2 LDLPE film resin

P2-	Ea, kJ/mol	Eta(0), Pas	Tau Relax, s	A
X1	36.4	8.01E+5	2.56	0.175
X2	35.4	7.81E+5	2.39	0.176
X3	36.5	9.05E+5	2.90	0.175
X4	36.7	1.27E+6	4.44	0.169
X5	35.5	1.03E+6	2.38	0.161
X6	37.4	1.38E+6	5.07	0.169
X7	38.4	1.19E+6	4.69	0.174

3. Blown Film Processability and Properties: The blown film properties for these seven samples are shown in Table 4. The processability of these seven samples was judged to be the same. However, considerable differences in the blown film properties can be seen in Table 4. Samples X1, X2, X5, and possibly X3 were judged to be equivalent in film properties. Samples X4, X6, and X7, however, are clearly lower in both impact as well as TD tear strengths and this is attributed to the higher LCB levels as discussed above.

Table 4. Effects of long chain branching on the blown film properties and processability of P140 resins

Resin description	Dart impact, g	Tear, MD, g	Tear, TD, g	Comments
P2, control	152	24	379	no melt fracture
P2 + 10 ppm peroxide	111	23	193	slight melt fracture
P2 + 25 ppm peroxide	63	24	50	moderate melt fracture
P2 + 50 ppm peroxide	-	-	-	poor drawdown, severe melt fracture
P2 + 100 ppm peroxide	-	-	-	poor drawdown, severe melt fracture

LCB VIA POLYMERIZATION CONDITIONS

Three resins viz. P3, P4, and P5 were prepared using different polymerization conditions which were intended to affect LCB. The goal was to make 10 HLMI, 0.950 g/cc density resins in each case.

  1.GPC: The GPC and HLMI data for the above three resins is also shown in Table 5. It is clear that all three polymers are quite similar and in fact resins P3 and P4 are virtually identical in these characteristics.

Table 5. Effects of long chain branching due to polymerization conditions on rheology and molecular weight data

Resin	P3	P4	P5
HLMI, dg/min	10.5	11.7	13.5
Eta(0), Pas	5.05E+5	2.59E+6	2.93E+6
Tau Relax, s	2.6	19.5	227
A	0.211	0.167	0.173
E_a, kJ/mol	32.7	34.5	34.9
M_w	217230	215670	227910
M_n	6800	7740	8020
M_w/M_n	31.9	27.9	28.4

2. Rheology: The rheological data for these resins, again represented by the CY parameters, are given in Table 5. It is noteworthy that resin P3 has a lower flow activation energy of 32.7 kJ/mol compared to resin P4 with an E_a of 34.5 kJ/mol.

3. Blown Film Processability and Properties: When blown into film at the 75 RPM conditions (see experimental section for further details), resin P3 processed very well while both the other resins exhibited very severe melt fracture.

Blown film properties were evaluated on films made using a lower screw speed (25 RPM) such that no melt fracture was observed for any of the resins and are shown in the following Table 6. Resin P3 has the highest dart impact strength and TD tear strength in this set and it is significant to note that the dart impact strength of P4 is almost half of P3.

Table 6. Effects of long chain branching via polymerization changes on the blown film properties of HDPE film resins

Resin	Dart impact, g	Tear, MD, g	Tear TD, g
P3	151	26	429
P4	78	23	322
P5	118	25	339

CONCLUSIONS

The effect of long chain branching on the processing-structure-property behavior of blown film resins is the same regardless of whether the LCB is introduced via peroxide addition or via finishing/stabilization or polymerization changes. The presence of very small amounts of LCB in a blown film resin will have very significant effects on both the processability and property characteristics. These small levels of LCB are not easily detectable in typical GPC or rheology data and use of other rheological indicators such as flow activation energy and melt elasticity must be used to account for their presence and relative levels in PE resins. An increase in the LCB level results in an increase in the zero shear viscosity, relaxation time, flow activation energy and melt elasticity and a decrease in the breadth A parameter. An increase in the LCB level results in lower impact strength and transverse direction, TD, tear strength of the blown films. The magnitude of the reduction in film properties increases with increasing LCB levels. The melt fracture potential (i.e., tendency to exhibit sharkskin melt fracture) of the resins in film blowing increases with an increase in the LCB level. This is attributed to an increase in the characteristic relaxation time and possibly higher extensional viscosity as a result of the higher LCB content. The drawdown or downgauging ability of the resins was observed to decrease considerably with very high levels of LCB and brittle failures in the stalk region of the bubble were observed at high drawdown ratios. Very high LCB levels are therefore expected to have another undesired effect on the processability viz. the inability to obtain thin films. Based on the above results taken together, it is concluded that any significant level of LCB in PE blown film resins is extremely undesirable and should be minimized as much as possible. Improved bubble stability (from high LCB) can only be obtained at the expense of the other processability and film property characteristics described above and this tradeoff should be well recognized and expected.

ACKNOWLEDGMENT

The author would like to thank all his coworkers in the polymerization, finishing and polymer physics groups for their help in various parts of this project. In addition, the author would like to thank Phillips Petroleum Company for support and permission to publish this work.

REFERENCES

1.   J. K. Hughes, *Proc. SPE ANTEC*, Chicago, May 2-5, 1983.
2.   W. W. Graessley, *Accounts of Chemical Research*, **10**, 332 (1977).
3.   L. Wild, R. Ranganath, and D. C. Knobeloch, *Polym. Eng. Sci.*, **16(12)** 811 (1976).
4.   P.A. Small, **Advances in Polymer Science**, Eds H. J. Cantow *et al.*, Vol. 18, New York, 1975.
5.   W. W. Graessley and E. S. Shinbach, *J. Appl. Polym. Sci., Polym. Phys. Ed.*, **12**, 2047 (1974).
6.   Y. Matsuo *et al.*, **U.S. Patent 4 508 878** (1985).
7.   A. Garcia-Rejon, O. Manero, and C. Rangel-Nafaile, *J. Appl. Polym. Sci., Polym. Sci. Ed.,* **30**, 2941 (1985).
8.   J. Schiers, S. W. Bigger, and N.C. Billingham, *J. Appl. Polym. Sci., Polym. Chem.*, **A30**, 1873 (1992).
9.   J. P. Hogan, C.T. Levett, and R.T. Werkman, *SPE J.*, 87 (1967).
10.   H. Yamane and J. L. White, *J. Soc. Rheology*, **15(3)** 131 (1987).
11.   K. K. Dohrer and D.H. Niemann, *Proc. SPE ANTEC*, 28 (1989).
12.   E. A. Colombo, T.H. Kwack, and T. Su, **U.S. Patent 4 614 764** (1986).
13.   M. Fleissner, *Intern. Polym. Proc.*, **II(3/4)** 229 (1988).
14.   J. D. Ferry, Ed., **Viscoelastic Properties of Polymers**, 3rd Edn., *John Wiley & Sons*, New York, 1980.
15.   C. A. Hieber and H. H. Chiang, *Rheol. Acta*, **28**, 321 (1989).
16.   C. A. Hieber and H. H. Chiang, *Polym. Eng. Sci.*, **32**, 931 (1992).
17.   A. M. Sukhadia, *J. Plast. Film and Sheeting*, **10(3)**, 213 (1994).
18.   B. H. Bersted, *J. Appl. Polym. Sci.*, **30**, 3751 (1985).
19.   D. Anzini, **U.S. Patent 5 073 598** (1991).
20.   L. E. Nielsen, **Polymer Rheology**, *Marcel Dekker*, New York, 1977.
21.   L. T. Pearson and M. Souza, *Polym. Eng. Sci.,* **32(7)** 475 (1992).
22.   A. Harlin and E. Heino, *J. Polym. Sci., Polym. Phys.* Ed., **33**, 479 (1995).
23.   R. Gachter and H. Mueller, Eds., **Plastics Additives Handbook**, 3rd Edn., *Hanser Publishers*, New York, 1990.
24.   M. S. Jacovic, D. Pollock and R. S. Porter, *J. Appl. Polym. Sci.*, **23**, 517 (1979).
25.   R. P Chartoff and B. Maxwell, *J. Appl. Polym. Sci.,* **(A2) 8**, 455 (1970).
26.   F. Pinaud, *J. Non-Newtonian Fluid Mech.*, **23**, 137 (1987).
27.   D. R. Michiels, *MetCon'94*, May, 1994.

Spherilene Process LLDPE for Conventionally Blown Shrink Film

T. J. Cowell

Montell Polyolefins Inc., Wilmington , Delaware, USA

ABSTRACT

Properties of blown shrink films produced from Spherilene Process LLDPE resins are detailed. These resins yield films with substantial transverse direction shrinkage and high resistance to pinholing, or burn through, during end use shrink tunnel processing. The current shrink film market is comprised largely of high pressure LDPE and LDPE-rich blends. Spherilene Process LLDPE may be used by replacing non-shrinking, conventional LLDPE in blends with LDPE. Further, these new resins may be used to decrease, or even eliminate, the LDPE content of shrink films. Examples are cited.

INTRODUCTION

Shrink film is used to bundle, collate, or over-wrap products for unitization, display esthetics, and/or tamper resistance. Once wrapped, the package is exposed to a thermal treatment (such as via a forced convection oven or heat gun) to tighten the film. Two general categories of shrink film exist-specialty shrink film produced via some extra-orientation process (such as tentering or double-bubble blown film fabrication) and industrial shrink film produced via conventional blown film fabrication. The latter is the focus of this paper.

The conventional blown shrink film market is dominated by high pressure low density polyethylene, LDPE. The most typical LDPEs used in this application are autoclave resins with melt indices around 0.4-0.7 g/10 min and tubular resins with melt indices around 0.2-0.3 g/10 min.

Films made from these LDPEs have very good shrink performance, give tight packages when unitizing goods, and heat seal at low temperatures. However, they don't downgauge well, have a tendency to form holes easily during the shrinking process, have relatively low physical properties (tensile strength, puncture, and tear resistance), and are generally more expensive than linear low density polyethylene, LLDPE. LDPE-rich blends with LLDPE, high density polyethylene, and ethylene vinyl acetate copolymers are used to overcome some of these deficiencies, as needed, in various premium, freezer, and high tensile film applications.

Shrink film manufacturers are currently looking for resins that will allow the production of a higher performance film (bringing value to their customers) or downgauging (bringing cost savings). Conventional LLDPE resins will yield film strength enhancements as well as an ability to downgauge, but films made from these resins will, instead of shrinking in the transverse direction, TD, expand upon heating. This expansion leads to loose packages that do not effectively unitize goods. Metallocene polyethylenes, mPEs, have been likewise explored as an alternative to LDPE in this market and have been found to shrink in the TD.[1] The cost of mPEs, however, still tends to be very high, lessening their impact on the conventional blown film shrink market. Spherilene Process High Performance LLDPEs[2] are shown to be a viable solution.

EQUIPMENT AND PROCESSES

BLOWN SHRINK FILM FABRICATION

A comparison was conducted of the shrink properties of three conventional LLDPEs (ethylene/1-butene, EB, ethylene/1-hexene, EH, and ethylene/1-octene, EO, copolymers) and Spherilene Process Shrink LLDPE A. The characteristic properties of the resins are detailed in Table 1. Films of 25.4 μm (1.0 mil) thickness were produced from these resins under the same conditions on an 80 mm (3.2 in) Gloucester blown film line with a 24 cm (9.5 in) die and 2.0 mm (80 mil) die gap at a 2.5:1 blow-up ratio, BUR.

Table 1. Resin characteristic properties

Resin	I_2,Melt Index g/10min	Base Density kg/m³
Ethylene/1-butene (EB) LLDPE	1.0	918
Ethylene/1-hexene (EH) LLDPE	1.0	918
Ethylene/1-octene (EO) LLDPE	1.0	920
Spherilene Process Shrink LLDPE A	0.95	909
Spherilene Process Shrink LLDPE B	0.60	909
Autoclave LDPE	0.9	920
Spherilene Process Shrink LLDPE C	0.50	912
Tubular LDPE	0.25	920
Ethylene/1-butene (EB) LLDPE 2	0.95	916

Spherilene Process Shrink LLDPE B and an autoclave LDPE (see Table 1) were fabricated into shrink films over a range of blend compositions. An 89 mm (3.5 in) Gloucester line with a 20 cm (8 in) die and 1.5 mm (60 mil) die gap was used. Fabrication conditions are listed in Table 2. The LDPE was downgauged, as much as was possible, to a 35.6 mm (1.4 mil) thickness and then made into a 66.0 mm (2.6 mil) film as well. The blends were fabricated at the thin gauge. All of these films were made at a 3.0:1 BUR.

Blown films made from Spherilene Process Shrink LLDPE C, a tubular LDPE, and EB LLDPE 2 (see Table 1) were fabricated at a 50.8 μm (2.0 mil) thickness-the lowest gauge possible with the LDPE-on the same equipment as the films described immediately above. Internal bubble cooling, IBC, and a 2.7:1 BUR were, however, utilized this time. The LDPE had to be run at a 15% lower output rate due to difficulties in downgauging. Fabrication conditions are detailed in Table 3.

Table 2. Blend study fabrication conditions

Shrink LLDPE B, %	0	25	50	75	100	0
Autoclave LDPE, %	100	75	50	25	0	100
Gauge, μm (mil)	35.6 (1.4)	35.6 (1.4)	35.6 (1.4)	35.6 (1.4)	35.6 (1.4)	66.0 (2.6)
Melt Temp., °C (°F)	193 (379)	229 (444)	234 (454)	237 (458)	237 (458)	193 (379)
Pressure, MPa (psi)	18.4 (2670)	21.0 (3050)	27.9 (4050)	32.1 (4650)	32.2 (4670)	18.0 (2610)
Amperage, A	66	68	78	94	114	67
Screw Speed, rpm	35.5	36.5	32.5	31.5	31.5	35.5
Line Speed, m/s (ft/min)	0.610 (120)	0.575 (113)	0.574 (113)	0.570 (112)	0.570 (112)	0.305 (60)
Output, g/s/cm (lb/hr/in) die circ	3.5 (10.9)	3.5 (10.9)	3.5 (10.9)	3.5 (10.9)	3.5 (10.9)	3.5 (10.9)

Table 3. Blown film fabrication conditions

	Shrink LLDPE C	Tubular LDPE	EB LLDPE 2
Gauge,μm (mil)	50.8 (2.0)	50.8 (2.0)	50.8 (2.0)
Melt Temp., °C (°F)	244 (471)	246 (475)	229 (444)
Pressure, MPa (psi)	29.9 (4340)	17.4 (2530)	32.5 (4710)
Amperage, A	103	60	110
Screw Speed, rpm	25.9	35.0	28.7
Line Speed, m/s (ft/min)	0.384 (75.5)	0.331 (65.1)	0.382 (75.1)
Output, g/s/cm (lb/hr/in) die circ.	3.2 (9.9)	2.7 (8.4)	3.1 (9.8)

FILM TESTING

The evaluation of film free shrinkage is accomplished by carrying out the following test. Square film samples, 10.0 cm x 10.0 cm (3.9 in x 3.9 in), are placed in metal trays containing a thin layer of Dow Corning 200 silicone oil. These are then put in a forced convection oven for 10 min at the set temperature. The samples are removed from the oven, cooled, and measured. The machine direction, MD, and TD free shrinkage of the samples are then computed and recorded.[3]

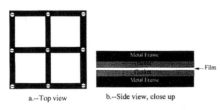

a.--Top view   b.--Side view, close up

Figure 1. Burn through test apparatus.

The resistance of film to pinhole, or burn through, upon shrinking is determined by fastening 27.9 cm x 27.9 cm (11.0 in x 11.0 in) film specimens in a metal frame containing four 10.2 cm x 10.2 cm (4.0 in x 4.0 in) holes. High temperature rubber gaskets are used to prevent slippage (Figure 1). The frame is placed in a forced convection oven at a set temperature for 10 min and then removed. Any hole, be it very small or very large, or number of holes in one of the four cells counts as a single failure. Therefore, at a given temperature the percentage failure could be 0, 25, 50, 75, or 100%.

Shrink tunnel testing of blown films was conducted by wrapping cardboard flats of goods, forming a lap seal on the bottom of the package, and sending the assembly through a forced convection oven on a wire mesh conveyor. Residence time was kept constant at 20 s as oven temperature was varied.

The Montell test method for puncture measures the resistance of a 10.2 cm (4.0 in) diameter film specimen to rupture by a 1.9 cm (0.75 in) diameter Teflon coated ball affixed to a rod moving through the plane of the sample at a rate of 25.4 cm/min (10 in/min). All other film tests were conducted under ASTM conditions.

RESULTS AND DISCUSSION

The free shrinkage data given in Table 4 show that all of the LLDPE resins, with the exception of Shrink LLDPE A, expand in the TD upon heating. Significant differences also exist in burn through resistance, as is depicted in Figure 2. From an applications viewpoint, the Spherilene Process resin would be the LLDPE of choice if blending with a LDPE because it shrinks more like LDPE and has greatly improved burn through resistance.

Table 4. LLDPE free shrinkage comparison

	EB LLDPE	EH LLDPE	EO LLDPE	Shrink LLDPE A
TD Free Shrinkage @ 135 °C, %	-5.8	-14.4	-15.2	25.5
MD Free Shrinkage @ 135 °C, %	79.0	66.3	77.1	50.8

The TD free shrinkage data presented in Table V and in Figure 3 shows that films made from blends rich in LDPE or in Shrink LLDPE perform well upon heating. Also, the films incorporating the Shrink LLDPE resin have considerably higher burn through resistance than does the pure autoclave LDPE film (Figure 4). The LLDPE film shrinks at temperatures much lower than its initiation of burn through; LDPE does not have this combination of properties.

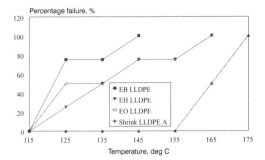

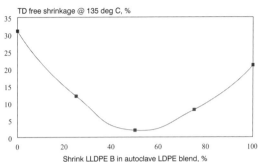

Figure 2. Burn through performance of LLDPE films.

Figure 3. TD free shrinkage LLDPE B/ autoclave LDPE blends.

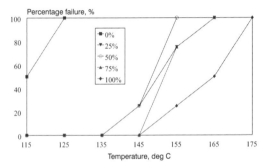

Figure 4. Burn through performance of shrink LLDPE B/ autoclave LDPE blends. Label gives percentages of shrinking LLDPE B.

When comparing the tensile, puncture, and tear properties of the 35.6 mm films, the LLDPE is considerably stronger (Table 5). Even when the gauge of the LDPE is increased by 85% to 66.0 mm, the physical properties of the LLDPE (with the exception of MD tear) are superior. This mixture of high physical properties and shrink performance allows a film fabricator to choose an appropriate LDPE/Shrink LLDPE blend for either downgauging or providing an enhanced strength shrink film.

Shrink bundling equipment was used to test the films described in Table 3. Trays of bottled drinks and canned goods were shrink wrapped. The EB LLDPE 2 film (with -15.3% TD free shrinkage) generally yielded a sloppy package not suited for unitizing goods (as would be expected for conventional LLDPEs). The tubular LDPE film (with 37.0% TD shrinkage) gave a tight package but had poor puncture resistance; at a shrink tunnel temperature of 177°C (350°F) some burn through was evident and at 204 °C (400 °F) huge holes resulted. The pure Shrink LLDPE C film (with 28.7% TD shrinkage) gave tight packages at the 177 and 204°C oven temperatures with much better puncture and no burn through. This demonstrates that a complete replacement of LDPE with Shrink LLDPE may be accomplished while obtaining a strong and tight package.

CONCLUSIONS

Improvements in strength or in downgauging conventionally blown LDPE shrink films may be obtained by using, either in blends or straight, resins from a new family of LLDPEs produced via the Spherilene Process.

Table 5. Physical properties of shrink LLDPE B/Autoclave LDPE Blends

Shrink LLDPE B, %	0	25	50	75	100	0
Autoclave LDPE, %	100	75	50	25	0	100
Gauge, μm (mil)	35.6 (1.4)	35.6 (1.4)	35.6 (1.4)	35.6 (1.4)	35.6 (1.4)	66.0 (2.6)
TD Free Shrinkage @ 135°C, %	30.9	12.2	0.5	8.2	21.8	34.8
MD Tensile Strength, MPa (psi)	28.7 (4,160)	31.5 (4,560)	34.3 (4,970)	39.5 (5,730)	48.2 (7,000)	24.2 (3,510)
TD Tensile Strength, MPa (psi)	22.9 (3,320)	27.1 (3,940)	33.0 (4,790)	35.4 (5,140)	38.2 (5,540)	19.4 (2,820)
MD 2% Secant Mod., MPa (psi)	217 (31,500)	196 (284,000)	202 (29,300)	188 (27,200)	174 (25,300)	181 (26,200)
TD 2% Secant Mod., MPa (psi)	252 (36,600)	194 (28,100)	226 (32,800)	208 (30,200)	224 (32,500)	181 (26,200)
MD Elmendorf Tear, g	250	127	42	83	294	345
TD Elmendorf Tear, g	122	273	466	598	443	347
Puncture, MPa (ft lb/in^3)	2.32 (28.0)	2.08 (25.2)	3.25 (39.3)	3.33 (40.3)	3.46 (41.9)	3.34 (40.3)

ACKNOWLEDGMENT

The author wishes to recognize the significant contributions of the following people to the work presented in this paper: Art D'Aloise, Bob Desmond, Jim Desmond, Jack Marvel, George Panagopoulos, Brad Rodgers, Dean Spencer, and Tucker Triolo.

REFERENCES

1.   K. L. Walton and R. M. Patel, *Proc. SPE ANTEC94,* 1430 (1994).
2.   G. Panagopoulos and R. D. Kamla, *Proc. TAPPI'96,* 241(1996).
3.   R. M. Patel and T. I. Butler, *Proc. SPE ANTEC'93,* 465 (1993).

Blown Film Characterization of Metallocene Resins Made in the Phillips Slurry Loop Process

A. M. Sukhadia

Phillips Petroleum Company, OK 74004, Bartlesville, USA

ABSTRACT

Metallocene catalyst polyolefins are widely regarded as a new generation of resins entering the marketplace. A number of metallocene resins, ranging in density from 0.905-0.945 g/cc and in melt index, MI, from 0.7-3 MI, were made using the Phillips slurry loop process. These resins were evaluated for blown film applications. The effects of density, molecular weight, slip/antiblock agents and film thickness on the processability, film property performance (impact and tear strengths) and clarity of blown films are reported in this work.

INTRODUCTION

Metallocene catalyst technology has recently gained considerable attention in the polyolefins industry. This is primarily due to the fact that these metallocene catalysts, also referred to as single-site catalysts, offer unique polyolefin structures characterized by narrow molecular weight distributions, MWD, and narrow i.e. homogeneous short chain branching distributions, SCBD. This SCBD is homogeneous at both the intramolecular (monomer sequence distribution along a polymer chain) and intermolecular (monomer distribution between different polymer chains) levels.[1] These unique structural features manifest themselves a number of advantageous properties e.g. excellent clarity, good toughness, low heat seal temperatures, low extractables (volatiles), etc.

To take advantage of these materials, Phillips has developed several metallocene catalysts that are compatible with the Phillips slurry loop process technology.[2] These catalysts were used to produce a wide range of polyethylene, PE, resins ranging in melt index, MI, values from 0.01-100 g/10 min and densities from 0.910 to 0.970 g/cc. Several of these resins were evaluated for blown film applications using typical linear low density polyethylene, LLDPE, processing conditions.

In this report, we focus on some of the processability and film property aspects of these resins. The melt fracture behavior and the role of slip/antiblock, SA, agents on processability are examined. The effects of density, MI or molecular weight, SA and film thickness on the film properties viz. dart impact and tear strengths are examined in detail. The clarity, as indicated by % haze, of the blown films is discussed in some detail with respect to the effects of density, SA and film thickness.

EXPERIMENTAL

RHEOLOGY

Melt index, MI, was measured as per ASTM D-1238, Condition F (190°C, 2.16 kg). High load melt index, HLMI, was measured as per ASTM D-1238, Condition E (190°C, 21.6 kg).

The rheological data was obtained on a Rheometrics Mechanical Spectrometer (model RMS-800) using a parallel plate geometry. Temperature-frequency sweeps were performed between 150-230°C in 20°C increments. This data was reduced to a single master curve at 190°C using the well known Williams-Landel-Ferry, WLF, time-temperature superposition method[3] through a proprietary algorithm to estimate the flow activation energy, E_a. The master curve was then fitted with the three parameter Carreau-Yasuda, CY, empirical model to obtain the CY parameters viz. zero shear viscosity, η_0, characteristic maximum relaxation time, τ_{relax} and A parameter. Details of the significance and interpretation of these three parameters may be found elsewhere.[4,5]

BLOWN FILM EVALUATION

The film blowing experiments were conducted on a laboratory scale and a commercial scale blown film line. Films of various gauges ranging from 0.0254 mm (1 mil) to 0.0762 mm (3 mil) were made depending on resin availability. The particulars of each line are as follows.

Laboratory Line: The line consists of a 38 mm (1.5 inch) diameter single screw Davis Standard extruder (L/D=24; 2.2:1 compression ratio) fitted with a barrier screw with a Maddock mixing section at the end. The die used was 102 mm (4 inch) in diameter and was fitted with a Dual Lip Air Ring using ambient cooling air. The die gap was 1.52 mm (0.060 inch). The film blowing was done at typical LLDPE conditions as follows: 2.5:1 blow up ratio, BUR, "in-pocket" bubble configuration and 190°C extruder and die set temperatures. The evaluations were made at an output rate of 27.2 kg/hr (60 lb/hr) using a screw speed setting of 115 RPM. These conditions were chosen since the film properties so obtained scale directly with those from larger commercial scale film blowing operations.[6]

Commercial Line: This line consists of a 89 mm (3.5 inch) diameter single screw extruder (L/D=24; 2.2:1 compression ratio) fitted with a LLDPE barrier screw. The die used was 203 mm (8 inch) in diameter and was fitted with a Dual Lip Air Ring using chilled air at 15°C (59°F). The die gap was 1.52 mm (0.060 inch). The film blowing was done at typical LLDPE conditions as follows: 2.5:1 blow up ratio, BUR, "in-pocket" bubble configuration and 190°C extruder and die set temperatures. The evaluations on this line were made at an output rate of 68 kg/hr (150 lb/hr) using a screw speed setting of 25 RPM.

BLOWN FILM PROPERTIES

The film properties were measured as follows:
1. Dart impact strength - ASTM D-1709 (method A)
2. Machine, MD, and transverse, TD, direction Elmendorf tear strengths - ASTM D-1922
3. % Haze was measured in accordance with ASTM D-1003 on a XL-211 Hazegard System.
4. Gloss (45°) was measured as per ASTM D-2457.

BLOWN FILM RESINS

A wide range of metallocene resins, deemed suitable for blown film applications, and ranging in density from 0.905 - 0.945 g/cc and in MI from 0.7 - 3 MI were evaluated at LLDPE type processing conditions. The reader is referred to ref.[2] for details on some of the more fundamental characteristics of these resins. Additional details relevant to this paper are presented in subsequent sections.

RESULTS AND DISCUSSION

RHEOLOGY

The rheology of a few representative metallocene resins, ranging from 0.7-3 MI, made using catalyst PA are shown in Figure 1 along with a chrome catalyst LLDPE type resin (PC-1) of broad MWD ($M_w/M_n = 24$) for comparison. The rheological particulars of these resins are shown in Table 1 along with the weight average molecular weight, M_w, data.

The metallocene resins exhibit viscosity curves, shown in Figure 1, that are typical of narrow MWD (ca. 2.25)[2] resins with no long chain branching, LCB, as reflected by low flow activation energies under 24 kJ/mol. Furthermore, it is clear from Table 1 that the MI and HLMI track quite well with the M_w values, both increasing as the M_w decreases. For the sake of brevity, it is simply mentioned here that all the metallocene resins exhibited rheological and molecular weight characteristics consistent with the above data and trends.

BLOWN FILM PROCESSABILITY

The processability of all the metallocene one resins was, in general, judged to be quite good. Resins ranging from 0.7-3 MI were readily converted into films on both the laboratory and commercial scale lines. Some processability limitations were encountered due to the lower densities and narrow MWD characteristics and they are discussed below.

Table 1. Basic rheological and molecular weight data for representative metallocene resins

Resin	Ea kJ/mol	HLMI dg/min	MI dg/min	M_w kg/mol
PA-71A	21.7	12.2	0.72	107.32
PA-74A	23.3	16.6	1.00	95.49
PA-76	23.7	25.1	1.36	87.09
PA-78A	23.7	32.4	1.89	80.37
PA-68	22.1	50.9	2.80	73.36
PC-1	33.9	24.2	0.24	215.68

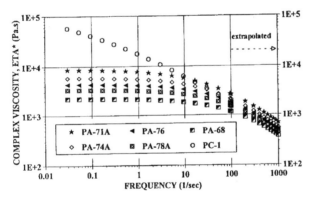

Figure 1. Rheology of metallocene LLDPE resins.

Blocking Behavior: It was observed that for metallocene resins lower than roughly 0.920 g/cc density, the bubble exhibited some friction with the wooden take-up tower. In addition, these films also tended to block or consolidate on the take-up rolls after being allowed to sit for 24-48 hours. Such blocking behavior is typical of LLDPE resins due to their higher coefficients of friction, COF, as has been noted in the literature.[7-10] Furthermore, the blocking behavior has been noted to be strongly dependent on the resin density, increasing as the density decreases,[7] which is very consistent with the observations made here. In subsequent trials, resins lower than 0.925 g/cc density were dry blended with 3 wt% (based on total polymer) of a commercially available LLDPE slip/antiblock, SA, concentrate

Table 2. Blown film properties of metallocene resins

Resin	Density	MI	HLMI	Dart^	MD^	TD^
				Impact	Tear	Tear
	g/cc	dg/min	dg/min	g	g	g
PA-						
186	0.9096	1.96	34.6	>1400*	238	390
180	0.9106	2.36	42.0	>1400*	266	395
79A	0.9179	1.06	18.4	388	200	398
183	0.9196	1.17	21.7	503	169	413
77	0.9217	1.65	27.8	275	275	499
178B	0.9220	1.02	18.5	158	232	485
78A	0.9222	1.89	32.4	145	174	453
73	0.9231	0.93	16.4	207	219	432
72A	0.9256	0.98	17.6	153	170	422
177	0.9306	1.92	45.3	65	65	267
178	0.9310	2.17	35.8	63	56	260
179	0.9313	1.67	29.0	68	46	278
99	0.9319	2.57	46.3	59	55	254
98	0.9326	1.56	28.1	54	43	226
97	0.9349	1.20	21.6	40	27	197
69A	0.9402	0.87	15.3	30	19	147

*no break with 1400 g maximum dart weight
^±95% confidence limits are with 10% of reported average

(Ampacet 10430 AB PE MB from AMPACET Corp.). The concentrate is silica based and the exact composition is proprietary. This, as expected, completely alleviated the friction and blocking problems noted earlier. It should be added here that the 3 wt% level was not optimized in any way but rather was thought to be more than sufficient to overcome the processing difficulties encountered in the density range studied between 0.900-0.925 g/cc.

Melt Fracture: Another common problem in the blown film processing of LLDPE type resins is the occurrence of sharkskin or surface melt fracture, SSMF, as noted in the literature.[11-13] The occurrence of SSMF is dependent on output rate, die geometry, gap, temperature and melt viscosity as well as other factors. It is well understood in the literature that SSMF occurs at a critical shear stress, although this stress level can vary with polymer structure (M_w, MWD, LCB, etc.). For the metallocenes resins of interest here, it was observed that the critical shear stress in the blown film die was of the order of 0.2 MPa. In resins which did exhibit melt fracture, addition of 0.07 wt% fluoropolymer processing aid (Dynamar FX-9613 from 3M Co.) completely eliminated the SSMF. It is important to clarify that the need for fluoropolymer to prevent SSMF in these resins will depend on a variety of processing factors mentioned above and thus cannot be generalized here.

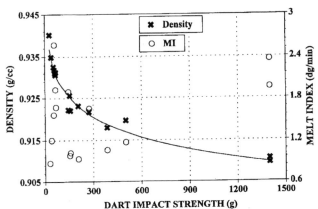

Figure 2. Effects of density and MI on the dart impact strength of metallocene blown films.

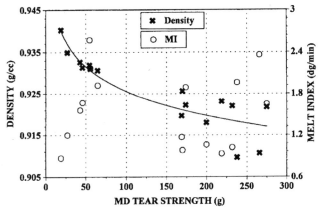

Figure 3. Effects of density and MI on the MD tear strength of metallocene blown films.

BLOWN FILM PROPERTIES

The dart impact and MD and TD tear strengths of the blown films are tabulated in Table 2 for a large number of representative resins, covering the range of 0.900-0.940 g/cc density and 0.7-3 MI. For ease of interpretation of this data, these three properties are plotted as functions of density and MI as shown in Figures 2, 3, and 4, respectively. All data is for 0.0254 mm (1 mil) films.

From Table 2 and Figure 2, it is clear that the dart impact strength is largely a function of the resin density, increasing as the density decreases, and independent of the MI or molecular weight. Furthermore, it can also be seen that the dart impact strength is most sensitive to the density changes in the lower density regime from 0.900 - 0.920 g/cc. A slight change of density in this range has a large effect on the dart impact strength. Lastly, the extremely high toughness of these resins especially at densities below 0.920 g/cc is noteworthy.

From Table 2 and Figure 3, it appears that the MD tear strength is also largely a function of density alone. For resin densities below 0.925 g/cc, the MD tear appears to be about 225 g on average (with exception PA-77, Table 2). At higher densities, the MD tear decreases substantially and appears to continually decrease with increasing density.

The trends in the TD tear strength data shown in Table 2 and Figure 4 appear similar to those observed for the MD tear strength discussed above. At densities below 0.925 g/cc, the TD tear strength is about 450 g on average. As the density increases above 0.925 g/cc, there is a sharp reduction in TD tear. The blown film properties (1 mil thick) and processability information for a few representative resins processed on the commercial line (see section on SA effects below) are shown in Table 3. While the absolute properties show some differences from those obtained on the laboratory line (Table 2), the

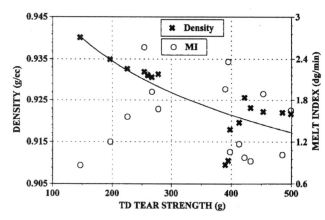

Figure 4. Effect of density and MI on the TD tear strength of metallocene blown films.

trends are quite consistent with the earlier observations. Extremely good dart impact strengths are observed with these resins even with the 3 wt % SA (see next section). The MD and TD tear strengths appear to be fairly constant through the range of resins evaluated on this line and exhibited excellent TD/MD tear balance. Haze values ranged from about 6-12 % with 3 wt% SA. In addition, the following processability characteristics were noted: good bubble stability, no blocking problems and no melt fracture for resins around 1 MI with fluoropolymer processing aid.

Table 3. Blown film properties and processability of metallocene resins on a commercial scale blown film line

Resin	PA-186	PA-4/3	PA-4/1	PA-4/2	PA-4/5	PA-4/5
MI^, dg/min	1.96	2.86	0.94	2.18	1.02	1.02
Density^, g/cc	0.9096	0.9169	0.9179	0.9213	0.9220	0.9220
Slip/Antiblock, wt%	3	3	3	3	0	3
Fluoropolymer, wt%	0	0	0.07	0	0.07	0.07
Dart Impact, g	614	350	444	120	242	270
MD Tear, g	346	349	278	299	280	253
TD Tear, %	491	533	541	528	565	531
% Haze	7.9	11.6	6.6	9.5	5.7	7.7
% Glass, 45°	83	79	98	86	111	88
Bubble Stability	fair	good	good	good	v. good	good
Blocking	none	none	none	none	some	none
Melt Fracture	none	none	none	none	none	none

^Density and MI are for neat resins, without SA concentrate or fluoropolymer

Table 4. Effects of thickness (gauge) on metallocene blown film properties

Resin	PA-186	PA-186 with SA	PA-4/2 with SA
Dart Impact, g			
0.0254 mm (1 mil)	>1400	685	140
0.0508 mm (2 mil)	>1400	>1400	269
0.0762 mm (3 mil)	>1400	>1400	396
MD Tear, g			
0.0254 mm (1 mil)	238	240	188
0.0508 mm (2 mil)	613	533	466
0.0762 mm (3 mil)	826	874	770
TD Tear, g			
0.0254 mm (1 mil)	390	390	405
0.0508 mm (2 mil)	701	677	739
0.0762 mm (3 mil)	982	1002	1069
Haze, %			
0.0254 mm (1 mil)	4.5	10.8	10.2
0.0508 mm (2 mil)	4.4	6.8	13.9
0.0762 mm (3 mil)	4.7	8.5	19.4

SA is 3 wt% Slip/Antiblock

Effects of Film Thickness: Higher gauge films (2 and 3 mil) were made from a few resins and the effects on film properties are shown in Table 4. Note that the properties for the higher gauges are reported as measured and not normalized per unit thickness. The dart impact strength increases roughly proportionally with thickness. The MD tear strength can be seen for all three cases to increase more than proportionally i.e. a doubling of thickness results in more than a doubling of the MD tear strength. In contrast, the TD tear strength increases but less than proportionally with thickness. The clarity shows an interesting trend in that the lowest density sample, PA-186, shows no change in haze with film thickness. Sample PA-4/2 with 3 wt% SA, however, increases in haze as film thickness increases. Other data (not presented here) appears to confirm this trend that at low densities (0.915 g/cc) the high clarity is maintained at higher gauges but at higher densities the clarity decreases as thickness increases.

Effects of Slip/Antiblock, SA: The effects of addition of 3 wt% SA on film properties are shown in Figure 5 for two representative cases. The dart impact strength for both cases reduces dramatically upon addition of the SA concentrate. However, and quite interestingly, there is absolutely no effect of the SA on either of the tear strengths. A few other points are worth clarifying here. First, the SA level added was probably in excess of that needed for good processability. Thus the potential to diminish the level of loss in the impact strength exists by simply lowering the SA level. Second, in cases where the resin is to be used in an inside layer of a multi-layer coextrusion type application, there would be no need to add any SA. Lastly, the clarity of the blown films as measured by haze did decrease (haze increased roughly by a factor of 2) upon addition of this level of SA. Therefore, some penalty in clarity should be expected with the addition of any particulate (silica here) agent, even at low levels.

Clarity of Metallocene Blown Films: The blown films of all the (neat) resins evaluated here, and ranging from 0.905-0.945 g/cc density showed excellent clarity as indicated by low haze and high gloss. These results are shown graphically in Figure 6 where the effects of both density and film thickness (gauge) on haze are shown. It is seen that the haze for the blown films ranged from about 3-8% for this entire density range of 0.905-0.945 g/cc. The excellent clarity of these films for densities above 0.925 g/cc was surprising and unexpected. This is currently thought to be a consequence of two main factors; an extremely homogeneous SCBD from this catalyst resulting in small and uniform crystallites and a very smooth film surface. Both factors are considered to be advantages over conventional LLDPE resins of equivalent density and molecular weight.

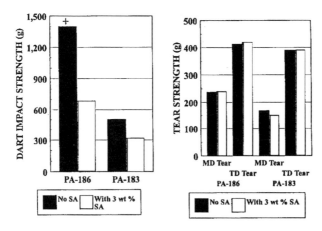

Figure 5. Effect of slip/antiblock on metallocene film properties.

CONCLUSIONS

A wide range of metallocene resins made in the Phillips slurry loop process were evaluated for blown film applications at typical LLDPE type processing conditions. It is seen that these resins offer an excellent balance of film properties such as dart impact and tear strengths, clarity and processability. The impact and tear strengths were observed to depend primarily on resin density, increasing as the density decreased. The clarity, as measured by haze, of all blown films ranging from 0.905-0.945 g/cc was found to be excellent and in the range of 3-8 % haze. Addition of 3 wt% of a slip/antiblock, SA, concentrate to reduce blocking resulted in a sharp drop in impact strength and a decrease in film clarity but no change in the tear properties. The properties of

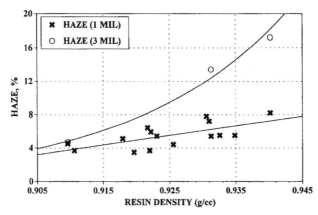

Figure 6. Effects of density and film thickness on % haze of metallocene blown films.

higher gauge films (2 and 3 mil) increased proportionally with thickness and TD/MD tear balance also improved. It is therefore concluded that metallocene resins produced in the Phillips slurry loop process will offer a very good balance of processability and performance characteristics, which can be optimized for various applications by appropriate choice of resin and processing conditions.

ACKNOWLEDGMENTS

The author would also like to sincerely thank Phillips Petroleum Company for support and permission to publish this work.

REFERENCES

1. Y. S. Kim *et al.*, *Proc. SPE ANTEC*, 1122 (1995).
2. M. B. Welch *et al.*, *MetCon 1995*, May 17-19 (1995).
3. J. D. Ferry, Ed., **Viscoelastic Properties of Polymers**, 3rd Edn., *John Wiley & Sons*, New York, 1980.
4. C. A. Hieber and H. H. Chiang, *Rheol. Acta*, **28**, 321 (1989).
5. C. A. Hieber and H. H. Chiang, *Polym. Eng. Sci.*, **32**, 931 (1992).
6. A. M. Sukhadia, *J. Plast. Film Sheeting*, **10(3)**, 213 (1994).
7. S. V. Karande, L. G. Hazlitt, and M. J. Castille, *Proc. SPE ANTEC*, 1262 (1993).
8. J. Radosta, *J. Plast. Film Sheeting.*, **7**, 181 (1991).
9. F. Ruiz, *Proc. SPE ANTEC*, 2766 (1994).
10. L. Hazlitt, S. Karande, and M. J. Castille, *J. Appl. Polym. Sci.*, **51**, 278 (1994).
11. T. J. Blong *et al.*, *Proc. SPE ANTEC*, 28 (1994).
12. S. E. Amos, *Proc. SPE ANTEC*, 2789 (1994).
13. G. W. Knight and S. Lai, *Proc. Polyolefins VIII*, Houston, TX, 227 (1993).

Comparison of the Fiber-Spinning Properties of Ziegler-Natta and Metallocene Catalyzed Polypropylene

E. Bryan Bond and J. E. Spruiell

*Center for Materials Processing, Department of Materials
Science and Engineering, Knoxville , TN 37996-2200, USA*

ABSTRACT

Two metallocene catalyzed isotactic polypropylene, miPP, resins and a standard Ziegler-Natta isotactic polypropylene resin, zniPP, were melt spun into fibers. The fibers were spun under similar conditions to allow for direct comparison of fiber properties between the various resins. Two of the miPP resins differed by molecular weight, with a third miPP resin containing a nucleating agent. The fiber properties were found to be dependent upon molecular weight, MW, as has been previously documented.[1] However, the tensile strengths of the metallocene resins were consistently higher in comparison with the conventional zniPP resin. The higher tensile strength can be attributed to the uniformity of the miPP chains in both length and in molecular microstructure that slightly retard crystallization and allow for a higher overall molecular orientation.

INTRODUCTION

Polypropylene has become a major engineering thermoplastic that has enjoyed widespread use due to its high strength to weight ratio and rheological properties. This paper will make a direct comparison between the fiber spinning properties of a standard Ziegler-Natta catalyzed and visbroken isotactic polypropylene and the new metallocene catalyzed polypropylene resins. This is the second part of a two part series; the first paper compares the melting and quiescent crystallization behavior of the metallocene and Ziegler-Natta resins.[1]

A substantial effort on polypropylene fiber spinning in recent years has been carried out both in our labs at the University of Tennessee-Knoxville,[2-6] and elsewhere.[7-9] Crystallization kinetics have been investigated in the spinline and it has been shown that molecular orientation is a major factor that enhances crystallization rates of iPP under stress.[6] A number of process parameters influence the molecular orientation present and, hence, they affect the structure and properties of the spun filaments. The filament structure and properties are also affected by the polypropylene resin characteristics such as molecular weight, molecular weight distribution, percent isotacticity, copolymer content and nucleating agent additives.[2,4,5] In particular, the high molecular weight tails in broader MWD resins are more effective in enhancing the crystallization rates in the spinline in comparison with narrow distribution resins with similar melt flow rate, mfr. The recently commercialized metallocene catalyzed resins have a different molecular microstructure and generally narrower molecular weight distribution than the Ziegler-Natta catalyzed resins. Thus the metallocene catalyzed resins would be expected to behave somewhat differently than the Ziegler-Natta catalyzed resins during the melt spinning process. The present paper describes some of the differences.

EXPERIMENTAL

MATERIALS

The basic characteristics of the four resins studied in this paper are listed in Table 1. The large, M, stands for metallocene catalyzed iPP and large, ZN, for Ziegler-Natta catalyzed iPP. The large, N, stands for the nucleated agent added to the 35 m resin, which is 1 %w/w. All the resins were unfractionated homopolymers that had a nominal 35 mfr, except for one metallocene resin that has an mfr of 22 g/10 min.

Table 1. Material characteristics

Sample Cod	FTIR % tacticity[a]	MFR	Polydispersity	M_n	M_w
M35	87.3	35.0	1.93	71,146	137,277
ZN35	89.5	35.0	2.56	59,577	152,332
M22	90.0	22.0	2.01	76,687	154,210

[a]IR method by Luango and Sundell[13,14]

These are the same resins whose quiescent crystallization behavior was detailed in paper.[1] It is important to understand the differences presented in the crystallization paper to understand what is taking place in the fiber spinning crystallization process.

FIBER PROCESSING

The fibers were melt spun by a Fourne' screw extruder and spinning head. The extruder screw was 13 mm in diameter and supplied resin to a constant mass throughput Zenith gear pump. The spinneret capillary was 3.81 mm long and 0.762 mm in diameter. The draw down was achieved by a pneumatic device that makes use of pressurized air to produce a suction force on the filament at the entrance to the device. Thus, the take-up velocity can be controlled by adjusting the applied air pressure. The take-up velocity is then calculated from the expression:

$$V = \frac{4Q}{\Pi D^2 \rho} \qquad [1]$$

where V is the velocity in m/min, Q is the mass throughput in g/min/hole, D is the final on-line filament diameter in meters and ρ is the density. The mass throughput ranged from 1.5-1.6 g/min and all extrusion temperatures were 210°C. References[2,3] should be consulted for complete details on the fiber spinning process and conditions.

RESULTS AND DISCUSSION

Several characteristics of the resins are shown in Table 1. The FTIR tacticities are shown for the resins, indicating the %isotactic content is approximately the same for all the materials, with the M35 resin having a slightly lower isotacticity. FTIR measures the average tacticity for the sample; it is sensitive to the crystallinity of the sample, so sample preparation is important.[10,11] The miPP resins have very low polydispersity values, equal to about 2.0 or less. The ZN35 resin has a low polydispersity, 2.56, but not as low as the miPP resins; this difference becomes important when these resins are spun into fibers.

Table 2. Thermal characteristics of resins*

Sample Code	Catalyst	Melting Temperature, °C		Crystallization Temperature, °C	
		Peak	Onset	Peak	Onset
M35	Metallocene	151.1	144.6	100.7	106.8
ZN35	Ziegler-Natta	159.7	152.8	104.8	110.6
M22	Metallocene	149.6	143.9	100.7	106.7

The quiescent melting and crystallization temperatures for the resins are given in Table 2. The differences in melting and crystallization temperatures arise from the differences in molecular architecture of the resins as explained in the crystallization paper. Since the defects are incorporated into each polymer chain, fractionation into atactic and isotactic chains is not possible. Since these defects probably cannot be incorporated into the growing lamellae, they likely limit the lamellar thickness and extent, producing lamellae that are not as stable as those grown from molecules containing less defects or in which longer isotactic sequences are present. Both possibilities lead to the same conclusion; zniPP and miPP of similar overall tacticity and average MW will have different melting and crystallization temperatures, with the miPP resin having the lower value in each case due to the defect distribution in each chain.

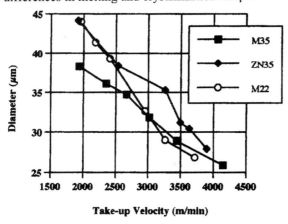

Figure 1. Diameter vs. take-up velocity.

Fiber diameter versus take-up velocity is shown for the various resins in Figure 1. Equation 1 shows that the observed diameter differences between the resins are a result of differences in the mass throughput, Q, and the sample density, ρ. They are basically unrelated to differences in rheology or molecular weight distribution, although the densities of the resins do vary slightly due to differences in the level of crystallinity developed (see below). However, the drawdown stress required to achieve a given spinning speed will be a function of the rheology and MWD of the resin.

The densities of the as-spun fibers are shown in Figure 2. The M22 resin had the highest densities followed closely by the ZN35 and M35 resins. The results agree with current theory, in that the higher MW resin should develop the higher density due to the higher amount of stress in the spin line at a given spinning speed. The greater density of the ZN35 resin compared to the M35 resin is consistent with a slightly lower tacticity of the M35 resin. Two other factors are also believed to contribute to this effect. The first is the narrower distribution of the M35 resin than the ZN35 resin. The second is the molecular architecture of the metallocene catalyzed resins compared to the Ziegler-Natta catalyzed resins. These factors are discussed further below.

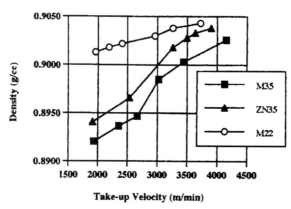

Figure 2. Density vs. take-up velocity.

The data in Figure 3 clearly show that the metallocene resins develop higher birefringence values at a given spinning speed. The M35 resin develops the highest birefringence values of all the resins studied, under all spinning speeds. And, somewhat surprisingly, the birefringence of the M22 resin is not as high as that of the M35 resin.

Figure 4 shows the birefringence measured on the running spinline as a function of the distance from the spinneret for filaments being spun at a pressure of 25 psig on the draw down device. These data show that the rapid rise in birefringence develops further down the spinline for the metallocene catalyzed resins than for the Ziegler-Natta catalyzed resins. However, once the birefringence begins to develop, the rate of increase and the final birefringence are higher for the metallo-

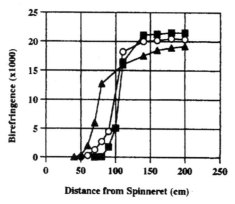

Figure 3. Birefringence vs. distance from spinneret.

cene catalyzed resins. This is especially true when comparing the two resins of similar mfr (M35 and ZN35). Since the rapid rise in birefringence occurs due to oriented nucleation and growth of crystals,

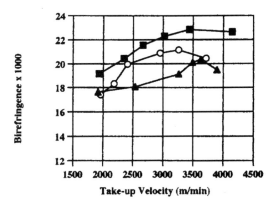

Figure 4. Birefringence vs. take-up velocity.

this result may be interpreted to indicate that crystallization occurs further from the spinneret and at a lower temperature in the metallocene catalyzed resins. This fact is consistent with the lower quiescent crystallization temperatures of the metallocene catalyzed resins as shown in Table 2. It is also consistent with a similar behavior observed in Ziegler-Natta catalyzed resins resulting from differences in polydispersity at similar mfr values.[5] Invariably, the broader MWD resin showed the rapid rise in birefringence closer to the spinneret, a result that was explained on the basis of the crystal nucleating effects of the high molecular weight tails in the broader MWD resins and the higher stresses and molecular orientation developed in the spinline. For the present resins the polydispersity is greater for the ZN35 resin than for the M35 (or M22) resin. Thus, independent of any influence of chain microstructure on quiescent crystallization, we would expect the broader MWD ZN35 resin to crystallize closer to the spinneret during melt spinning than the narrower MWD M35 resin.

It is also worth noting that the earlier work[5] showed that the broader MWD resin would develop lower final birefringence than the narrower MWD resin, as is the case in the present results. The higher final orientation of the narrower MWD resin was attributed to a tendency to produce higher orientation in the final fiber because the delay in crystallization allows greater amorphous orientation to develop before crystallization begins. The lower crystallization temperature for the metallocene catalyzed resin is also consistent with the lower crystallinity developed in this resin.

There is one troubling result from the data of Figure 4. The M35 resin has a higher birefringence than does the M22 resin. This result contradicts much of the work that has been done on MW effects on iPP fiber spinning of Ziegler-Natta catalyzed resins.[2-4] Previous work has shown that the higher MW resins should develop a higher birefringence due to a greater stress in the spinline promoting a higher overall molecular orientation. The reason for the present behavior is as yet unresolved, and we will need to examine it more fully in order to understand it. It would seem that the molecular structure, i.e., the uniformity of the molecular chains, plays an important and a heretofore unencountered role in the structure development in the spinline as a function of average molecular weight.

The next three figures, Figures 5, 6 and 7, show the results from mechanical property testing of the as-spun fibers. Figure 5, tensile strength versus take-up velocity, shows that the non-nucleated M35 and M22 resins have much higher tensile strengths than the ZN35 resin at similar spinning speeds. At moderately low spinning speeds the M35 resin has a tensile strength that is nearly twice that of the ZN35 resin. The M35N resin has a low tensile strength compared to the other miPP resins, especially at the lower spinning speeds. The M35N resin closely resembles the ZN35 resin up until the highest spinning speed, where it increases sharply. The differences in tensile strength among the resins tends to reflect the differences in the overall molecular orientation as measured by the birefringence data of

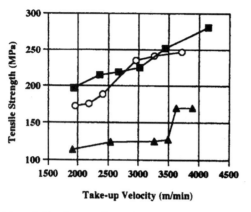

Figure 5. Tensile strength vs. take-up velocity.

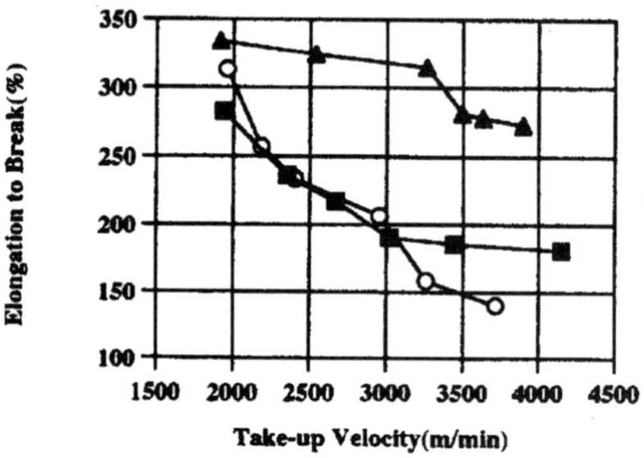

Figure 6. Elongation to break vs. take-up velocity.

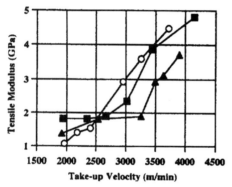

Figure 7. Tensile modulus vs. take-up velocity.

Figure 3. Again, the conventional logic has been that the higher MW resin will develop a higher tensile strength than the lower MW resin[2,4] due to higher molecular orientation. However, the orientation of the higher molecular weight M22 resin is lower, not higher than that of the M35 resin, a fact that is largely reflected in the tensile strength data.

The elongation-to-break data in Figure 6 are basically the reverse of the tensile strength data.

The modulus data in Figure 7 increase with an increase in spinning speed for all resins, indicating the dependence of modulus on molecular orientation. It is also known that increasing crystallinity increases modulus, all other things being equal. The fact that the M35 resin has a higher modulus than the ZN35 resin suggests that the M35 resins lower crystallinity is offset by its higher orientation at a given spinning speed, compare Figures 2 and 3.

CONCLUSIONS

The molecular architecture of the miPP resins has a profound impact on the mechanical and physical properties of melt spun isotactic polypropylene resins. The narrow molecular weight distribution, lower melting point and slower crystallization kinetics combine to delay the onset of crystallization in the spinline compared to conventional Ziegler-Natta catalyzed resins with similar mfr. This delay allows for higher molecular orientation to be developed. This results in higher birefringence, tensile strength and modulus, but lower elongation-to-break. The effects of adding a nucleating agent increases the temperature at which the structure develops, and reduces the molecular orientation, birefringence, tensile strength and modulus while increasing the elongation-to-break.

ACKNOWLEDGMENTS

The authors wish to thank Exxon Chemical Company for supplying the resins and for supporting this research.

REFERENCES

1. E. B. Bond and J.E. Spruiell, *ANTEC*, 1997.
2. F.-M. Lu and J. E. Spruiell, *J. Appl. Polym. Sci.*, **34**, 1521 (1987).
3. F.-M. Lu and J. E. Spruiell, *J. Appl. Polym. Sci.*, **34**, 1541 (1987).
4. S. Misra, F.-M. Lu , J. E. Spruiell, and G. C. Richeson, *J. Appl. Polym. Sci.*, **56**, 1761 (1995).
5. J. E. Spruiell, F.-M. Lu, Z. Ding, and G. C. Richeson, *J. Appl. Polym. Sci.*, (1996).
6. F.-M. Lu and J. E. Spruiell, *J. Appl. Polym. Sci.*, **49**, 623 (1993).
7. Q. Fan, D. Xu, D. Zhao, and R. Qian, *J. Polym. Eng.*, **5**, 95 (1985).
8. F. Kloos, *4th International Conference, Polypropylene Fibres and Textiles*, Nottingham, U.K., 1987.
9. E. Andreassen, O. J. Myhre, E. L. Hinrichsen, and K. Gronstad, *J. Appl. Polym. Sci.*, **57**, 1075 (1995).
10. J. P. Luongo, *J. Appl. Polym. Sci.*, **3**, 302 (1960).
11. T. Sundell, H. Fagerholm, H. Crozier, *Polymer*, **37**, 15 (1996).

Metallocene Linear Low Density Polyethylene Blends for Pouch Materials in High-Speed Hot-Liquid Filling Process

T. Tomatsuri, N. Sekine, and N. Furusawa

Toppan Printing Co., Ltd, Saitama , Japan

ABSTRACT

Pouch materials for liquid fillers (e.g., sauce and dressing) are usually laminated films consisting of nylon substrate and extrusion-laminated heat-sealable materials such as linear low density polyethylene, LLDPE, and ethylene-vinyl acetate copolymer, EVAc. Recently the use of metallocene linear low density polyethylene is on the increase because of the superior high-speed sealability, however, the heat resistance of metallocene-LLDPE is not enough for sterilizing hot-filling process. The blend of conventional LLDPE with metallocene-LLDPE has improved the heat resistance of the pouch material, and this blend material can be satisfactorily used for high-speed hot-filling process.

INTRODUCTION

Small pouch type packages are widely used for a variety of products such as ketchup, jams, mayonnaise, and salad dressings. These small pouch packages are usually made from a laminated film consisting of substrate (e.g., biaxially oriented PET or nylon film, aluminum leaf) and heat-sealable material (e.g., LDPE, LLDPE, EVAc). The web roll of the laminated film is fed to a vertical form/fill/seal machine and continuously converted into a number of different bag styles such as three- or four- sided seal package, pillow style bag and gusseted bag. Metallocene-LLDPE, which has superior low temperature heat-sealability and hot-tack property, has been recently preferred for the heat-sealable material because of the demand for the high-speed filling.[1-3] However, the heat resistance of metallocene-LLDPE is not enough for high-speed hot-liquid filling process necessary for product pasteurization. This often causes a crack in the pouches when filled into a cardboard box while they are still hot. In order to improve the heat resistance for hot-liquid filling, conventional LLDPE was blended into metallocene-LLDPE. In this paper, we discuss the characteristics of the blend of metallocene-LLDPE and conventional LLDPE especially in terms of filling and sealing properties.

EXPERIMENTAL

MATERIALS

Metallocene-LLDPE from Mitsubishi Chemical (57L, extrusion/lamination grade) was used due to the good low-temperature sealability and hot-tack property. For the heat resistance improvement, one LDPE and three conventional LLDPE materials (Table 1), having relatively low melting temperature, were chosen as the counter blend component because of their solubility with metallocene-LLDPE.

Table 1. Materials used in this study

Type	Comonomer	Density g/cm³	Malt Index g/10min	Melting Point °C
Metallocene-LLDPE	Hexene	0.905	11	98
HP-LDPE	-	0.917	9.5	106
Conventional-LLDPE-1	Butene	0.910	10	100, 108, 121
Conventional-LLDPE-2	Butene	0.914	10	101
Conventional-LLDPE-3	Octene	0.910	8	100, 112, 123
-LLDPE-C		0.906	9	104

BLEND PREPARATION

Metallocene-LLDPE/LDPE and metallocene-LLDPE/conventional LLDPE blends (Table 2) were prepared using an intermeshing corotating twin-screw extruder (Toshiba TEM-35B) operating at 200 rpm, output rate of 200 g/min, and temperature of 200°C. The premixed pellet materials at the desired proportions were melt-blended in the twin-screw extruder, then the strand-shaped extrudates were cooled in a water bath and pelletized.

Table 2. Heat of fusion of the blends Hf, J/g

Blends	100/0*	90/10*	80/20*	50/50*	20/80*	10/90*	0/100*
Meta-LLDPE/HP-LDPE	71.5	74.0	78.4	86.9	95.3	100.0	107.5
Meta-LLDPE/LLDPE-1	71.5	-	70.8	76.1	80.3	-	82.8
Meta-LLDPE/LLDPE-2	71.5	-	71.7	79.3	86.9	-	91.4
Meta-LLDPE/LLDPE-3	71.5	70.2	71.6	79.3	86.0	87.3	91.0

*blend ratio

THERMAL PROPERTIES

Thermal properties were analyzed using a differential scanning calorimeter (Perkin-Elmer DSC-7). In the first run the heating scan up to 200°C was followed by rapid cooling-down at a rate of 500°C. The thermal properties of the blend pellet samples were recorded in the second run up to 200°C at a heating rate of 10°C/min.

FILM LAMINATION

Laminated films for heat-sealability study were prepared using a two-extruder tandem system. The 25 μm thickness of conventional LLDPE with density of 0.910 g/cm^3 and MI =10 was first extrusion-laminated at 270°C with ozone treatment onto an oriented nylon film which was previously coated with an adhesive. Then the metallocene-LLDPE blends were extrusion-laminated onto the LLDPE/nylon laminated film at 270°C.

SEALABILITY TEST

Heat seal strength and hot tack strength were measured for the sealability study of the metallocene-LLDPE blends. The heat seal strength was evaluated using a heat-sealed laminated film with 15 mm width. The heat-sealing was carried out with a sealing time of 1 s and sealing pressure of 0.2 MPa at various temperatures. The hot tack strength was measured using a hot-tack tester (Theller Engineering). The laminated film was heat-sealed with a sealing time of 0.5 s and sealing pressure of 0.5 MPa at various temperatures, then the sealed film was peeled off immediately at 200 cm/min. The hot-tack strength was determined after 0.25 s period.

FORM/FILL/SEAL TEST

Form/fill/seal, FFS, test was carried out using the laminated film of the metallocene-LLDPE blends at various operating speeds. The seal quality was evaluated by the appearance observation, and the pressure resistance of the sealed pouches was evaluated by loading 980 N for 3 min. The test conditions are as follows:

FFS machine	Super Komakku (Komatsu Machine)
Filler	Soy sauce
Pouch size	65 mm (W) x 90 mm (H)
Filling temp.	90°C
V. seal. Temp.	190°C
H. seal. Temp.	140°C
Line speed	20, 24, 27 m/min

RESULTS AND DISCUSSION

BLEND PREPARATION

Using a twin-screw extruder metallocene-LLDPE/LDPE and metallocene-LLDPE/conventional LLDPE blends of various blend ratios (90/10, 80/20, 50/50, 20/80, 10/90) were prepared. All blend samples could be pelletized without difficulty and had good uniformity. Single screw extruder may be enough for the blend preparation.

THERMAL PROPERTIES

Table 2 shows the melting point and heat of fusion of the blends. As seen in DSC thermograms of the blends (Figures 1-4), metallocene-LLDPE shows a single broad fusion peak with the melting temperature of 98.2°C. LDPE also has a single endotherm peak with the melting temperature of 107.5°C, on the other hand 3 conventional LLDPEs show multi endotherm peaks. DSC thermograms of the blends were observed as the overlap of each component. Since LLDPE-3 has large endotherm peak area in high temperature region (around 100 to 125°C), it is expected to improve the heat resistance of the

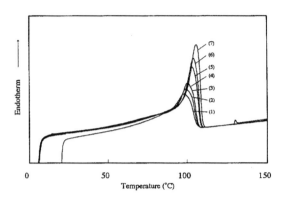

Figure 1. DSC thermograms of meta-LLDPE/HP-LDPE blends at heating rate of 10°C/min. Blends ratio: (1) 100/0, (2) 90/10, (3) 80/20, (4) 50/50, (5) 20/80, (6) 10/90, (7) 0/100.

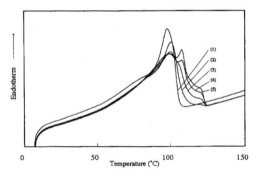

Figure 2. DSC thermograms of meta-LLDPE/LLDPE-1 blends at heating rate 10°C/min. Blends ratio: (1) 100/0, (2) 80/20, (3) 50/50, (4) 20/80, (5) 0/100.

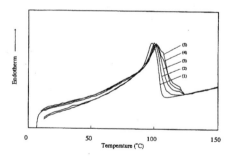

Figure 3. DSC thermograms of meta-LLDPE/LLDPE-2 blends at heating rate 10°C/min. Blends ratio: (1) 100/0, (2) 80/20, (3) 50/50, (4) 20/80, (5) 0/100.

metallocene-LLDPE. Consequently, LLDPE-3 has been chosen for the blend component. Figure 5 shows the heat of fusion of metallocene-LLDPE/LLDPE-3 blends as a function of blend composition. Since lower heat of fusion of a sealant is required for high-speed heat sealing, the blend ratios of metallocene-LLDPE/LLDPE-3 = 80/20 or 70/30 were suggested as the optimum blend composition for the sealant material.

SEALABILITY TEST

Figure 6 shows the heat-seal strength of metallocene-LLDPE, metallocene-LLDPE/LLDPE-3 = 80/20 and 70/30 blends, and LLDPE-C most commonly used for sealant material. Metallocene-LLDPE showed larger heat-seal strength in low sealing temperature range and more rapid rise in heat-seal strength than the LLDPE-C. Metallocene-LLDPE/LLDPE-3 = 80/20 and 70/30 blends also had more rapid rise in heat-seal strength, thus it was expected that these blend materials could be used for high-speed heat sealing.

Figure 7 shows the hot-tack strength of the same samples as in the heat-seal strength test. Metallocene-LLDPE gives a sharp hot-tack strength curve with a peak at 95°C around the melting point; its hot-tack strength above 100°C is considerably small, thus the heat resistance of this material is not enough for hot-filling process. On the other hand metallocene-LLDPE/LLDPE-3 = 70/30 blend and LLDPE-C give broad hot-tack strength curves with a peak at 100 and 110°C respectively. Metallocene-LLDPE/ LLDPE-3 = 80/20 blend shows a moderate hot-tack strength curve with a peak at around 105°C, which indicates that the heat resis-

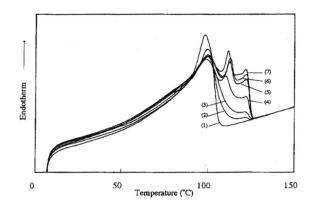

Figure 4. DSC thermograms of meta-LLDPE/LDPE-3 blends at heating rate of 10°C/min. Blends ratio: (1) 100/0, (2) 90/10, (3) 80/20, (4) 50/50, (5) 20/80, (6) 10/90, (7) 0/100.

tance of this blend is better than that of metallocene-LLDPE. And this blend material maintains enough hot-tack strength in a wide range of sealing temperature. Therefore, the metallocene-LLDPE/LLDPE-3 = 80/20 blend is expected to be a good heat-resistant sealant for hot-filling process.

FORM/FILL/SEAL TEST

Table 3 shows the result of the form/fill/seal, FFS, test. The pouches containing metallocene-LLDPE sealant made at an operating speed of 20 m/min had good seal quality and enough pressure resistance, however, those made at an operating speed of 24 m/min gave broken samples in the pressure resistance test. The pouches containing LLDPE-C and metallocene-LLDPE/LLDPE-3=70/30 blend gave broken ones even at an operating speed of 20 m/min. The pouches containing metallocene-LLDPE/LLDPE-3=80/20 blend could be made at any operating speed without trouble. It has been found that this blend material had highest ability for high-speed hot-liquid filling process.

Table 3. Form / Fill / Seal test

Sealant	Line Speed 20 m/min Quality	Line Speed 20 m/min Press. Resist.	Line Speed 24 m/min Quality	Line Speed 24 m/min Press. Resist.	Line Speed 27 m/min Quality	Line Speed 27 m/min Press. Resist.
Meta-LLDPE/LLDPE-3 (80/20)	o	o	o	o	o	o
Meta-LLDPE/LLDPE-3 (70/30)	o	Δ	x	x	x	x
Meta-LLDPE	o	o	o	Δ	o	Δ
LLDPE-C	o	x	x	x	x	x

CONCLUSIONS

Blending of conventional LLDPE with metallocene-LLDPE improved the heat resistance and the ability for high-speed hot filling process. Based on the DSC analysis, the best appropriate LLDPE material and blend composition have been determined as metallocene-LLDPE/LLDPE-3 = 80/20 blend. Measurements of the heat-seal strength and hot-tack strength for various metallocene-LLDPE blends showed that this metallocene-LLDPE/LLDPE-3 = 80/20 blend material had good sealing ability in hot filling process. The form/fill/seal test confirmed that this blend material had highest ability for high-speed hot-liquid filling process. It is prospected that this method can be applied in many of similar systems to improve the ability for high-speed hot filling process.

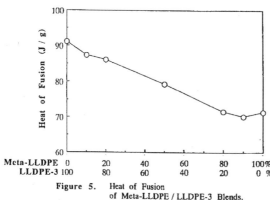

Figure 5.   Heat of Fusion
            of Meta-LLDPE / LLDPE-3 Blends.

Figure 5. Heat of fusion of meta-LLDPE/LLDPE-3 blends.

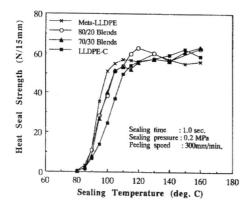

Figure 6. Heat seal strength of sealant materials.

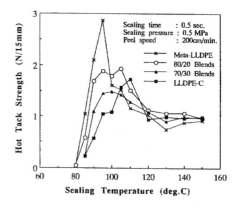

Figure 7. Hot tack strength of sealant materials.

REFERENCES

1.   D. Van der Sanden and R. W. Halle, *TAPPI*, 1991.
2.   D. Van der Sanden and R. W. Halle, *TAPPI*, 1992.
3.   D. Van der Sanden and R. W. Halle, *TAPPI*, 1993.

Blends of SIS Block Copolymers with Ultralow Density Polyethylenes

D. M. Bigg
R. G. Barry Corporation, Columbus , OH

ABSTRACT

A series of blends of styrene-isoprene-styrene block copolymers and ultralow density polyethylenes were evaluated for mechanical properties. Two types of block copolymer were investigated; one with a substantial concentration of diblock content, another with only triblock construction. Two classes of ultralow density polyethylene were examined; one made from Ziegler-Natta catalysis, the other from metallocene catalysis. The metallocene based polymer exhibited greater interfacial adhesion with the SIS triblock copolymer than any of the other combinations investigated. Elastomeric blends were produced which maintained a high level of softness and extensibility, while exhibiting a very low degree of surface tack. These blends were based on a matrix of SIS copolymer with at least 20% polyethylene.

INTRODUCTION

Thermoplastic elastomers have a number of advantages over vulcanizable rubber. The primary advantage is that, since no cure is required, processing is faster. As a result TPE's have made significant inroads into markets traditionally served by rubbers. Still, natural rubber is dominant in articles that require very low values of hardness. Such products are characterized by very high degrees of flexibility and softness. Examples include baby nipples, condoms, medical gloves, and electrical worker's gloves. While there are several TPE's that have hardness values as low as the natural rubber used in the above products, those TPE's have an unacceptable degree of surface tack. It was recently discovered that the addition of polyethylene to low hardness styrene-isoprene block copolymers will eliminate the surface tack under certain conditions.[1] Those conditions require the polyethylene to be dispersed in the elastomer at a concentrations of at least 20%. Maintenance of a low degree of surface hardness in the blend also requires the polyethylene material to have a low degree of hardness. This condition is met with the recently commercialized ultralow density polyethylenes.

There are two types of styrene-isoprene block copolymer available; those containing only SIS triblock construction, and those with significant degrees of SI diblock character. There are also a number of ultralow density polyethylene materials available. In this project they were characterized according to the synthesis technique used to produce them; Ziegler-Natta, or metallocene. This paper describes the properties of several blends made using triblock and diblock/triblock styrene-isoprene with ultralow density polyethylene made from Ziegler-Natta and metallocene catalysis.

In this investigation the key technical issues related to the effect of polyethylene on the properties of the styrene-isoprene elastomer. Numerous articles have been published on the effect of constituent concentration on the properties of polymer blends.[2] The most elementary method of analyzing the effect of the dispersed phase on the matrix phase properties is according to how well the properties adhere to the simple rule-of-mixtures relationship:

$$P_b = P_A \phi_A + P_B \phi_B$$ [1]

where P is some property and ϕ is the volume fraction of material in the mixture. The subscript, b, refers to the blend, while, A, and, B, refer to the two materials which make up the mixture. Most polymer blend properties deviate from the rule-of-mixtures in a negative manner, unless the degree of adhesion between the two components is very good. Nicolais and Nicodemo presented a lower bound relationship for tensile strength, σ, when there is no adhesion between the two phases.[3] That relationship is:

$$\sigma_b = \sigma_m (1 - 1.21 \phi_d)^{2/3}$$ [2]

where, b, refers to the blend, m, the matrix, and, d, the dispersed phase. In the case of the tensile strength, the value of the data relative to the rule-of-mixtures and lower bound values gives a measure of the interfacial adhesion.

MATERIALS
STYRENE-ISOPRENE BLOCK COPOLYMERS

The styrene-isoprene block copolymers investigated are described in Table 1. These materials were provided by Shell and Dexco, a joint venture company between Exxon and Dow Chemical Company. Shell (under the Kraton™ tradename) and Dexco (under the Vector™ tradename) provided different types of styrene-isoprene block copolymers. These block co-polymers are elastomeric materials having mechanical properties very similar to vulcanized natural rubber. Styrene-isoprene block copolymers have this characteristic because the styrene blocks form a dispersed, hard, glassy phase at temperatures below the glass transition temperature of polystyrene (~100°C). The hard, glassy phase segments physically link the rubbery isoprene blocks, which constitute the matrix phase in the copolymer.

Table 1. Description of styrene-isoprene block copolymers

Grade	Kraton 1107	Vector 4113	Vector 4100D
Supplier	Shell	Dexco	Dexco
Density, g/cc	0.92	0.92	0.92
Tensile strength, MPa	18.6	18.8	23.2
Elongation-at-break, %	1350	1500	1400
Set at break, %	10	-	-
200% Stress, MPa	1.0	0.4	0.6
Hardness, Shore A	36	35	38

The ratio of styrene to isoprene and the molecular length of each segment determines the properties of the elastomer. The styrene-isoprene block copolymers provided by Dexco were made by a different chemical synthesis procedure than the more established Kraton block copolymers. The synthesis technique used to produce the Vector copolymers resulted in an optically clear material, as opposed to the cloudy appearance of Kraton copolymers. The Dexco copolymers also have less residual salt residue from the catalyst, a factor which should increase the electrical breakdown strength of these materials. Two grades of Dexco copolymer were evaluated, Vector 4113 and Vector 4100-D. Vector 4113 is offered as direct substitute for Kraton 1107. Both Kraton 1107 and Vector 4113 contain 15% styrene and 85% isoprene. Vector 4100-D is a high strength developmental grade that contains 15% styrene. Vector 4110-D has the characteristic of having an almost pure styrene-isoprene-styrene triblock construction. The Kraton copolymer and Vector 4113 have a significant concentration of styrene-isoprene diblock molecules.

ULTRALOW DENSITY POLYETHYLENES

Table 2. Description of ultralow-density polyethylenes

Grade	MQF0	Exact™ 4041
Supplier	Enichem	Exxon
Density, g/cc	0.915	0.878
Tensile Strength, MPa	6.8	28.7
Elongation-at-break, %	900	800
200% Modulus, MPa	3.7	3.4
Hardness, Shore A	80	76

The major shortcoming of very low hardness styrene-isoprene block copolymers is that they have an unacceptable degree of surface tack, or stickiness. This problem also exists to some degree in natural rubber, where it is eliminated by surface chlorination. While that approach can also be used to modify the surface of the styrene-isoprene block copolymers, an alternative solution is to add an ultralow density polyethylene to the block copolymer. This has the advantage of eliminating tackiness in a molded part. The addition of ultralow density polyethylene to the styrene-isoprene block copolymer was found to be an effective means of eliminating surface tack from the block copolymer.

Ultralow density polyethylene materials were provided by Enichem and Exxon. The properties of these materials are summarized in Table 2. There is a wide range among the properties of these materials. Exxon's Exact™ 4041 grade was of interest because it has a particularly high tensile strength. Exact 4041 ultralow density polyethylenes were made by metallocene catalysis, while the Enichem MQF0 was made by Ziegler-Natta catalysis. The metallocene polymer exhibited a lower value of Shore A hardness than ultralow density polyethylenes made by Ziegler-Natta catalysis. The Enichem material had the lowest hardness of any polyethylene made by Ziegler-Natta catalysis.

PREPARATION OF BLENDS

Blends were prepared by melt mixing the polymers on a two-roll mill at 135°C for 10 minutes. Enichem MQF0 polyethylene was blended into the Kraton 1107, Vector 4113, and Vector 4100-D elastomers. Exxon's Exact 4140 polyethylene was blended into Vector 4113 and 4100-D block copolymers. The compounded materials were compression molded into 38 cm x 38 cm x 1 mm sheets. Compression molding was performed at 150°C under 1.7 MPa of pressure for 5 minutes.

EVALUATION PROCEDURES

As discussed previously, one important attribute of polymer blends is that their properties are not usually proportional averages of the properties of the two constituents, since the degree of compatibility between constituents cannot be predicted *a priori*. This necessitates experimental evaluation of all of the key properties. The key properties investigated included tensile strength, Barcol A hardness, stress at 200% elongation, elongation-to-break, tear resistance, and puncture resistance. An Instron testing unit was used to make most of these measurements.

The range over which these materials maintain elastomeric properties was determined by Dynamic Mechanical Analysis, DMA, performed on a Rheometrics Mechanical Spectrometer, model 605. The samples were analyzed in the torsional mode from -125°C to 125°C at a frequency of 6.28 rad/sec.

RESULTS

MECHANICAL PROPERTIES

The mechanical properties of the blends of the Kraton 1107 styrene-isoprene copolymer and MQF0 polyethylene are summarized in Table 3. The relative tensile strength (that of the blends divided by that of the elastomer) is plotted as a function of polyethylene concentration in Figure 1. Clearly, the tensile strength data do not follow the straight line rule-of-mixtures relationship, nor do they fall on the curves following the lower bound model. This suggests that there was some degree of adhesion between the elastomer and polyethylene for polyethylene concentrations below 28% and above 40%. The blends consisted of polyethylene dispersed in the copolymer at polyethylene concentrations below 28% polyethylene. Above 40% polyethylene, polyethylene became the matrix phase. The phase inversion is quite noticeable in data in Figure 1.

Table 3. Mechanical properties of (styrene-isoprene-styrene/ultralow-density polyethylene) blends - Kraton 1107 block copolymer - MQF0 polyethylene[a]

Property	100/0	90/10	80/20	76/24	70/30	60/40	50/50	40/60	30/70	20/80	10/90	0/100
Hardness, Shore A	36	40	45	47	53	64	67	71	79	82	84	86
Tensile strength, MPa	18.6	16.8	13.5	13.2	9.3	3.2	3.3	3.6	4.1	4.5	4.6	6.6
200% Stress, MPa	1.0	0.8	1.2	1.3	1.3	1.9	2.0	2.2	2.9	3.0	3.4	3.7
Tensile set, %	8	7	12	-	54	148	-	-	-	-	-	-
Ultimate elongation, %	1350	1450	1350	1920	1375	1150	900	900	800	850	750	900
Tear resistance, kN/m	31	27	25	23	26	25	30	30	32	36	35	44

[a]styrene-isoprene-styrene/ultralow-density polyethylene ratios

The Shore A hardness of the Enichem MQF0-Shell Kraton 1107 blends are plotted as a function of polyethylene concentration in Figure 2. At polyethylene volume fractions up to 0.28 the Shore A hard-

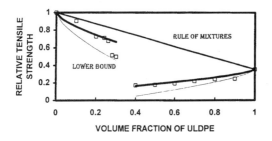

Figure 1. Effect of polyethylene concentration on the relative tensile strength of Kraton 1107/Enichem MQF0 blends.

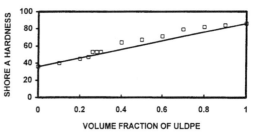

Figure. 2 Effect of polyethylene concentration on Shore A hardness of Kraton 1107/Enichem MQF0 blends.

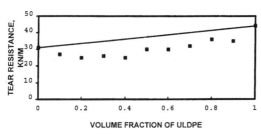

Figure 3. Effect of polyethylene concentration on the tear resistance of Kraton 1107/Enichem MQF0 blends.

Table 4. Mechanical properties of (styrene-isoprene-styrene/ultralow-density polyethylene) blends - Dexco 4113 copolymer-MQF0 polyethylene

Property	4113/MQF0 100/0	4113/MQF0 90/10	4113/MQF0 80/20
Hardness, Shore A	35	39	45
Tensile Strength, MPa	18.8	18.4	11.0
200 % Stress, MPa	0.4	1.1	0.7
Ultimate Elongation, %	1500	1200	1200
Puncture Resistance, kN/m	77	69	68
Tear Resistance, kN/m	22	27	27

ness followed the rule-of-mixtures quite closely. Above this polyethylene concentration, where polyethylene became the matrix phase, the hardness of the blends was slightly greater than the simple proportional average of the values of the two materials.

Figure 3 is a graph of the tear strength of the MQF0/1107 blends as a function of increasing polyethylene concentration. The presence of polyethylene reduced the tear strength of the elastomer at all concentrations in which the block copolymer was the matrix phase. Similarly, the presence of the dispersed elastomer reduced the tear strength at all concentrations in which polyethylene was the matrix. This behavior is a good indication that, while there was some adhesion between phases indicated by the tensile strength data, it was not sufficient to enhance the tear strength.

The mechanical properties of blends of MQF0 polyethylene in Vector 4113 styrene-isoprene copolymer are summarized in Table 4. It appears that Vector 4113 is an acceptable alternative to Kraton 1107. Vector 4113 is less tacky than Kraton 1107 and has comparable mechanical properties. The tear resistance values of the MQF0/4113 blends were comparable to those of the MQF0/1107 blends. Finally, the phase inversion from elastomer matrix to polyethylene matrix occurred in the same concentration range for these blends as for the MQF0/1107 blends.

Table 5. Mechanical properties of (styrene-isoprene-styrene/ultralow-density polyethylene) blends - Dexco 4100D copolymer - MQF0 polyethylene

Property	4100D/MQF0 100/0	4100D/MQF0 95/05	4100D/MQF0 90/10	4100D/MQF0 80/20
Hardness, Shore A	38	41	41	46
Tensile Strength, MPa	19.3	23.9	19.2	19.7
200 % Stress, MPa	0.7	0.7	0.6	1.7
Ultimate Elongation, %	1400	1350	1300	1500
Puncture Resistance, kN/m	106	84	53	55
Tear Resistance, kN/m	30	30	29	25

The mechanical properties of blends of MQF0 polyethylene with Vector 4100-D styrene-isoprene elastomer are summarized in Table 5. The properties of these blends reflected the increased strength of this elastomer relative to those elastomers in which there was a significant level of diblock concentration. Of particular note are the improvements in the puncture and tear resistance provided by this elastomer.

Table 6. Mechanical properties of (styrene-isoprene-styrene/ultralow-density polyethylene) blends - Dexco 4113 copolymer - 4041 polyethylene

Property	4113/4041 100/0	4113/4041 80/20	4113/4041 75/25	4113/4041 70/30
Hardness, Shore A	35	43	47	45
Tensile Strength, MPa	18.8	18.7	11.2	16.8
200 % Stress, MPa	0.4	0.6	1.0	0.7
Ultimate Elongation, %	1500	1300	1100	1200
Puncture Resistance, kN/m	77	83	83	78
Tear Resistance, kN/m	22	30	25	24

The mechanical properties of the blends of Exact 4041 polyethylene and Vector 4113 styrene-isoprene copolymer are summarized in Table 6. Note that the tensile strength and puncture resistance are dramatically higher than those of previous blends.

Table 7. Mechanical properties of (styrene-isoprene-styrene/ultralow-density polyethylene) blends - Dexco 4100-D block copolymer-4041 polyethylene[a]

Property	100/0	90/10	80/20	70/30	60/40	50/50	40/60	30/70	20/80	10/90	0/100
Hardness, Shore A	38	39	46	50	55	59	64	67	71	75	78
Tensile strength, MPa	23.2	27.6	24.8	24.80	18.6	20.4	21.5	23.3	26.2	27.0	28.7
200% Stress, MPa	0.6	0.9	1.2	1.4	2.1	2.0	2.3	2.8	2.8	3.1	3.4
Ultimate elongation, %	1400	1300	1100	1100	1000	900	900	900	700	800	800
Puncture resistance, kN/m	106	117	122	144	112	132	124	131	126	105	100
Tear resistance, kN/m	30	38	37	27	28	33	33	38	42	43	46

[a]styrene-isoprene-styrene/ultralow-density polyethylene ratios

Table 7 summarizes the properties of blends of Exact 4041 ultralow density polyethylene in Vector 4100-D styrene-isoprene block copolymer. Figure 4 shows the relative tensile strength of the 4100- D/4041 blends. As before, the light solid line was generated from Equation 2. Clearly, the data do not follow this interpretation of zero interfacial adhesion. When the elastomer formed the matrix phase the tensile strength behavior closely followed the rule-of-mixtures. This implies that the adhesion between the two phases was quite good. The tensile data suggest that the polyethylene interacted more strongly with the elastomer when the polyethylene was the dispersed phase, than the elastomer did with the polyethylene when the elastomer was the dispersed phase.

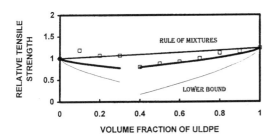

Figure 4. Effect of polyethylene concentration on the relative tensile strength of Vector 4100-D/Exact 4041 blends.

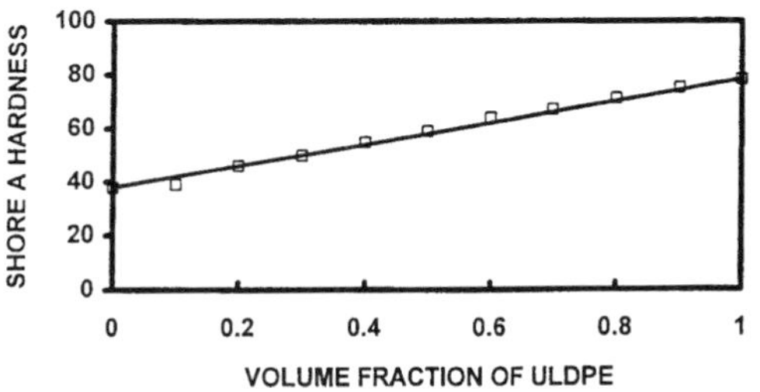

Figure 5. Effect of polyethylene concentration on the Shore A hardness of Vector 4100-D/Exact 4041 blends.

Figure 5 shows that the rule-of-mixtures was closely adhered to when determining the Shore A hardness of these blends. Figure 6 shows that the tear resistance follows the same general trend observed for the tensile strength; namely, that the tear strength of the elastomer is enhanced at low polyethylene concentrations, while the tear strength of the polyethylene is diminished when it is the matrix

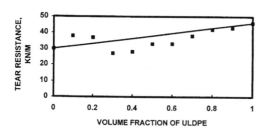

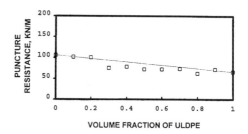

Figure 6. Effect of polyethylene concentration on the tear resistance of Vector 4100-D/Exact 4041 blends.

Figure 7. Effect of polyethylene concentration on the puncture resistance of Vector 4100-D/Exact 4041 blends.

phase. Contrast this with the data shown in Figure 3 where the presence of the Ziegler-Natta based Enichem MQF0 polyethylene in the Kraton 1107 elastomer resulted in a drop in tear strength. The metallocene based polyethylene enhanced the tear strength of the blend when the elastomer was the matrix phase.

Interestingly, the puncture resistance, Figure 7, is higher for all blends than it is for the two pure materials. The disparity of results among these various tests reflects not only the degree of interfacial adhesion, but the various ways in which the blends respond to the different stress mechanisms imposed in each of these testing procedures.

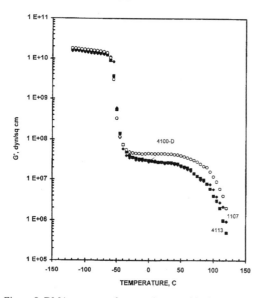

Figure 8. DMA response of styrene-isoprene block copolymers.

DYNAMIC MECHANICAL PROPERTIES

DMA was used to measure the dynamic shear modulus, G', of Vector 4100-D, Vector 4113, and Kraton 1107 styrene-isoprene block copolymers; Exact 4041, and Enichem MQF0 ultralow density polyethylenes; and the 80/20 blends of Vector 4100-D with Exact 4041 and Enichem MQF0. The measurements were made over the temperature range of -120°C and 120°C. The results are shown in Figures 8 through and 10. The responses of the three grades of block copolymer investigated, shown in Figure 8 are very similar. They all have an isoprene phase glass transition temperature at -50°C, and a styrene phase glass transition temperature at 110°C. Each has a long plateau region between the two transition temperatures, marking the rubbery region of the polymer. The responses of the Kraton 1107 and Vector 4113 are identical within the limits of the test equipment. Since they are sold as equivalent materials

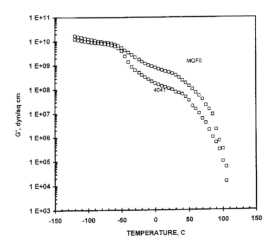
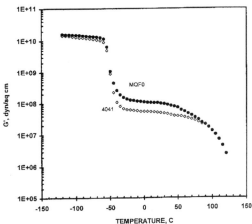

Figure 9. DMA response of ultralow density polyethylene.   Figure 10. DMA response of 80/20 blends of Vector 4100-D with MQF0 and 4041.

this is not surprising. Vector 4110-D exhibits a slightly higher modulus and flatter plateau region that the other two elastomers. Because all three polymers have the same concentration of styrene, this difference can be attributed to the triblock nature of the 4100-D grade.[4] The other two grades have a significant level of diblock fraction.

The responses of the ultralow density polyethylene materials is shown in Figure 9. To varying degree they all exhibit a glass transition temperature at -45°C, they all have a plateau region between the glass transition temperature and approximately 40°C. Between 40°C and 100°C the modulus drops off in a steadily increasing manner. This drop off is attributed to the fact that these polymers have a very broad melting range.[5] The Enichem MQF0 has the highest modulus at all temperatures above the glass transition temperature. The metallocene based polymers have similar responses, with minor differences exhibited around the glass transition temperature. The level of modulus in the plateau region is a measure of the relative degree of crystallinity in the polymers. Since the MQF0 grade has the highest density of the group it has the highest degree of crystallinity, and would be expected to have the highest plateau modulus.

Figure 10 shows the DMA response of the 80/20 blends of the four polyethylenes with the Exact 4100-D elastomer. The only observable response is that the plateau modulus of the blend utilizing the MQF0 polyethylene is appreciably higher than those of the blends using the metallocene based polyethylenes. 80/20 SIS/ULDPE blends were of interest because blends with at least 20% polyethylene exhibited little surface tack, yet retained the basic characteristics of the elastomeric phase.

SUMMARY

The mechanical properties of several poly(styrene-isoprene) block copolymer/ultralow density polyethylene blends were examined. Blends based on Dexco's Vector 4100-D block copolymer and Exxon's Exact 4041 ultralow density polyethylene were of particular interest. The presence of the polyethylene had the important feature of eliminating surface tack at concentrations above 20%. Polyethylene was found to become the matrix phase at concentrations above 30%. Of the properties exceeding the specifications, tensile strength, tear resistance, and puncture resistance were the most significant, and most highly desired. Analysis of the data showed that there was a significant degree of adhesion between the two phases of these materials when the polyethylene had been produced by metallocene catalysis.

REFERENCE

1.   D. M. Bigg, *U. S. Patent 5 451 439*, 1995.
2.   L. A. Utracki, **Polymer Blends and Alloys**, *Hanser Verlag*, Munich, 1989.
3.   L. Nicolais and L. Nicodemo, *Polym. Eng. Sci.*, **13**, 469 (1973).
4.   C. A. Berglund and K. W. McKay, *Polym. Eng. Sci.*, **33**, 1195 (1993).
5.   D. E. Turek, B. G. Landes, and J. M. Winter, SPE Conference, 1995.

Preparation of Metallocene Plastomer Modified High Flow Thermoplastic Olefins

Thomas C. Yu

Exxon Chemical Company, Baytown, TX 77522, USA

ABSTRACT

Metallocene plastomers are ethylene-α-olefin copolymers with a density range from 0.91 to 0.86 and a melt index ranging from less than 1 to 125. The high efficiency single site catalysts provide uniform comonomer insertion, so that at a relatively low comonomer level, the copolymer exhibits both plastics and elastomeric characteristics. In this study, both batch (Banbury) and continuous (extruder) mixing devices were used to melt blend the plastomer into a high flow (35 MFR) homopolymer polypropylene. A mixing device of choice needs to provide both distributive and dispersive mixing as well as adequate mixing time. Examples of good mixing devices are a twin screw extruder or a single screw extruder equipped with a mixing section.

INTRODUCTION

The recently introduced metallocene plastomers have been found to possess performance properties similar to ethylene-propylene rubber in many thermoplastic olefin compositions. In comparison with pelletized rubber, the higher density plastomers are produced as free flowing pellets, and can therefore be used in either batch or continuous mixers. Another advantage of the plastomers is the plastic-like molecular weights (high melt index) which allow for ease of dispersion in polypropylene.[1,2] For many years, high molecular weight ethylene-propylene rubber bales have been processed together with the polypropylene base resin of choice in an intensive batch mixer to produce thermoplastic olefins, TPOs. To allow for continuous compounding, the rubber bales had to be pelletized ahead of time in an intensive mixer. Also, an exterior dusting agent, such as polyethylene powder, is required to improve pellet stability. The free flowing plastomer pellets, on the other hand, can be processed via both batch and continuous mixers. With the rapid development of thermoplastic olefins as a preferred automotive plastic, high flow TPOs are especially desirable in order to produce large complex parts. For this reason, the lower molecular weight plastomer is often selected over the higher molecular weight ethylene-propylene rubber to minimize melt flow rate suppression. In this study, two series of experiments were conducted in order to select the best type mixing device for high flow thermoplastic olefin production. In the first experiment, both commercial batch and continuous mixers were used to compound an ethylene-butene plastomer into a high flow isotactic polypropylene based resin. A mixing study of a typical automotive bumper composition, i.e., a three component blend of the same plastomer in combination with a mineral filler and a high flow polypropylene impact copolymer was next investigated. Based on these two experimental results, guidelines for a mixer device selection was derived.

EXPERIMENTAL

MIXING DEVICE SELECTION

Both commercial size batch and continuous compounders were used for this study. As shown in Table 1, the batch mixer used was a Stewart Bolling #10 intensive mixer. Several continuous mixers were used; they varied from a 2-1/2" single screw extruder to a 2-1/2" single screw extruder with a mixing segment, a Farrel CP500 continuous mixer, and finally a 57 mm ZSK twin screw extruder. The produced compounds, together with the original dry blend, were then injection molded into 3-1/2" discs and 1/8" thickness on a 75 ton Van Dorn equipped with a polypropylene screw without a mixing section.

Table 1. Mixing device selection

- Dry Blend (in an injection molding machine)
  - Metering screw with no mixing segment
- Single screw extruders
  - 2-1/2" no mixing segment
  - 2-1/2" with distributive mixing segment (Mattox mixer)
- D57 mm ZSK Twin screw extruder
- Stewart Bolling #10 intensive mixer
- CP-500 Farrel continuous mixer

POLYPROPYLENE AND PLASTOMER SELECTIONS

There are several types of polypropylene available commercially, i.e., homopolymer, random copolymer, and impact copolymers. Among them the high flow homopolymer was found to be most resistant to impact improvement. We purposely selected a high flow homopolymer for the first experiment together with EXACT™ 4033, an ethylene-butene plastomer, as an impact modifier. As shown in Table 2, 70 wt% of Escorene PP1105(35 MFR) and 30 wt% EXACT™ 4033 (0.8 MI/0.880 density)

Table 2. Mixing study for plastomer polypropylene blends

| Escorene PP 1105 E-1 (35 MFR) | 70 Wt% |
| EXACT 4033 (0.8MI/0.88D, C4) | 30 Wt% |

- Dry blend (in an injection molding machine)
  - Metering screw with no mixing segment
- Single screw extruders
  - 2-1/2" no mixing segment
  - 2-1/2" with distributive mixing segment (Mattox mixer)
- 57 mm ZSK Twin screw extruder
- Stewart Bolling #10 intensive mixer
- CP-500 Farrel continuous mixer

was compounded on all mixing devices as shown in Table 1. A high flow polypropylene impact copolymer (Escorene PD 7565, 45 MFR) was selected for the second experiment. The mineral filler used in the second experiment was a 1 μm average particle size talc (Cimpact 710); and EXACT 4033 again served as the impact modifier. The composition and mixing devices used for this experiment are shown in Table 3.

Table 3. Mixing study for talc filled plastomer polypropylene blend

Escorene PD 7565 (45 MFR, 8% C2)   60 Wt%
EXACT™4033(0.8MI/0.88D,C4)   30 Wt%
Cimpact 710 (1 micron talc)   10 Wt%
- D57 mm ZSK Twin screw extruder
- Stewart Bolling #10 intensive mixer
- CP-500 Farrel continuous mixer

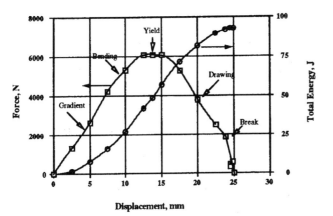

Ductility Index DI = [(Total Energy - Energy at Yield)/Yield Energy] × 100

Figure 1. High speed falling weight instrumental impact testing.

TEST CRITERION

Preliminary testing showed that most of the common mechanical properties were insensitive to the mixing device selection. Only impact properties such as a falling weight dart impact could be used to distinguish the performance of each type of mixing device. The impact tester employed was a Ceast Fractovis using a 0.5" hemisphere test dart and a 3" circular holder. A single test speed of 4 m/sec, and test temperatures ranging from 23 to -40°C were used. A typical force displacement tracing generated by the Ceast instrument is shown in Figure 1 for a test specimen exhibiting ductile failure. Note that for ductile failure, a substantial amount of energy is absorbed after the yield point due to dart induced sample drawing. Integration of the force-displacement curve generates an energy absorption versus displacement curve also shown in Figure 1. A ductility index DI can be defined as:

Ductility Index DI =[(Total Energy-Yield Energy/Yield Energy)] X 100

The typical failure modes are shown in Figure 2. Escorene PD-7194, a medium impact polypropylene, was found to be brittle at -20°C and its ductility index was calculated to be zero. When the same polypropylene was modified with 20 wt% of EXACT 4033, its ductility was drastically improved from zero to 100. The right side of Figure 2 shows the force displacement response of EXACT 4033 at -20°C. The elastomeric nature of EXACT 4033 is clearly illustrated in that the sample shows extensive drawing after its yield point, with a ductility index of over 100.

RESULTS AND DISCUSSION

BINARY PLASTOMER ISOTACTIC POLYPROPYLENE BLEND

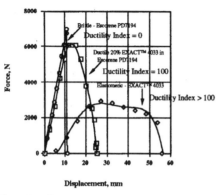

Figure 2. Failure mode for instrumented impact. Test temperature @ -20°C.

The total energy absorbed by all the test blends as a function of test temperature is shown in Figure 3. It is interesting to observe that from 23 to -10°C the total energy absorbed is about the same for all melt blends and the control dry blend. When the test temperature was lowered further, several blends started to show brittle failure as reflected by the lower energy absorbed. At -40°C test temperature, only the twin screw compounded blend and the 2-1/2" single screw with mixing segment compounded blend still remained ductile. The conclusion is supported by the ductility index calculated for these blends as shown in Figure 4. Again, only the twin screw and single screw with mixing segment blends were found to be ductile at -40°C. Both the low shear batch mixer, and the short resi-

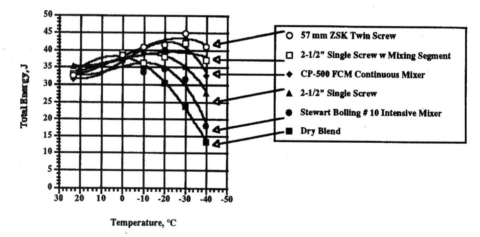

Figure 3. Effect of mixing devices on total impact energy for plastomer polypropylene blend.

dence time continuous mixer failed to produce acceptable compounds. It was necessary to provide some degree of dispersive mixing, as demonstrated by the fact that without a mixing segment, the single screw extruder also failed to produce a compound with cold temperature impact resistance.

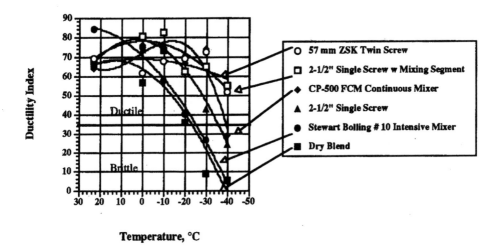

Temperature, °C

Ductility Index DI = [(Total Energy - Energy at Yield)/Yield Energy] × 100

Figure 4. Effect of mixing devices on ductility index for plastomer polypropylene blend.

TERNARY PLASTOMER/POLYPROPYLENE/TALC BLEND

A second series of experiments based on a typical automotive bumper composition was compounded using a batch mixer (Stewart Bolling), and two continuous mixers (Farrel CP500, and 57 mm ZSK twin screw extruder). Figure 5 shows that all three compounders produced final products with identical total impact energy from 23 to -40°C test temperature. The ductility index as shown in Figure 6 also reinforced our observation that all three types of compounders are equally effective in producing impact resistance compounds. We may conclude that the addition of a mineral filler increases the melt viscosity of the polypropylene matrix resin, thus reducing the viscosity disparity between EXACT plastomer and polypropylene. This allows even the low shear batch mixer and extremely short residence time continuous mixer to be effective compounding device.

CONCLUSIONS

Metallocene plastomers are a new family of modifiers for polypropylene. They are especially suited as impact modifiers in high flow thermoplastic olefins in the rapidly growing market of automotive parts. Adequate mixing time in a mixing device is needed to provide distributive and dispersive mixings. Examples of good mixing devices are a twin screw extruder, or a single screw extruder equipped with a mixing section. When a mineral filler is added to the TPO compound, the choice of mixing devices broadens to include both the batch and continuous type of compounders.

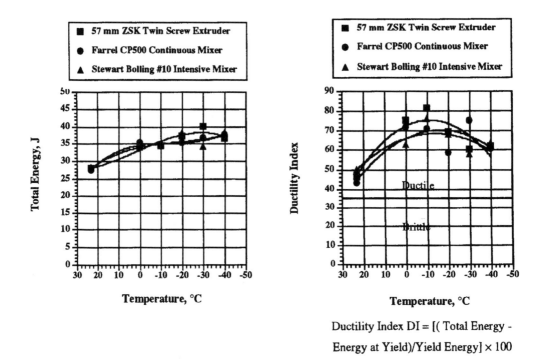

Figure 5. Effect of mixing devices on total impact energy for talc filled plastomer polypropylene blend.

Ductility Index DI = [(Total Energy - Energy at Yield)/Yield Energy] × 100

Figure 6. Effect of mixing devices on ductility index for talc filled plastomer polypropylene blend.

ACKNOWLEDGMENT

The author would like to thank Kelli Brightwell, Ruth Spilman, Tony Flores and the Polymer Application Laboratory in Baytown for their support of this study. D. S.Davis and L. G. Kaufman reviewed the manuscript and made several valuable suggestions.

REFERENCES

1. S. C. Speed, B. C. Trudell, A. K. Mehta, and F. C. Stehling, *SPE RETEC Polyolefins VII*, 1991.
2. T. C. Yu and G. J. Wagner, *SPE RETEC Polyolefins VIII*, 1993.

Metallocene Capabilities Drive Success of Innovative Packaging Manufacturer

D.C. Fischer
Cypress-Cryovac, USA

INTRODUCTION

The last decade of the twentieth century might well become known as the "Metallocene Age." A study by Business Communications Co. Inc. of Norwalk, Connecticut says that the market for metallocene plastomers and elastomers was 671 million pounds in 1995, and is expected to increase to 941 million pounds by 2000, an astounding rise given that metallocene catalysts were only developed in the 1980's and didn't come to market as viable commercial products until the early 1990's.

The unique features of metallocene catalysts, coupled with the ability to modify the metallocene structure itself, have fostered the development of innovative products, new businesses, and unique ways of achieving results more economically. The end result is that today we are able to provide our customers with amazing new capabilities and significant benefits that were not available just a few years ago.

Through our imaginative utilization of metallocene technology, our company transformed itself from a manufacturer of commodity products to one of value-added, higher margin business. It is from this entrepreneurial perspective that I have been asked to address you today.

In 1989 my partner and I mortgaged all our worldly possessions to purchase a failed commodity bread bag manufacturing plant in upstate New York. Armed with an ambitious business plan we set out to conquer the poly bread bag world and Cypress Packaging was born. Buffeted by losses in our first year of operations and terrorized by occasional gut wrenching feelings of "what the hell have we got ourselves into," we turned our attention to our California agricultural roots.

As young boys growing up in the fertile Salinas Valley, we spent our summers harvesting strawberries and lettuce. Little did we know that we would someday manufacture polyethylene bags in which lettuce, cauliflower, carrots, and celery would be packed. Cypress quickly became a major supplier of commodity polybags to the produce industry. Soon hundreds of millions of produce polybags were shipped all over the country from our plant.

While we were profitable, our product line lacked imagination and we were just another commodity polybag manufacturer. We yearned for innovation, new product development and something new and exciting. Opportunity came knocking one day in early 1991 when one of our major California based produce customers asked our assistance in solving a serious packaging problem plaguing their new fresh-cut produce division.

It seems their polyethylene rollstock, supplied by one of the world's largest flexible packaging conglomerates, had a narrow heat seal range resulting in defective bags coming off their vertical form and fill equipment. The launch of this company's new fresh-cut salad mixes was ruined by bags breaking open in the supermarket or unacceptable numbers of bags with broken seals causing the product to spoil.

Opportunity needed only knock once at our door and within days we provided our customer with a new test film made of a unique blend of polyethylene resins including ultra linear lows. Our formula worked and our customer immediately placed all their rollstock orders with Cypress. We were an overnight success in the foundling fresh-cut produce industry because we had just been named the supplier to Dole.

Since those pioneering days we have introduced various monolayer, coextruded and laminate films to this and other traditional markets. Our sales quadrupled in four years and profits soared. Industry analysts soon proclaimed us the leader of fresh-cut packaging. Several months ago we sold our company to W.R. Grace and today are part of one of the most highly respected packaging companies in the world — Cryovac. Cryovac, a leader in fresh-cut food service packaging, abandoned plans to build a new plant dedicated to the fresh-cut industry opting instead to purchase our company.

The growth, development, and success of Cypress over the past five years is directly attributable to the fact that we recognized very early that metallocene resins would provide distinct advantages in the vertical form, fill, and seal fresh produce markets, and ultimately in more traditional markets. These advantages included better clarity, superior oxygen transmission, and better sealing characteristics than any other product on the market. We have not used traditional LLDPE and LDPE EVA's for over four years. All have been replaced by new metallocene based resins.

Working with Exxon Chemical and other resin suppliers, we became early adopters of this new technology. It is most interesting that neither Dow or Exxon had targeted produce packaging as an application for their metallocene resins. This market ended up being one of the largest consumers of these resins. Not only did Cypress take the first rail car of metallocene, but in 1995 we were the world's largest producer of blown film using metallocene resins.

MODIFIED ATMOSPHERE PACKAGING

The fastest growing segment of the produce industry is "pre-cut" or "fresh-cut" produce. This includes ready-to-eat salads, cole slaw, carrots, and other fruits and vegetables delivered to the consumer in an attractive, modified atmosphere package that dramatically extends shelf life. Properly formulated fresh-cut packages reduce spoilage waste and provide a wide variety of pre-washed, pre-cut produce that can last for two weeks or longer in the consumer's refrigerator.

The Wall Street Journal reported that sales of retail fresh-cut produce in 1992 approached $200 million. Retail sales today are almost $900 million a year and, it is estimated that by 2000, 35% of all fresh produce sold will be in modified atmosphere packages.

Unlike barrier films, in which most food is packaged, modified atmosphere packaging is specifically designed to match the respiration rates of each individual product. A living plant takes up CO_2 and gives off oxygen. Once harvested, however, a plant goes into reverse photosynthesis and consumes oxygen and gives off CO_2. All vegetables and fruits have different respiration rates.

Modified atmosphere packaging must be permeable enough to take in the exact amount of required oxygen and exhale the excess carbon dioxide.

Long product shelf life is dependent on production of a film that can be modified to match the exact oxygen transmission rates, OTR, of the target product. OTR is defined as the rate at which oxygen, in cubic centimeters, penetrates a one hundred square inch area of film in a 24-hour period. OTR rates for specific products are computed from a formula that takes into account the O_2/CO_2 needs of the product, weight of the product, area of film comprising the bag, and other factors.

NEW FILMS DEVELOPED

Cypress' entrance into this challenging fresh-cut market was a direct result of our quick response to a major customer's critical need. We were able to respond because we did a better job than our competition in delivering a film that met the customer's requirements, not just supplying an existing film from our product line. Listening and responding quickly to customer needs — that has been the hallmark of our success. Very shortly thereafter we were asked to develop laminated films because of their greater consumer appeal. Laminated films have better optics, a stiffer feel which consumers prefer, a higher quality appearance, and better color and graphics integrity with reverse printing.

We discovered very quickly that there are many facets to designing the right package for a product. Supplementing physical testing, we developed mathematical models to determine the OTR effect that various resin combinations would yield. We had to develop these models ourselves because they simply did not exist. Our research and development efforts lead to our invention of CP601, a film that is a tri-laminate structure consisting of blown film layers of EXACT™ and EXCEED™ metallocene resins supplied by Exxon Chemical Company married with a core of butadiene styrene copolymer. Coextrusion capability was added to our plant in order to produce an entire new family of produce films.

CP601 film was unique enough to receive a patent. This film was patentable because by modifying certain elements of the structure, we are able to dial in a wide range of OTR requirements. By modifying the structure, we can provide the exact OTR levels needed by the customer to match the size, weight and type of product being packaged. To our knowledge, there is not currently another film available that has CP601's broad range of capabilities.

New entrants into the fresh-cut packaging market have discovered how technically sophisticated a film manufacturer must be to properly service this unique industry. Unlike many barrier film applications found in the snack food and candy industries, one film formulation cannot be used for a customer's wide range of fresh-cut items. For one of our major customers we have over two dozen different films, each one manufactured to meet exacting customer performance standards.

Today it is not uncommon for unscrupulous converters to provide seriously out of spec film to unsuspecting fresh-cut operators. Few converters or their fresh-cut customers have invested in sophisticated, modern laboratory testing equipment or hired professional laboratory technicians to implement proper quality assurance programs that are desperately needed for this industry.

Fortunately times are rapidly changing and an organized fresh-cut industry, as well as suppliers like Cypress, are educating the produce industry on the critical nature of selecting the proper film for specific food applications. Our company has long advocated that it is the responsibility of the flexible packaging supplier to certify that its films meet customer specifications. Several of our customers now demand film certification from their suppliers.

CUSTOMER BENEFITS WITH METALLOCENE

The use of metallocene resins in the vertical form, fill, and seal markets provide our customers with substantial benefits over traditional EVA blends of sealants. First, metallocene films have superior sealing characteristics, or hot tack. We found that customers using EVA films were running their form, fill, and seal lines at between 35 and 40 packages per minute. The limiting factor was time required for good seal integrity. However, with the superior sealing characteristics of metallocene resin, our customers were able to increase line speed up to about 60 packages a minute and still maintain absolute seal integrity. Moreover, vertical form and fill equipment can be run at much lower temperatures reducing wear on that equipment.

With no modifications to their equipment, customers could dramatically improve productivity simply by changing to metallocene films. This was a significant advantage and, coupled with metallocene's greater clarity, superior oxygen transmission characteristics, and improved puncture resistance, was responsible for the industry's acceptance of the advantages of metallocene over more traditional blends of resins.

TRADITIONAL PACKAGING MARKETS

We recognized that the application of metallocene technology had provided significant benefits to our customers in the fresh produce markets. We realized also that many of these same benefits were transferable to more traditional packaging markets including meat, cheese, industrial and medical. We employed the same strategy to enter these markets, although, instead of competing against traditional EVAs, we were competing against ionomer resins.

Barrier films developed for these markets use metallocene in the sealing layers, and combinations of additional resins in the interior layers. Although OTR is not a factor in these markets, metallocene still provides significant customer benefits. These benefits include superior hot tack that provides hermetic seals, great optical clarity for better product display, and high strength and puncture resistance. Here again customers can run equipment at lower temperatures that was possible with traditional resins.

An additional benefit is cost. Ionomer resins are much more expensive than metallocene. So, in addition to providing our customers with superior barrier films, we can also provide these films at a significant reduction in cost. Also, because of its superior strength and puncture resistance, metallocenes offer significant down-gauging opportunities in traditional markets.

In early 1994 opportunity once again knocked on our door when we were briefed by Kodak packaging engineers about the impending launch of their new single use camera. The two major flexible packaging suppliers to Kodak were given one year to develop a new clear laminate film with exacting performance standards including a very low MVTR (moisture vapor transmission) number. Nine months passed and neither supplier had solved Kodak's film challenge. Meanwhile, single use cameras were being manufactured and placed in warehousing awaiting the film overwrap.

While Kodak had never heard of Cypress Packaging, both companies were headquartered in Rochester, New York. We were invited to compete for their business. Searching the world for barrier films to be used in our laminate process, we discovered just the right candidate in Japan. Metallocene resins were incorporated into our laminate film. Within one month we supplied Kodak with a test roll of laminate film for machinability trials and laboratory tests. Our film passed all the stringent tests and Cypress became the supplier to Kodak for this major new product line.

Cypress has enjoyed great success in introducing metallocene based films in other traditional packaging markets such as meat, cheese and medical applications. We recently opened a pouch making department and have successfully introduced standup pouches.

Once again Cypress successfully competed against the best in the business and won the race. Our dedication to solving customer and prospective customer problems and our partnershipping with suppliers have lead to phenomenal successes in very divergent packaging markets. Particularly interesting is the fact that we have achieved great technological breakthroughs without a single research scientist or packaging engineer on staff.

While we couldn't afford the luxury of research and development facilities and staff we quickly learned the value of partnershipping with our suppliers — the inventors of new and unique resins, films, adhesives, additives, coatings and inks. Eager to explore new markets for their products, these suppliers found in Cypress a manufacturing facility that would open its doors to experimentation, allowing them to trial new materials on our equipment.

Without these multiple partnership arrangements, Cypress would not have succeeded as well as we did. Our successes in permeable and barrier film technology are directly attributable to the magnificent working relationships initiated years ago with our suppliers. Large competitors of ours have traditionally been too reluctant to allow suppliers into their plants, much less to run experiments on manufacturing equipment. They often limit supplier contact to new product briefings in a conference room and keep them away from manufacturing.

Giving certain suppliers almost unlimited access to our staff and plant yielded fantastic results. Few companies have invented so many new film structures in divergent markets in such a short time as Cypress. Our successes lead to offers from many multinational companies and our recent sale to W.R. Grace.

CONCLUSION

The packaging industry has changed dramatically over the past six years. Much of that change has come as a direct result of the development and marketing of new metallocene resins. These resins can provide new capabilities and benefits to our customers.

At Cypress, we believe that we have been one of the companies responsible for recognizing the enormous advantages of these new resins, and of bringing them to market.

Certainly, our company has seen enormous growth and success over the past few years. We believe that this success has come about as a direct result of the fact that we recognized metallocene's capabilities early and because we were an entrepreneurial, flexible company, we were able to make the decisions and take the actions required to develop this technology. As new grades of metallocene resins are developed we will find innovative applications and new markets to introduce this amazing technology.

Impact of Technology on Global Polyethylene Technology

D. F. Bari

Chem Systems Inc., Tarrytown, NY, USA

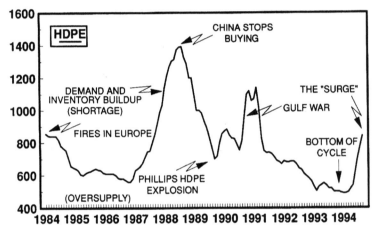

Figure 1. In addition to traditional business cycles global events have impact.

The polyolefins industry is truly cyclical (Figure 1). Business cycles are driven downward by periods of oversupply and economic recession. Increasing market demand and tight supply conditions have the opposite effect in driving up profitability. In addition to the normal industry cycles, the polyolefins business is strongly impacted by unpredicted global events such as import swings, plant outages, war, etc. The bottom of the current cycle was apparently reached in mid-1993 to March 1994, when polymer prices and export demand began to increase.

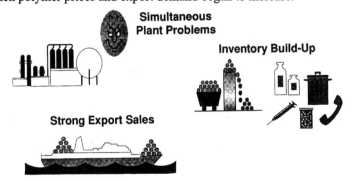

Figure 2. Current feedstock/polymer shortages have resulted in global "surge" but how long will it last?

Most recently, an unusually high number of simultaneous problems have created an unforeseen surge (Figure 2). These have included:

* Plant accidents in the United States and Europe
* Weather related shutdowns and production cutbacks in Asia
* Power supply/mechanical problems in Asia, Latin America and the United States

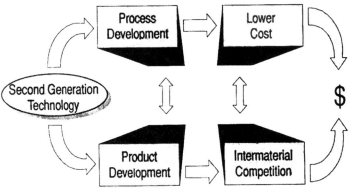

Figure 3. Technology development is aimed at improving long term profitability.

The resulting global polymer shortage has driven up the export sales. Some of this is replacing the output from shutdown plants and some is building inventory. The main question now is how long and strong this surge will last.

Without regard to the current market situation, companies have looked to technology development for product enhancement/differentiation and cost improvements to increase demand in order to improve profitability (Figure 3). In fact, the current commitment to second generation technology development has been strong.

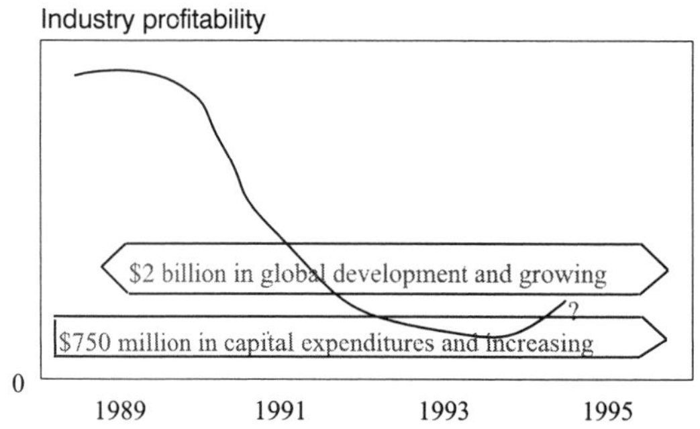

To date, it is estimated that $2 billion has been spent globally on research and development and market development, in spite of the poor industry performance over the last few years (Figure 4). In addition, Chem Systems estimates that from 1991 to 1995, $750 million in capital expenditures will have been made globally. Therefore, it is apparent that the marketing of products and/or licensing technology will be critical to the long term success of a company's second generation technology.

Figure 4. In spite of poor profitability, the commitment to second generation technology is strong.

As such, there have been several types of second generation technologies developed globally (Figure 5). In addition to metallocene-based polyethylene and polypropylene resins, key non-metallocene-based second generation resins include Catalloy, super hexene LLDPE, broad molecular weight distribution LLDPE and bimodal highly processable LLDPE. Each of the second generation materials offer an improved family of properties relative to conventional resins and are targeted to specific segments of the polymer business.

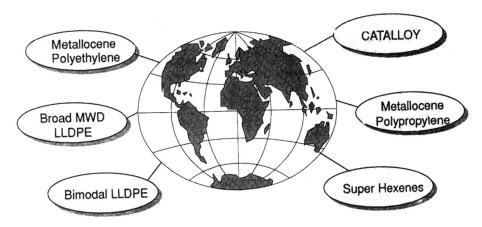

Figure 5. Second generation developments have been diverse and global.

Figure 6. Second generation polyethylene capability in the U.S.

Company	Type	Capacity, kTA	Year (start-up)
Dow	Metallocene	120	1993/1995
Exxon	Metallocene	280	1991/1994
(Montell) Himont	Bimodal	200	1994
Mobil	Super hexene	200	1993
Phillips	Broad MDW	160	1993
Union Carbide	Bimodal	300	1995
Total		**1,260**	

The success and growth of each of these types of resin will be dependent on their ability to perform as advertised and the producers' strategies; they will not, however, be capacity limited. By the end of 1995 there will be at least 1.2 million metric tons of second generation capacity in the United States alone (Figure 6).

Metallocene-based polyethylene technology has received a great deal of attention. These resins are being targeted to several markets. For example, because of improved thermal, mechanical, environmental (purity, low extractables) and/or handling properties compared to conventional materials, metallocene-based polyethylenes are being targeted to a variety of LDPE, LLDPE and non-polyethylene markets and initially at high value materials such as the traditionally insulated markets

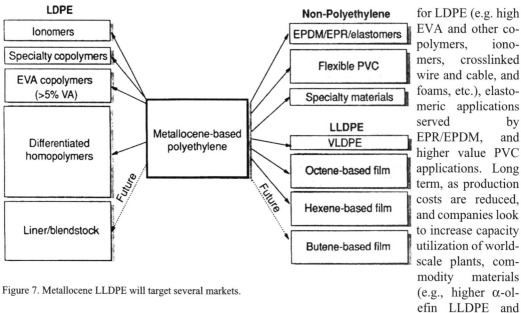

Figure 7. Metallocene LLDPE will target several markets.

for LDPE (e.g. high EVA and other co-polymers, iono-mers, crosslinked wire and cable, and foams, etc.), elasto-meric applications served by EPR/EPDM, and higher value PVC applications. Long term, as production costs are reduced, and companies look to increase capacity utilization of world-scale plants, com-modity materials (e.g., higher α-ol-efin LLDPE and LDPE blendstock material) will be targeted by metallocene-based polyethylene as well (Figure 7).

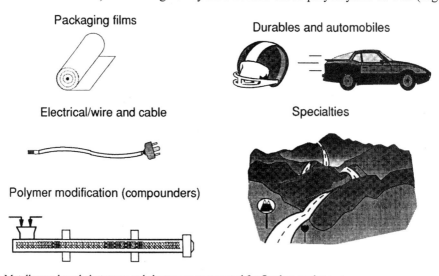

Figure 8. Metallocene-based plastomers and elastomers are targeted for five key market sectors.

The initial metallocene LLDPE resins have been plastomer (i.e., density of 0.885-0.910 g/cm^3) and elastomer (i.e., density of 0.860-0.885g/cm^3) materials which are targeted to five key market sectors (Figure 8), including:

- Packaging films:
    Modified atmospheric packaging
    Seal layers
    Tough films
- Electrical/wire and cable:
    Insulation
    Jacketing
    Semiconductive shields
    Molded accessories
- Polymer modification for compounders and molders:
    Improved impact
    Improved sealing performance
    Improved flexibility (asphalt modification)
- Durables and automotive:
    Under-the-hood automotive and TPOs (gaskets, hoses)
    Sporting goods (foams)
    Molded goods (medical, etc.)
    Housewares
- Specialties (much is still developmental)
    Foams
    Adhesives
    Coatings

While metallocene-based resins represent a significant advancement in polyolefin technology, other second generation developments are also expected to have a significant and potentially more immediate impact on the business. More specifically, the bimodal and broad MWD-based polyethylenes are being targeted at LDPE commodity and higher α-olefin LLDPE film applications (Figure 9).

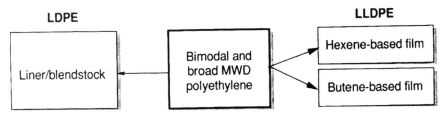

Figure 9. Other second generation resins are expected to have a significant impact on the business with larger volumes in 1995.

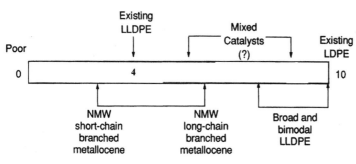

Figure 10. Processability will be critical for some second generation resins but other properties are also critical.

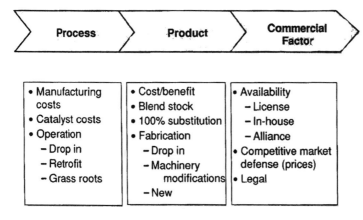

Figure 11. Several key issues will determine the rate of market penetration.

The principal objective is to produce a highly processable resin using a low pressure process to target LDPE applications where clarity and melt strength are not critical. However, some of these resins are also being developed with enhanced clarity and strength properties. Therefore, while improved processability will be critical for the success of some new resins, other properties are more critical for other resin types (Figure 10). Initial commercial testing has shown that some of the bimodal LLDPEs do appear to be "drop-ins" on existing LDPE blown film equipment.

As indicated above, there are a whole host of resin types in the early stages of commercialization. The market success of these resins, that is, those with different processing and handling characteristics, and enhanced performance properties, will depend on a variety of process, product and commercial factors (Figure 11).

Globally, Chem Systems is forecasting annual demand for second generation polyethylene resins to reach almost seven million metric tons (15.3 billion pounds) by 2010, with the largest segment, bimodal/broad molecular weight distribution at 3.3 million metric tons (7.3 billion pounds), accounting for 48 percent of the total. Metallocene-based polyethylenes are projected to account for 36% of the total. Super-hexene-based products are forecast to have a 16% share (Figure 12). Super-hexene resins are improved Ziegler-Natta based LLDPE gas-phase resins with improved strength properties relative to

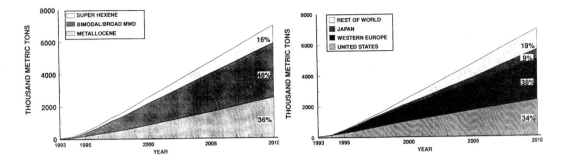

Figure 12. Global demand for second generation polyethylene.

Figure 13. Europe and the U.S. will dominate the global demand.

their conventional counterparts. Overall, the United States and Western Europe are expected to dominate the demand (Figure 13).

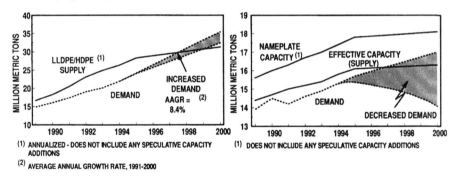

Figure 14. The impact on the various polyolefin segments will be significant.

The impact of these new resins is expected to be significant. More specifically, these resins will enhance the demand of the overall LLDPE market and improve the supply/demand situation, effectively reducing projection by two years. On the other hand, the growth will come primarily at the expense of the LDPE market, forcing some long term rationalization (Figure 14).

In conclusion, the competitive dynamics of second generation polyolefins will be complex as several of these new materials will be targeting similar markets (Figure 15). The results will most likely be declining prices in many market segments.

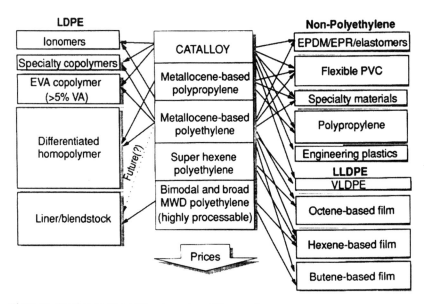

Figure 15. Conclusion: Competitive dynamics will be complex.

Impact of Metallocene Based Polypropylene on Existing Polypropylene Technologies and Markets

E. Shamshoum
Fina Oil and Chemical Co., R&D, LaPorte, TX, USA
J. Schardl
Market Development, Fina Oil and Chemical Co., Dallas, TX, USA

ABSTRACT

Metallocene or single-site catalyzed polyolefins represent the most recent breakthrough in this field. These catalysts have been applied in the development of new or improved versions of polyethylene, polycycloolefins, EPR, EPDM, syndiotactic and isotactic polypropylene and other specialty polyolefins.

The marketability of the new polypropylene homo- and copolymer resins as a direct replacement of the existing resins will depend on whether such resins introduce additional benefits that outweigh the additional cost. Another is the use of polypropylene resins in totally new markets. It has been reported in the past that metallocene catalysts are more expensive than the conventional Ziegler-Natta catalysts which in turn lead to higher price for the resins. This paper will attempt to match process, catalyst chemistry and polypropylene resin properties with applications that exhibit balance between properties and cost. The attributes of metallocene resins will be used to compare to the needs of the different market segments in the major geographical areas such as U.S., Europe and Asia Pacific (Japan in particular).

INTRODUCTION

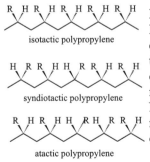

isotactic polypropylene

syndiotactic polypropylene

atactic polypropylene

Figure 1. Basic Structures of Polypropylene Resin.

Metallocene catalyzed polyolefins have been in existence since 1954. Breslow and Chien in 1957 and 1958[1] have shown that Cp_2TiCl_2 was capable of catalyzing the formation of polyethylene. The activity, at that time, was extremely low which discouraged the use of such catalysts in commercial scale applications. From that time until mid to late seventies, metallocenes were only a curiosity and were used for modelling Ziegler-Natta catalysis. In 1976, Sinn and Kaminsky[2] showed that methylaluminoxane, MAO, is the ideal cocatalyst to be used in order to commercialize metallocenes. Since that time, the potential of metallocene catalysts have been explored in many different applications including the polymerization of olefins.

Polypropylene polymer consists of three basic structures: isotactic, syndiotactic, and atactic. The isotactic polypropylene, iPP, (Figure 1) is where all the methyl groups are located on the same side of the polymer chain. Syndiotactic polypropylene, sPP, is where the methyl groups alternate above and below the polymer chain while atactic polypropylene, aPP, has the methyl groups in a random/statistical distribution.

DISCUSSION

The majority of commercial isotactic polypropylene resin, found by Natta in 1954, is produced via conventional Ziegler-Natta type catalysts. The resins produced with such catalysts are used in injection molding, fibers, and filaments, film blow molding extruded sheets and others. This resin is produced in various types of processes which are listed below:

1. *Low Pressure Slurry Process*: This process includes the use of diluents. such as hexane, heptane, butane/isobutane or propane. This diluent acts as a carrier of the PP powder until reaching the drying section. Depending on the thermal properties of the diluent, either centrifugation or flashing can be applied to separate the liquid from powder.

2. *Low Pressure Bulk Propylene Process*: The only diluent/carrier used here is liquid propylene. The separation of the powder here is completed via flashing of the propylene. Two types of processes exist here: a) loop or liquid full reactor and b) stirred reactor with head space containing hydrogen, propylene and possibly another comonomer.

3. *Low Pressure Gas Phase Process*: Two types of processes a) vertical reactor where the polypropylene powder is fluidized in gaseous propylene (or with a comonomer) and b) horizontal reactor where the powder is agitated with a mechanical agitator.

4. *Low Pressure Liquid-gas phase Process*: This process consists of one or two liquid propylene reactors to produce homopolymer followed by one or two gas phase reactors to produce ethylene-propylene rubber. The homopolymer reactors are either loop type or CSTR. The critical step in this type of process is the liquid/powder separation before entering the gas phase reactors. This combination of reactors is designed to produce impact copolymers.

The majority of polypropylene today is produced via the low pressure bulk polypropylene process.

POLYPROPYLENE CATALYSTS

Using the Ziegler $TiCl_4 + Et_3Al$ catalyst preparation, Natta in 1954 produced a mixture of amorphous and crystalline polypropylene which were later separated by using solvent extraction methods. At a similar time frame, Phillips Petroleum used the chromium on silica catalyst to produce polypropylene. Later, they isolated and characterized isotactic polypropylene in the form known today. Titanium containing Ziegler-Natta catalysts used to produce polyolefins and more particularly polypropylene can be placed into three categories:

1. *First Generation Catalysts*: At least two types of catalysts belonging to this category were prepared by employing a precipitation step resulting from the reaction of $TiCl_4$ with aluminum alkyl component. Precipitation temperature, type of solvent, type of aluminum alkyl and Ti form ($\alpha, \beta, \gamma, \delta$) are important parameters controlled during the preparation. Many other methods were developed to achieve the active Ti^{3+} oxidation state. The activity of this catalyst is extremely low at 1000 g-PP/g catalyst/hr. The stereospecificity of the catalyst was low as reflected in the low isotactic index of the resin. This generation of catalyst produced polypropylene resins with 91-92% isotactic index and broad molecular weight distribution with M_w/M_n of 11 - 13 (molecular weight distribution is a relative value).

2. *Second Generation Catalysts*: Due to the very low activity and stereospecificity of the first generation catalyst, researchers continued to find ways for improving such a catalyst. It was found that improvements could be made by the addition of Lewis base during the synthesis which in turn will act as a stereocontrol reagent. Another important finding is the presence of aluminum byproducts (EtAlCl$_2$) that act as catalyst poisons. With the removal of such compounds, catalyst activities increase to 5000-7000 g-PP/g catalyst/hr and the isotactic index increased to 96%. The molecular weight distribution decreased (M_w/M_n = 9-10).

3. *Third Generation Catalysts*: These catalysts are supported in nature. The most common support used is MgCl$_2$. Magnesium has been described as the optimum support for this catalyst because its ionic radius (Mg^{2+} ionic radius is 0.066 nm) is very similar to that of Ti^{4+} (ionic radius of Ti^{4+} is 0.068 nm). In addition, a Lewis base (also defined as internal electron donor) such as mono- or di-esters are also present in the catalyst matrix to increase the stereospecific properties of the catalyst. This donor is incorporated during the catalyst synthesis and more particularly during the titanation step. Another isotacticity control reagent is added during the polymerization process to further increase the isotactic index. Fina recently developed proprietary methods to commercially produce isotactic polypropylene with an isotactic index of 99.7% and %mmmm of >97.5%. More recently, reports were issued on the use of diethers as internal donors. These reports also claim that external electron donors are not necessary to achieve a high isotactic index. Important features for this generation are the very high activity (>25,000 g-PP/g catalyst/hr), high isotactic in-

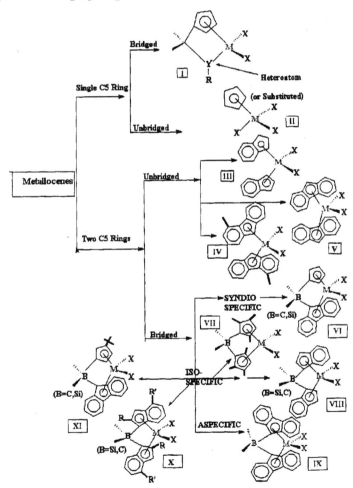

Figure 2. Key metallocene structures.

dex (>97%), low ash content and narrow molecular weight distribution (M_w/M_n = 6-7.5).

Recent breakthroughs in polyolefin catalysts led to the finding of new families of homogeneous catalyst precursors called metallocenes. These also known as "Single Site" or "Constrained Geometry" catalyst precursors. These types of molecules gained commercial and economic popularity after Kaminsky and Sinn, in 1976, demonstrated that Cp_2TiCl_2 in the presence of methyl aluminoxane, MAO, produce polyethylene at very high rates. Since that point, a new revolution began, resulting in the findings of other metallocenes that were used to produce special grades of polyolefins such as LLDPE, mDPE, polycycloolefins, elastomers, and iso, syndio, and atactic polypropylenes. Some of the key metallocene structures are shown in the Figure 2. With regard to the polypropylenes, the bridged (VI, VII, VIII, X, and XI)[3] and unbridged (IV) metallocenes are used in combination with MAO or perfluoro-tetra-phenyl borate and conventional aluminum alkyls to produce the isotactic structure. Metallocene VI produces syndiotactic while the rest produce atactic structures. The activities of most of these metallocenes are very high and range from 100,000 to 1,000,000 g-PP/g catalyst/hr compared to 5,000-50,000 for conventional Ziegler-Natta catalysts. Consequently, the ash content in the resin decreases dramatically. Economically speaking, the preferred approach for using metallocenes in producing polypropylene, is a "drop in" into one of the processes described above. However, one of the drawbacks for using homogeneous metallocenes is the very low bulk density polymer they produce (<0.12 g PP/liter) which generates problems in the circulation, separation and extrusion of powder. As a result, research efforts continued in order to develop catalysts that are compatible with current Ziegler-Natta processes and can be used with minor process modifications. One of the reasons for Fina's success in this area is the ability to use isospecific and syndiospecific metallocenes with minor changes.

MARKETS

The approximate breakdown for the consumption of polypropylene in the U.S., Europe, and Japan since 1985 in various applications is shown in the Table 1.

Table 1. Polypropylene consumption (%)

	Injection molding	Fibers & filaments	Film	Blow molding	Extruded sheet	Others	Total demand 1995
US	44	33	10	2.5	2.5	8	9.8 B lbs
Western Europe	43	27	16	1	-	13	10.5 B lbs
Japan	56	7.5	22	2	-	13	4.8 B lbs

The Table 1 shows some similarities in consumption breakdown of polypropylene between western Europe and the U.S. There appears to be a good balance between fibers and filaments and injection molding applications. With regard to the Japanese markets, injection molding share is much higher than the U.S. and Europe (56 vs. 44%). The fibers and filaments, however, are very low (7.5% for Japan and 33% and 27% for U.S. and Europe, respectively).

Worldwide capacity/utilization of polypropylene resin was reported by K. Scheidl.[4] This information is summarized in the Table 2.

Table 2. Worldwide capacity utilization of polypropylene

Year	Consumption Ktons/yr (MMlbs/yr)	Capacity Ktons/yr (MMlbs/yr)	% Utilization (US)	AAGR%
1990	12,500 (27,558)	14,028 (30,926)	89, 90 (87)	
1991	13,841 (30,514)	15,599 (34,390)	89, 83 (87)	10.7
1992	14,977 (33,018)	16,774 (36,980)	89, 81 (87)	8.2
1993	16,065 (35,417)	18,091 (39,883)	89, 81 (90)	7.3
1994	17,690 (38,999)	18,996 (41,879)	93, 87 (98)	10.1
1995	18,577 (40,955)	20,447 (45,077)	91, 89 (94)	5.0
1996	19,633 (43,327)	22,589 (49,780)	87, 89 (97)	5.8
1997	21,691 (47,820)	26,390 (58,199)	82, 86 (89)*	10.4
1998	23,456 (51,711)	29,332 (64,655)	80, 86 (88)*	8.1
1999	24,991 (55,095)	31,039 (68,429)	81, 86 (86)*	6.5
2000	26,655 (58,737)	31,994 (70,534)	83, 85 (86)*	6.7
2005	34,019 (74,842)		109*	5.0
2010	43,416 (95,515)			5.0

* estimated values; US % utilization represents the average of SPI, Chem Data, Chem System, Fina

The cumulative global growth in PP demand between 1990 and 1995 was 48.6% and average utilization was 90%. With such a level of growth it is very difficult to justify the introduction of a new technology and new products into the market. The average annual growth is also listed in the Table 2. When considering growth rates of 7% for the years 1999 through 2010 combined with a low % utilization (up to 2000), new technology developments and commercialization will be more probable than today.

Metallocene produced polypropylenes, m-PP, represent a new grade of polypropylenes that must contain additional/improved properties to warrant the worth of the effort spent for commercialization and marketing. For the rest of the paper, the syndiotactic and isotactic polypropylenes will be discussed separately for each of the injection molding, film and fibers applications

SYNDIOTACTIC POLYPROPYLENE, sPP

Syndiotactic polypropylene was characterized by Natta in the late fifties. This sample was obtained by successive extraction methods from isotactic polypropylene. The early catalysts used to produce sPP with low yield and syndiotacticity were vanadium chlorides. In 1987 Fina Oil and Chemical Company developed a new metallocene (structure VII) that was capable of producing sPP with very high tacticity (>75%) and activity. Some of the key property differences between sPP and conventional iPP were reported in the past by Fina and Mitsui Toatsu and summarized in the Table 3.[5]

Table 3. Key property differences between sPP and conventional iPP

Property	sPP
Impact strength	Higher (14 vs <0.5 ft.lb/inch @ room temperature)
Melting point	Lower 127 vs. 160°C comparable to random copolymer
Clarity	Higher
Crystallization rate	Slower
Heat seal temperature	Lower (comparable to Super Z polymer)
Flexural modulus	Lower
Molecular weight distribution	Narrower (2-3 vs 7-8) broad MWD sPP was produced ($M_w/M_n = 4.5$-5.0)
Melt strength	Lower
Tear resistance	Higher
Xylene extractables	Lower for comparable T_m material (RCP)
Gamma radiation stability	Higher
MVTR	Higher for sPP

With the results summarized in the Table 3, it is unlikely that sPP will replace isotactic polypropylene due mainly to its lower melting point and slower crystallization rate. The higher melting point of iPP makes it a very useful plastic. The slower crystallization rate of sPP reduces plant throughput thus making it difficult to use as neat material. However, sPP has been found to improve the properties of polypropylene homo and copolymers when blended in small quantities. The major improvements found in film applications, injection molding, and fibers are shown in the Table 4.

Through several plant trials (the first on April 27, 1993), Fina today is capable of producing sPP resins with melt flows of 2-10 and molecular weight distributions M_w/M_n of 2-5. The higher M_w/M_n is a bimodal product which was first produced on a large scale during a plant trial conducted in Jan. 1995. This product offers an improvement in processability and crystallization rate. The $t_{1/2}$ for this product is 1/3 of the narrow MWD sPP. Another successful attempt was made in 1996 to improve the crystallization (setup time) behavior of sPP. This improvement made it easier for Fina to continue the efforts towards its commercialization of this product.

Table 4. Improvements in film, injection molding, and fibers

Application	Concerns	Improvements
Film (little sPP blend)	Stickiness when used at large concentration	Clarity Tear resistance Heat seal Impact resistance Developed clear impact sPP (CI-sPP)
Injection molding	Longer cycle time	Impact resistance Clarity
Fibers		Loft and hand Elasticity Crimp stability

ISOTACTIC POLYPROPYLENE

After the successful sPP plant trials, Fina decided to commercialize m-iPP. The first plant trial was in Sept., 1995 followed by others in March and Nov., 1996. During these trials several products were produced. Another major trial is planned to be conducted in the first quarter of 1997. The Table 5 summarizes these products.

Table 5. Comparison of products from different trials

Trial date	Melt flow	T_m, °C	Type	%mmmm
September 95	26 26	148 143	homo copolymer	91.5 91.5
March 96	15 180 290	146 146 145	special homo homo	91 91 91
November 96	2.5 18 30	152 152 151	hom homo homo	97 97 97

Metallocene iPP resins continue to be evaluated in the areas of fibers, film and injection molding applications. McAlpin from Exxon on many occasions, reported on the evaluations of m-iPP resins in the areas of fibers (SPO-94,95) and film (BOPP, Shrink, and Cast films-SPO-96). He reported that single site, NMWD, resins introduce some attractive attributes during processing and in the properties of the spun fibers. In the BOPP film areas (using laboratory TM Long machine), he reported that broad MWD (at least two metallocenes) resins are preferable when considering the broad operating window of conventional iPP resins. He also reported that m-iPP can be run at very high processing speeds which in turn could favor the economics of using metallocene type resins. Random copolymers produced with metallocenes were reported to have excellent balance of shrink performance and processing speed.

When comparing the performance of m-iPP produced with Fina developed technology and con-

Table 6. The results of melt spinning

	Conditions	Conventional Z-N resin	Fina m-iPP (Finacene)
Melting point, °C		163	148
MFI		17	25
Elongation @ break, %	2000 m/min 2500 m/min 3000 m/min 4000 m/min	119.5 112.5 85.1	158.4 149.0 135.5 132.0
Tenacity @ max., g/den	2000 m/min 2500 m/min 3000 m/min 4000 m/min	3.88 4.17 4.54	3.20 3.35 3.47 4.21
Free shrinkage, %	2000 m/min 2500 m/min 3000 m/min 4000 m/min	14.8 13.9 15.6	13.5 11.1 10.2 11.3

ventional Z-N resin in melt spinning application, the results are given in Table 6.

At constant draw ratio of 3:1, the tenacity of m-iPP at max is within the acceptable range while the % free shrinkage is lower than what was observed for conventional products used primarily in melt spinning applications.

m-iPP resins developed by Fina were evaluated in at least two commercial BOPP lines. The Table 7 below summarizes some of the conditions and the results for one experiment.

Table 7. Conditions and results of experiment

	Conventional Z-N resin	m-iPP
Melting point, °C	160	152
MFI	2.8	2.8
$T_{extrusion}$, °F	450-470	450-470
T_{oven}, °F	330	330
Gauge, µm	18	18
MVTR	0.2	0.17
Haze, %	0.2	0.9
See through clarity, %	70	60
Gloss, %	99	97
Tensile strength, MD, psi	20,000	23,000

CONCLUDING REMARKS

With all these results in hand, one can easily assess the economic value of metallocene produced polypropylenes. It is clear so far that sPP will not compete with isotactic polypropylene homopolymer. It is possible that it will capture some markets to replace high ethylene (>4%) random copolymers. It can also be used in blends with other polymers especially iPP homopolymer. Assuming sPP will be consumed at 2-5% of the fibers and film markets through replacement as neat products or in blends or for new applications, the maximum capacity worldwide will be 160-400 Ktons/yr (calculated based on 1995 results) starting the year 2010. The U.S. capacity will be 50-100 Ktons/year. This is approximately the size of one or two world scale lines. Fina is taking the necessary steps to prepare for this speculative new capacity by adding another world scale homo and impact copolymer line of 550 MM lbs/yr at its PP facility in LaPorte, TX. The additional capacity should free the smaller lines to produce metallocene based propylene resins.

The m-iPP is another story. With the attributes observed so far in melt spinning and film applications, we estimate that 10-15% of the world consumption will be produced with metallocene technologies. Assuming the price of m-iPP becomes competitive with that of conventional Z-N type resins, the expected world wide quantity of m-iPP resin that will be consumed by the year 2010 is 1500-2200 Ktons/yr. The U.S. consumption will be 600-1000 Ktons/yr. It is the authors opinion that the acceptance of this resin will be most favorable in fiber applications.

One of the key points to keep in mind is that 10-15 years ago high activity catalysts were developed and implemented in the majority plants. However, some producers still use the low activity second generation catalysts. Metallocene products are different. The commercialization path for metallocene based polymers will be different than what was experienced in the past. The difference in properties compared to conventional polypropylene resins and the potential for new and niche markets will require a longer maturity in the market.

REFERENCES

1.   D. S. Breslow, N. R. New Guy, *JACS*, **79**, 5072 (1957).
2.   J. H. Sinn, W. Kaminsky *et al.*, *Angew. Chem. Int. Ed. Eng.*, **15**, 630 (1976).
3.   a. Fina Technology, **U.S. Patent 4 892 851**.
      b. Fina Technology, **U.S. Patent 5 334 677**.
      c. J. Ewen, A. Razavi *et al.*, *JACS*, **110**, 6255 (1988).
      d. Spaleck *et al.*, *Angew. Chem. Int. Ed. Eng.*, **31(11)** 1347 (1992).
      e. Spaleck *et al.*, *Organometallics*, **13**, 954 (1994).
      f. S. Miya *et al.*, **Catalytic Olefin Polymerization**, Ed. Keii, Vol S6, 531, *Elsevier*, N.Y., 1990.
      g. J. C. Stevens *et al.*, **European Patent 416 815**.
      h. J. C. Stevens, *MetCon'93*, 1993.
      i. J. Canich, **U.S. Patent 5 055 438**.
4.   K. Scheidl, *MAACK Conference*, Zurich, Sept 18-20, 1996.
5.   a. T. Shiomura *et al.*, *Macromol. Symp.*, **101**, 289 (1996).
      b. T. Shiomura *et al.*, *Macromol. Rapid Commun.*, **17**, 9 (1995).
      c. E. Shamshoum *et al.*, *Proc. Pofyolefins IX Conference*, 1995.
      d. E. Shamshoum *et al.*, *Proc. SPO 93, 94, 95, and 96*.
      e. E. Shamshoum *et al.*, Metallocenes 96, Dusseldorf, 1996.
      f. E. Shamshoum *et al.*, *MetCon 93 and 94*, Houston, TX.
      g. T. Shiomura, E. Shamshoum *et al.*, *Proc. SPO'96*, 1996.

New Polyethylenes for Targeted Performance Based Applications

Pradeep Jain, P. J. Lonnie, G. Hazlitt, and J. A. DeGroot

The Dow Chemical Company, 2301 Brazosport Boulevard, B-1607 Freeport, TX 77541 USA

ABSTRACT

Homogeneous polymers produced from "single-site" catalyst technology have induced significant changes on the future of industrial production of ethylene/α-olefin copolymers. In combination with the appropriate polymerization process, the impact could be dramatic. For example, the unique characteristics of Dow's solution process, when combined with constrained geometry catalyst, CGC, technology lead naturally to unprecedented control of both the molecular weight distribution, MWD, and the short chain branching distribution, SCBD, of the copolymers. In addition, Dow's INSITE[TM] catalyst technology allows the controlled incorporation of long chain branching, LCB, into the copolymerization products. Product designs can therefore rely upon all the known molecular design strategies (MW, MWD, SCBD, and LCB) to obtain a high degree of structural specificity for the desired polymer product. Dow will soon commercialize a broad range of new polyethylene products, or Enhanced Polyethylenes, EPE, that rely on INSITE[TM] catalyst technology and are a natural extension of Dow's existing solution process technology. Enhanced Polyethylenes are precisely designed and produced with a high degree of reproducibility for specific applications. The complete process/product technology package behind Enhanced Polyethylenes leads naturally to a few key polymer design concepts. These design concepts have been developed from a materials science understanding of the performance needs behind each targeted application, while giving careful consideration to the manufacturing capabilities of EPE solution process technology. Consequently, Enhanced Polyethylenes are a targeted performance-based portfolio of new products for a wide range of traditional LLDPE applications as well as new applications once thought to be outside of the range of polyethylene capability.

INTRODUCTION

Enhanced Polyethylenes, EPE, are a new polyethylene product portfolio, providing unique combinations of processability and finished product performance not available in traditional polyethylenes. These polymers are produced using Dow's existing solution process in combination with INSITE[TM] Technology. The solution process has been the basis of Dow's highly successful DOWLEX[TM] franchise, which is widely recognized as the world leader in terms of LLDPE product performance. The recent introduction of INSITE[TM] process technology, which includes the unique constrained geometry catalyst, CGC, and related new process technology, resulted in the commercial successes of AFFINITY[TM] Polyolefin Plastomers, POP, and ENGAGE[TM] Polyolefin Elastomers, POE. In contrast to the specialty applications of AFFINITY[TM] and ENGAGE[TM], Enhanced Polyethylenes are targeted for more traditional LLDPE markets, but with emphasis on higher value applications where commercial LLDPEs are not currently considered to be the best choice either because of poor product performance or difficulties in fabrication. We have selected three examples from industrial packaging films, wire and cable jacketing, and food and specialty packaging films, to illustrate the capability of Dow's new EPE technology. The methodology we have chosen will be to elucidate how the solution process capabilities can be used to leverage specific product design goals to achieve ultimate performance characteristics.

THE MOLECULAR DESIGN AND MATERIALS SCIENCE PRINCIPLES

The new molecular design principles that EPE technology can leverage to achieve the desired product performance arise from the understanding of materials science. The three principles we have chosen to discuss are tie-chains for tougher finished parts, long chain branching, LCB, and Molecular Weight Distribution, MWD, control for melt processability, and SCBD/MWD control to improve sealing performance in many food packaging applications. Where possible, we have emphasized the importance of solution process principles in achieving the desired polymer structures indicated by the materials science. The methodology employed to elucidate these principles will be to provide three examples of product applications where an EPE was specifically designed to achieve performance advantages by providing the desired combinations of molecular structure(s).

Materials Science Principle #1: Tie chains have been shown to have a dramatic effect on the tensile behavior LLDPE. Specifically, the strain hardening behavior that occurs just prior to failure in a typical stress/strain curve has been ascribed to the presence of tie chains. Tie chains have been shown to arise through the control of comonomer content (especially branching frequency) and molecular weight.[1] The probability for tie chain formation as a function of both molecular weight and density is shown in Figure 1.[1] The homogeneous nature of comonomer incorporation by CGC leads to the ability to increase the relative population in all SCBD of key fractions of a given polymer, while simultaneously controlling that particular fraction's molecular weight. EPE products with a high tendency for tie chain formation are designed by carefully incorporating the right fraction of co-polymer, controlling its molecular weight and co-monomer content of that fraction, and ultimately producing the molecular weight and density desired for the product. Thus an otherwise heterogeneous MWD and SCBD produced from a heterogeneous catalyst, is modified to produce a unique SCBD and MWD. The result is, for example, blown film produced from an EPE product with a

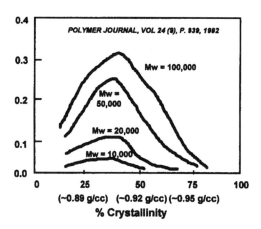

Figure 1. Probability of tie molecule formation. Homogeneous polymer and LLDPE polymer fractions.

density of 0.925 g/cc that has dart impact over twice that of blown film produced from a 0.920 g/cc DOWLEX™ product.[2]

Materials Science Principle #2: LCB and MWD are structural attributes of EPEs that can significantly impact melt processability. EPE technology is capable of introducing LCB into the polymer structure via the contribution of INSITE™ technology. In tandem with the use of MWD control, unique combinations can be achieved to deliver a wide range of extrusion performance where lower high shear viscosity at the same or higher zero shear viscosity can result in better processability, higher melt strength, and improved resistance to melt fracture. The result is an EPE product designed for wire and cable jacketing having increased flexibility, low notch sensitivity, good abrasion resistance, and no melt fracture during the extrusion process.[3]

Materials Science Principle #3: SCBD and MWD are structural attributes of EPEs that can affect the sealing performance in many food packaging applications.[4] Sealing performance of conventional Ziegler-Natta based Ultra Low Density Polyethylene, ULDPE, can be greatly improved through proper copolymer design and the application of EPE technology. The control of specific regions of the SCBD/MWD distributions can improve seal strengths and provide unprecedented control of the initiation temperature. The materials science basis for these improvements are not easily ascertained, but it is clear that by controlling the relative mass of specific comonomer fractions in conjunction with their molecular weight, a great deal of specificity can be brought to bear on the sealing performance of the product.

THE SOLUTION PROCESS PRINCIPLES

The solution process is capable of excellent control and is inherently capable of achieving high levels of comonomer incorporation. Also, it is easily adapted to achieve multiple reactor schemes[5] using more than one catalyst. Differences in catalyst reactivity toward monomers as well as the ability to independently control molecular weight are important factors to be considered in any product design. Dow's solution process, when coupled with multiple choices of catalyst, is extremely adaptable to the molecular design needs. One reason for this versatility is the absence of diffusional resistances that are often present in other commercial polymerization reactor systems and processes. In addition, Dow's solution process provides a great deal of flexibility and allows precise feed forward control for producing a desired overall molecular architecture.

HIGH MODULUS/HIGH DART BLOWN FILMS

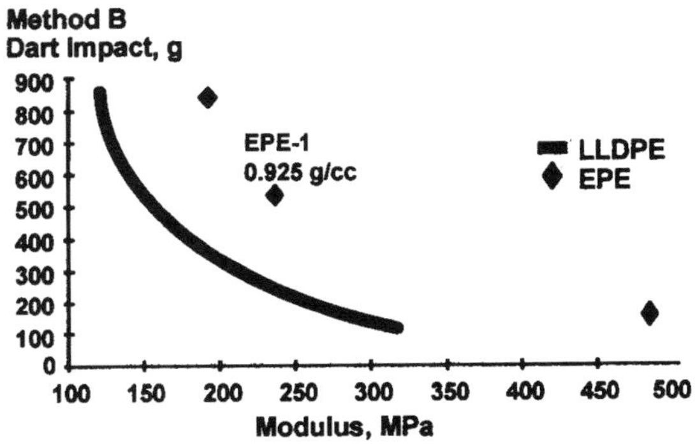

Film convertors have reached limits in their ability to downgauge using conventional LLDPE products. In order to design a product with improved performance, consideration has been given to the molecular basis of the performance improvement that is desired. One major contribution is that of tie molecules. The materials science dictates that these be of a specific branching frequency and molecular weight. EPE technology allows for the facile incorporation of substantial amounts of such a tie molecule fraction over a broad polymer density range. In Table 1 and Figure 2, the EPE product (EPE-1) at a 0.925 g/cc density has much

Figure 2. New performance combination. High dart impact and high modulus.

Table 1. High Modulus/High Dart Impact Blown Film Data

	DOWLEX 2049*	EPE-1	DOWLEX 2045*
Pellet Data			
Melt Index, dg/min	1.0	0.85	0.95
Density, gm/cc	0.926	0.925	0.921
I_{10}/I_2	8.1	7.2	8.3
Fabrication Data**			
Melt Temperature, °C	462	471	462
Pressure, psi	5160	6170	5040
Amps	71	81	69
Screw Speed, RPM	66.7	68.8	66.2
Output, lbs/hr	120	120	120
BUR	2.3	2.3	2.3
Film Results			
MD Yield, psi	2,045	2,130	1,665
CD Yield, psi	2,200	2,280	1,730
MD 2% Modulus, psi	36,470	33,960	26,360
CD 2% Modulus, psi	44,600	39,890	31,210
MD Break Strength, psi	6,975	7,475	6,525
CD Break Strength, psi	4,800	7,390	5,480
MD Elmendorf Tear, gm	477	430	691
CD Elmendorf Tear, gm	718	749	819
Dart Impact, gm	177	532	236

*Tradmark of The Dow Chemical Company; **Screw Type-Single flight mixing screw; Die Type-Gloucester die with 70 mil gap.

higher dart impact, B, strength than a comparable 0.920 g/cc; conventional Ziegler-Natta product.

Table 2. Wire and Cable Jacketing*

	EPE-2	Gas Phase HDPE
Resin Properties		
I_2, dg/min	0.89	0.12
I_{10}/I_2	10.7	29
Density, g/cc	0.957	0.958
Flexural Modulus, MPa	1,147	1,523
Abrasion, g lost/1000 cycles	0.29	0.29
Compression Molded Notch Sensitivity Tensile Test		
Notch 1 Tensile Strength, MPa	20.9	10.8
Notch 1 Tensile Elongation, %	529	53
Finished Cable Properties[1]		
Flexibility: Force at 5 mm, Deflection, kg	6.8	8.0
Flexibility: Force at 20 mm, Deflection, kg	16.8	19.7
Cable Tensile Elongation at the Overlap, %	220	40
Resin Processability[2]		
Amps, A	35	34
Pressure, bar	75	101
Specific Output, kg/hp-hr	4.30	4.56
Surface Quality Rating[3], Melt Fracture	100	80
Cable Uniformity[3], Melt Strength	pass	pass

*All resins contained 2.7 wt% carbon black
[1]For specific test methods see Ref. 3
[2]The cable jacketing line was equipped with a 6.35 cm, 20:1 L/D extruder; 5-turn metering screw, 3.66:1 compression ratio. The die and tip inside diameters of the crosshead die were 2.04 and 1.73 cm, respectively, 0.318 cm die gap and zero land. Screw speed=55 rpm; cable line speed=7.62 m/min; melt temperature=232-241°C.
[3]These test were done by visual inspection. The surface quality was evaluated using a magnifying glass (10x) and was given a rating from 0-100; 100 representing the "best" or the smoothest surface.

WIRE AND CABLE JACKETING

Wire and cable jacketing manufacturers want to produce jackets with greater flexibility, while maintaining abuse resistance. Improved flexibility allows for easier cable installation. Abuse resistance is often measured by abrasion resistance and notched tensiles. Additional key performance requirements

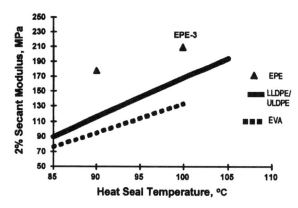

Figure 3. New performance combination. Low heat seal initiation and high modulus.

for this application are maintaining extrusion processibility, a uniform cable coating, and a smooth surface (lack of melt fracture).

Enhanced Polyethylene Technology is a means of implementing a molecular design for these performance requirements because INSITE™ technology can deliver LCB in addition to MWD control. The wire and cable product(s) produced via the EPE technology redefine the relationship between flexibility and abrasion resistance that is currently provided by gas phase polyethylenes. EPE-2 provides a significant reduction in compression molded flexural modulus at the same level of abrasion resistance (Table 2). Other enhancements versus the conventional technology resin are described in Table 2. For example, EPE-2 exhibits dramatically higher notched tensile strength and elongation relative to the gas phase product. The flex modulus and notched sensitivity testing of a compression molded plaque are a result of materials science principles #1 and #2. These performance improvements were accomplished with no sacrifice in processibility and a low tendency for melt fracture.

BLOWN FILMS FOR FOOD AND SPECIALTY PACKAGING APPLICATIONS

Blends and coextrusions of LLDPE, ULDPE, ethylene vinyl acetate resins, EVA, are often used in food and specialty packaging. LLDPE provides good toughness and stiffness, but packaging line speeds are usually limited due to higher sealing temperatures. ULDPE and EVA's provide the low sealing temperature, but have low modulus and higher blocking tendencies that can result in poor convertibility in bag and pouch making processes. Enhanced Polyethylene Technology is well suited for the development of products that provide low temperature sealability, high abuse resistance, low blocking tendency, good convertibility, and processibility.

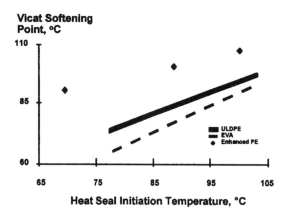

Figure 4. New performance combination. Low seal initiation and temperature resistance.

Materials science principle #3 can be used to design the products in this market segment. The combination of seal initiation temperature and heat seal strength at targeted densities allow for unique combinations of heat seal performance and modulus as shown in Figures 3 and 4. At a comparable heat seal initiation temperature to ULDPE or EVA (9% VA),

Table 3. Sealing Performance. Comparative Data.

	EPE-3	LLDPE	ULDPE
Sealing Property			
Melt Index, g/10 min	0.85	1.0	1.0
Density, g/cc	0.920	0.920	0.912
Vicat Softening Temperature, °C	109	105	95
Sealing Performance			
Heat Seal Initiation, °C, at 1 lb/in	100	109	95
Ultimate Heat Seal, lb/in	12.4	11.8	12.0
Hot Tack Range*, 4 N/in	50	37	45
Blown Film Properties**			
Dart Impact B, g	>850	266	680
Block, g	20	92	134
Clarity	68	60	77
45° Glass	61	61	72
Haze, %	12	11.3	7.6

*Structure failure prevented testing beyond 150°C 1 mil nylon/1 mil EAA/1.5 sealant blown coextrusion
**2 mil monolayer blown films, 2.5:1 BUR

EPE-3 provides the stiffness of a LLDPE. The data in Table 3 illustrates the overall physical performance of EPE-3 relative to a traditional LLDPE, and ULDPE.

The ability of EPE technology to produce such a product is enhanced greatly by the solution process for many of the same reasons as were mentioned above. In this case, the necessary level of control is extremely high since small variations in SCBD and MWD will effect the heat sealing behavior.

CONCLUSIONS

The production of these products using EPE technology signals a new era in the manufacture of LLDPE products. EPE technology will provide the ability to obtain targeted performance improvements or wholly new products while maintaining or even increasing product consistency. Precise knowledge of kinetics, mass and energy balances coupled with the high degree of control capability provided by the solution process are prerequisites to manufacturing EPE products. A good indicator of the potential success of the technology might be the ability to produce products like EPE-3. Since heat sealing performance is determined by the SCBD and MWD, the ability to target a narrow sealing "window" and produce a product consistently is extremely challenging for any modern production facility. EPE technology includes the necessary process control capability and discipline to achieve such a goal. The detailed connections between the process fundamentals and the polymer molecular structure give both the product chemist and the process engineer the means to develop and produce new products.

Extension of the INSITE™ Technology platform has led to the introduction of a new class of polyethylene resins called Enhanced Polyethylene. These unique polymers live up to their name in large part due to the control of the structures necessary to achieve the performance desired in any particular application. The adaptability of the solution process to the implementation of EPE Technology will insure that these products can be produced in a very cost competitive manner and provide performance enhancements to both large volume, "commodity" and small volume, "specialty" applications.

ACKNOWLEDGMENTS

The authors wish to especially acknowledge the work of Nicole Whiteman, Laura Mergenhagen, Kenny Stewart, Larry Kale, Trudy Iaccino, Ken Bow, Ron Markovich, and Steve Chum. In addition, there are numerous folks within Dow who work tirelessly to improve and maintain the solution process, to identify new markets, to provide analytical and fabrication data, and to provide customer support; and they are realty the reason EPE technology is even possible. We would also like to acknowledge the advice and support of Che-I Kao, Tony Torres, Steve Ellebracht, Brian Kolthammer, Ian Robson, and Kaelyn Koch.

REFERENCES

1.   *Polymer Journal*, **24(9)** 939 (1992).
2.   J. A. DeGroot, *SPO Conference*, September, 1996.
3.   L.T. Kale, T. L. Iaccino, and K. E. Bow, *ANTEC'96*, May, 1996.
4.   N. F. Whiteman, J. A. DeGroot, L. K. Mergenhagen, and K. B. Stewart, *RETEC'95*, February, 1995.
5.   Mitchell, **U.S. Patent 3 914 342**.

Polyvinyl Chloride Replacement for Flexible Medical Tubing

J. H. Ko and L. Odegeard
3M Company, St. Paul, MN 55144, USA

ABSTRACT

The research involved studies of blending two classes of polyolefin resins for the development of chlorine-free medical tubing. The first resin used is a soft, flexible metallocene polyolefin thermoplastic elastomer with excellent clarity. The second resin is a tough, ionic polymer that has good compatibility with the first resin. Performance advantages of the blend tubing as compared with tubing made from the individual resins and polyvinyl chloride, PVC, tubing were reported.

INTRODUCTION

PVC based tubing and film are used in numerous medical products. The material is easy to formulate and process. It also is relatively inexpensive and performs well in most applications. Typical requirements for tubing used in medical devices, such as intravenous sets (IV), include clarity, flexibility, kink resistance, toughness, scratch resistance, ease of bonding with some common solvents or adhesives, and suitable for gamma sterilization. However, PVC is viewed as hazardous to personal health[1] environment,[2] and inadequate for some drug delivery application due to PVC drug interaction.[3]

While other thermoplastic polymer has been used to replace flexible medical tubing made with PVC, none to date has been able to match the advantages provided by PVC materials. For example, polyurethane or silicone is relatively expensive. Some commercial offerings of non-PVC medical tubing are available on the market. The Braun tubing made of polybutadiene-based material is translucent and kinks easily. The Baxter tubing has a three-layer construction with the outer layer consisting of a plasticized PVC and the inner layer polyolefin material. None has been used successfully as environmentally-compatible replacements for PVC-based material. In this study, we utilized two chlorine-free thermoplastics to mimic the key properties of flexible PVC tubing.

EXPERIMENTAL

MATERIALS

The polymer blends were made from a non-ionic thermoplastic ethylene/butane copolymer designated as E-1 and an ionic copolymer designated as I-1 and I-2, respectively. The I-1 is a copolymer of polyethylene, methacrylic acid neutralized with zinc base. The I-2 is a terpolymer made of ethylene, soft acrylate and methacrylic acid neutralized with zinc base. After extrusion, the tubing was aged in the room conditions for more than one month prior to testing. PVC tubing (trade name Unichem) with diethylhexyl phthalate plasticizer, DOP, was supplied by Colorite Plastics Company.

EXTRUSION

Two sizes of tubing commonly used for IV sets were extruded using a two-inch extruder (Model 200 from Welex Co. of Blue Bell, PA). The dimensions were: Inside diameter, ID, 0.24 cm and outside diameter, OD, of 0.35 cm; ID: 0.27 cm and OD 0.41 cm.

γ-IRRADIATION

Some of the tubing were γ-irradiated at 25, 50, 75 and 100 kiloGray (KGray) and aged for a month prior to testing.

KINK MEASUREMENT

Tubing was bent until kinked. The distance between the two parallel bent tubing was reported as kink number. The kink point of the tubing was marked and set straight for 30 seconds. Rekink property at the first kink point of the tubing was measured and reported as rekink number. The testing device was a "Kink-O-Meter" (made in-house).

MECHANICAL PROPERTIES

The tensile, elongation and Young's modulus were tested using an Instron tester (Model 2122). initial length was set at 2"/min with a head speed at 20"/min.

TEM

Uniformity of the blend was analyzed using a transmission electron microscope. Cross-section (outer, inner and middle areas) of a blend tubing were examined.

TMA ANALYSIS

Heat stability of the tubing was measured using a thermal mechanical analyzer, TMA, equipped with 50 mN penetration probe and operated over a temperature sweep from 0 to 100°C at a rate of 10°C/min.

RESULTS

Tables 1 and 2 show the kink and mechanical properties of two sizes of blend tubing consisting of the ionomers and poly(ethylene-butene) copolymer. Table 3 tabulated the same properties of tubing made with individual resins in Tables 1 and 2.

Table 1. Evaluations of extruded tubing with blend of poly(ethylene butane) copolymer E-1 and ionomer I-1

OD cm	ID cm	Kink cm	Rekink cm	Young's Modulus MPa	Ultimate Tensile Strength MPa	Elongation %
0.40	0.27	2.29	2.67	22.5	18.1	898
0.35	0.24	2.16	2.79	21.5	14.9	830

The blend tubing has much better kink and rekink properties compared with E-1 tubing or a stiff tubing made with I-1. The tubing made from the E-1 resin had a poor rekink property, a tacky surface that easily attracted dirt, a strong cold memory and an unacceptable pull out force when adhesively bonded with connec-

Table 2. Evaluations of extruded tubing with blend of poly(ethylene butane) copolymer E-1 and ionomer I-2

OD cm	ID cm	Kink cm	Rekink cm	Young's Modulus MPa	Ultimate Tensile Strength MPa	Elongation %
0.41	0.28	2.41	2.79	20.3	17.9	850
0.35	0.24	2.03	2.54	22.3	18.8	826

Table 3. Evaluations of extruded tubing with individual resins

Resin	OD cm	ID cm	Kink cm	Rekink cm	Young's Modulus MPa	Ultimate Tensile Strength MPa	Elongation %
E-1	0.41	0.27	2.67	3.56	14.7	19.5	1039
I-1	0.41	0.27	2.41	3.56	46.0	31.0	435
I-2	0.41	0.27	2.41	2.92	23.4	19.4	700
PVC	0.41	0.31	2.29	2.79	23.6	17.4	302

tors for IV set assembly. The tubing made from the I-1 material is too stiff to be suitable for the end use, such as IV sets. Tubing of the E-1/I-1 blend has comparable stiffness to PVC tubing. The blend is about 22 MPa that matches well with the Young's modulus of PVC tubing.

Kink properties of the blend tubing are comparable with some soft version of ionomer, such as I-2. However, I-2 has relatively low heat stability (softening temperature of approximately 42°C) as studied using a TMA. The thermogram of the TMA shows that blend of I-2 and E-1 exhibited a higher softening tempera-

ture at approximately 65°C; almost the same as the softening point of E-1. The result suggests that adding E-1 in the ionomer improves heat stability of the blend tubing over tubing with I-2 alone. The overall properties of the blend tubing with E-1 and I-2, including cost saving, are greatly improved compared with tube samples extruded with individual resins such as I-2.

Mechanical properties of γ-irradiated tubing made with the blends and individual resins were tabulated in Table 4. There is little color change for the blend tubing after exposed to various levels of irradiation. The results also show that mechanical properties such as Young's modulus, ultimate tensile strength and elongation of the blend, I-2 and E-1 changed only slightly with a higher dose of γ–irradiation. However, in a separate study,[4] value of melt index decreased markedly from 2.72 dg/min to 0.05 dg/min for a tubing made of poly(ethylene butene) copolymer tubing exposed to 50 KGray irradiation. There was only 24.6% soluble extracted in decahydro naphthalene boiling solvent (194.6°C) when exposed to 100 KGray cobalt source. This indicated that some grafting and crosslinking occurred on the blend tubing when exposed to higher dose of γ-sterilization.

Table 4. Effect of γ-irradiation on tubing properties

Resin	Radiation KGray	Young's Modulus MPa	Tensile Strenght MPa	Elongation %
E-1	0	14.7	19.5	1039
	25	14.8	16.0	1057
	50	12.4	17.3	1109
	75	13.0	17.2	1096
	100	12.8	16.9	1109
I-2	0	23.4	19.4	700
	25	32.2	19.9	713
	50	26.6	20.0	716
	75	26.8	21.0	736
	100	26.8	21.6	718
E-1/I-2	0	21.4	18.3	840
	25	20.5	18.4	856
	50	23.8	18.5	873
	75	21.2	17.4	841
	100	18.9	17.0	843
PVC	0	23.6	17.4	302
	25	25.5	18.2	338
	50	20.2	17.4	340
	75	22.9	17.9	338
	100	28.1	18.5	338

PVC tubing became brownish after exposure to 25 KGray or higher dose of cobalt irradiation. This suggested that dehydrohalogenation had taken place in the PVC chains after γ-irradiation.

Cross-section of a blend tubing was examined with TEM. The result shows the blend tubing to be more uniform in the middle portion than in the outer and inner layers. Further investigation is needed. The result, however, affects very little the clarity of the tubing. Increased extrusion temperatures did not appear to physically affect the clarity of the tubing.

CONCLUSIONS

The blend tubing made of the two types of thermoplastics satisfies various performance needs for medical tubing application that tubings made with individual resins fail to do.

Compared with flexible PVC tubings, the blend tubings have advantages in clinical application. And, they provide a viable alternative to solve environmental issue when PVC material is used.

ACKNOWLEDGMENT

The authors would like to acknowledge S. Larson for drug adsorption studies and J. Tucker for Instron measurement of irradiated samples. Support from Medical Specialties and Infusion Therapy Departments at 3M is also appreciated.

REFERENCES

1.   A. Kaul *et al.*, *Drug Intelligency and Chemical Pharmacy*, **9** (1982).
2.   B. Switzer, **Report 01-5099**, *Southwest Research Institute* (9/92).
3.   C. Larson, *Internal report on evaluation of drug adsorption with PVC and blend tubing.*
4.   D. Domine, *private communication.*

INDEX